MULTIPHASE
FLOWS
with DROPLETS
and PARTICLES

MULTIPHASE FLOWS
with DROPLETS
and PARTICLES

Clayton Crowe
Martin Sommerfeld
Yutaka Tsuji

CRC Press

Boca Raton   Boston   New York   Washington   London

Library of Congress Cataloging-in-Publication Data

Crowe, C. T. (Clayton T.)
    Multiphase flows with droplets and particles / Clayton T. Crowe,
Martin Sommerfeld, Yutaka Tsuji.
       p. cm.
    Includes bibliographical references and index.
    ISBN 0-8493-9469-4 (alk. paper)
    1. Multiphase flow. 2. Drops. 3. Particles. I. Sommerfeld,
Martin. II. Tsuji, Yutaka, 1943– . III. Title.
TA357.5.M84C76 1997
620.1′.064—dc21

                                             97-24341
                                                CIP

© 1998 by CRC Press LLC

No claim to original U.S. Government works
International Standard Book Number 0-8493-9469-4
Library of Congress Card Number 97-24341
Printed in the United States of America 1 2 3 4 5 6 7 8 9 0
Printed on acid-free paper

Preface

Multiphase flows with particles and droplets cover a wide range of applications from the flow of mud to the flow field in a gas turbine engine. There have been significant advances in the science and technology of multiphase flows in the past few years because of enhanced computational and experimental capabilities. The field is evolving from one in which empiricism played a major role to one in which analysis and modeling can now be used to complement design and control. Still, the technology is far from mature and many challenges remain.

Because of the wide variety and importance of multiphase flows in industrial processes, courses on multiphase flows have been developed at several universities and have become the subject of short courses for industry. A course on multiphase flows was initiated at Washington State University in the mid seventies and has been taught on a two-year sequence ever since. The material for the book has evolved largely from this course. Material on particle-particle interaction (YT) and experimental methods (MS) has been added to complete the book.

Scope

The purpose of this book is to provide an organized, pedagogical approach to the study of multiphase flows with droplets and particles. The book is designed to complement a first-year graduate class in multiphase flows or to serve as a reference for engineers and scientists working in this general area. Problems are provided at the end of each chapter for instructional purposes and a numerical program for quasi-one-dimensional flow is available for student-oriented numerical projects.

The first chapter of the book provides an overview of multiphase flows and provides examples of industrial applications of multiphase flows with droplets and particles.

In the second chapter definitions and parameters specific to dispersed phase flows with particles and droplets are introduced. The idea of coupling and the qualitative effects of coupling are described. The purpose of this chapter is to give the reader some physical insights into the mechanics of flows with particles and droplets.

The third chapter reviews the parameters to quantify particle size such as mass median diameter and Sauter mean diameter. Commonly used distributions are also presented.

In the fourth chapter, the mass, momentum and energy transfer between individual droplets and particles and the carrier phase are addressed. Current information on the mass transfer, drag and heat transfer is included. The equations for the motion and thermal history of droplets and particles are derived in detail in Appendix A.

Chapter 5 summarizes particle-particle and particle-wall collisions which are important in dense phase flows. The hard sphere and soft sphere models are presented including the spring-damper representations for numerical simulations.

The fundamental equations for the carrier phase with suspended droplets and particles are developed in Chapter 6 and derived in detail in Appendix B using the volume averaging approach. The equations for quasi-one-dimensional flows are developed first to illustrate the basic principles and to support the numerical model described in Chapter 8. The equations are then presented as differential equations for a multidimensional flow. Radiative heat transfer is also addressed in this chapter.

The equations for a cloud of particles and droplets are developed in Chapter 7. Both the discrete element and two-fluid approaches are discussed as well as the dispersion and effective shear of the dispersed phase. Appendix C presents the equations for particle dispersion due to Brownian motion.

In Chapter 8 the fundamental ideas underlying numerical models for single phase flows are discussed first. These concepts are then extended to flows with suspended droplets and particles. Numerical models for both dilute and dense phase flows are described. A numerical program for fluid-particle flow in a quasi-one-dimensional duct is documented in Appendix D. This program can be used for the numerical projects suggested at the end of Chapter 8. This program, written in FORTRAN, can be obtained by sending a 3.5 inch floppy disc to Professor Crowe at Washington State University. It is also available through e-mail.

In Chapter 9 the various experimental techniques available for measuring particle size, velocity and concentrations are reviewed. These include laser-Doppler anemometry, phase-Doppler anemometry as well as particle-image velocimetry. The information provided in this chapter is adequate to carry out a preliminary design of a measurement system.

Chapter 10 provides an assessment of the state-of-the-art of multiphase flow with particles and droplets and outlines future needs.

Several books have appeared over the years on multiphase flows with droplets and particles. The first book to address the specific subject of multiphase flows was written in 1967 by S.L. Soo, *Fluid Dynamics of Multiphase Systems*. Since that time several books have appeared with material on multiphase flows with particles and droplets. These books include:

1969 - *One-Dimensional Two-Phase Flows* - G. Wallis

1971 - *Flowing Gas-Solids Suspensions* - R.G. Boothroyd

1978 - *Bubbles, Drops and Particles* - R. Clift, J.R. Grace & M.E. Weber

1980 - *Fundamentals of Gas-Particle Flows* - G. Rudinger

1981 - *Gas-Solid Transport* - G. Klinzing

1982 - *Handbook of Multiphase Systems* - G. Hetsroni (editor)

1989 - *Particulates and Continuum* - S.L. Soo

1991 - *Slurry Flows: Principles and Practice* - C.A. Shook & M.C. Roco

1993 - *Theory and Numerical Modeling of Turbulent Gas-Particle Flows and Combustion* - L. Zhou

1993 - *Multiphase Flow and Fluidization* - D. Gidaspow

These publications provide considerable detail on a wide spectrum of fluid-particle flows and are valuable sources of information. Some of these publications have been written to complement instruction, some to provide a source of information for a specific area and others to serve as a source of general information for multiphase flows.

Authors:

Dr. Crowe is Professor of Mechanical and Materials Engineering at Washington State University. He received his BS degree from the University of Washington in 1956, his MS degree from MIT in 1957 and his Ph.D. degree from the University of Michigan in 1962. His area of emphasis was aeronautical and astronautical engineering. From 1962 to 1969 he was employed in the rocket industry where he worked on gas-particle flow in rocket nozzles. In 1969 he joined Washington State University where he has taught thermal fluid sciences, numerical modeling and two-phase flows. He has coauthored a text book on fundamental fluid mechanics which is in its sixth edition. He has also received several teaching awards including the ASEE Pacific Northwest Outstanding Teaching Award in 1980.

While working in the rocket industry Dr. Crowe was involved in measuring the drag coefficient of particles in rarefied flows. After moving to Washington State University, he became involved in numerical modeling of multiphase flows and developed the PSI-Cell (Particle-Source-in-Cell) model for dispersed phase flows which has been used widely. He has also published works on particle dispersion in turbulence and on the modulation of carrier phase turbulence due to the presence of particles. He has organized numerous symposia on multiphase flows through ASME and other organizations. In 1995 he received the ASME Fluids Engineering Award for his contributions to multiphase flows.

Dr. Sommerfeld is Professor of Mechanical Process Engineering at the Martin Luther University of Halle-Wittenberg in Germany. He received his Dipl.-Ing. degree in 1981 and his Dr.-Ing. in 1984 from the Technical University of Aachen. The work for his doctoral degree focused on shock wave propagation through gas-particle mixtures. In 1984 he spent a year at Kyoto University in Japan on a research fellowship from the Japan Society for Promotion of Science and the Alexander Humboldt Foundation. From 1986 to 1994, Dr. Sommerfeld was at the Institute of Fluid Mechanics at the University of Erlangen where he lead a research group on two-phase flows. In 1994 he was promoted to Professor of Mechanical Engineering at Martin Luther University and Director of the Institute for Mechanical Processes and Environmental Protection.

The research activities of Professor Sommerfeld have focused on detailed experimental analysis of multiphase flows using modern optical instrumentation such as digital image analysis and phase-Doppler anemometry. He has also developed numerical methods for design and optimization of industrial processes involving multiphase flows such as stirred vessels, spraying systems and pneumatic conveying. He has organized a continuing series of workshops on two-phase flow predictions, ASME symposia and several other international conferences. He is currently the coordinator for the ERCOFTAC Special Interest Group on dispersed turbulent two-phase flows and chairman of the IAHR Section on industrial two-phase flows.

Dr. Y. Tsuji is Professor of Mechanical Engineering at Osaka University in Japan. He received the B.E. (1966) and M.E. (1968) degrees in Aeronautical

Engineering at the University of Osaka Prefecture. His early research work centered on the study of fluid turbulence and transition. After receiving the D.E from Osaka University in 1974, he directed his attention to fluid-solid multiphase flows. His research interests are numerical analyses and optical measurements of fluid-solid flows. His work has led to a number of publications.

He is a recipient of the Jotaki Award (in 1985) from The Society of Powder Technology, Japan and the JSME Medal (in 1992) from The Japan Society of Mechanical Engineers. In addition to his contributions to research and education, Dr. Tsuji has been active in many professional services including editorships for several international and domestic journals on multiphase flows, powder technology and fluids engineering. He has also been involved with many international conferences and symposia as an organizer, chairman and/or committee member. He is currently the Coordinator of the Committee on Simulation for The Association of the Powder Process Industry & Engineering in Japan.

Acknowledgments:

Many people have contributed to the development of this book. Dr. W. Comfort of LLL provided the initial encouragement to write the book. Other input resulted from professional leaves at Energy International (Dr. S. Bernstein) and IWT of the University of Bremen (U. Fritsching and K. Bauckhage). Professor Crowe acknowledges the contributions of his students, who have offered many suggestions on improving the manuscript. These students include A. Bakkom, K. Eichmann, J. Martinez, R. Moehrle, D. Peterson, M. Rank, T. Swanson and C. Wark who edited and constructively criticized the most recent version of the manuscript. He also is appreciative of Dr. Wang of the University of Delaware and Dr. R. Berry of INEEL who used an early draft of the book in classes they teach. Professor Crowe is also indebted to his colleagues, J. Chung and T. Troutt, for the many years of scintillating discussions. The assistance of Brenda Syre and David Cunningham in drawing and scanning figures is appreciated. A sincere depth of gratitude is extended to Danielle Bishop who played an indispensable role in rescuing the manuscript from abysmal depths of software and hardware failure. Dr. Crowe acknowledges the continual support and encouragement from his wife, Linda, and sons, Kevin and Chad.

Professor Tsuji acknowledges the assistance and contributing research of former students in his laboratory. These people include Dr. Toshitsugu Tanaka, Associate Professor of Osaka University, Mr. Toshihiro Kawaguchi, Research Associate of Osaka University and Dr. Neng Yao Shen who is an engineer with Sanko Air Plant, Ltd.

Clayton T. Crowe
Martin Sommerfeld
Yutaka Tsuji

This book is dedicated to

Thomas and Goldie Crowe
Linda, Kevin and Chad

and

Megumi

and

Martina,
Julia, Isabel and Katharina

Contents

Chapter 1

Introduction

The flow of particles and droplets in fluids is a subcategory of multicomponent, multiphase flows. The flow of multicomponent, multiphase mixtures covers a wide spectrum of flow conditions and applications. A *component* is a chemical species such as nitrogen, oxygen, water or Freon. A *phase* refers to the solid, liquid or vapor state of the matter. Examples of single and multicomponent, multiphase flows are provided in Table 1.1.

	Single component	Multicomponent
Single-phase	Water flow Nitrogen flow	Air flow Flow of emulsions
Multiphase	Steam-water flow Freon-Freon vapor flow	Air-water flow Slurry flow

Table 1.1. Examples of single and multicomponent, multiphase flows.

The flow of air, which is composed of a mixture of gases (nitrogen, oxygen, etc.), is the best example of a single-phase multicomponent flow. It is common practice to treat these types of flows as the flow of a single component with a viscosity and thermal conductivity which represents the mixture. Such an approach is practical unless the major constituents of the component gases have significantly different molecular weights. In this case the momentum associated with the diffusional velocities may be important. Also, the multicomponent nature of air will be important at high temperatures where dissociation occurs, or at very low temperatures where some species may condense out.

The flow of mixtures of liquids is also an important industrial application. For example, water is sometimes used to flush oil from a well which gives rise to a multicomponent single-phase flow. If the two liquids are miscible, then the mixture will be treated as a single-phase with modified properties. If the liquids are immiscible, then the liquid cannot be regarded as homogeneous and

1

treatment of the flow problem becomes much more complex. In this situation one may have "globs" of oil in the water or for high oil content, globs of water carried by the oil. The mixtures of two liquids are generally referred to as emulsions.

Single component multiphase flows are typically the flow of a liquid with its vapor. The most common example is steam-water flows which are found in a wide variety of industries. Another example of single component, multiphase flows are refrigerants in a refrigeration system.

The flow of fluids of a single phase has occupied the attention of scientists and engineers for many years. The equations for the motion and thermal properties of single-phase fluids are well accepted (Navier-Stokes equations) and closed form solutions for specific cases are well documented. The major difficulty is the modeling and quantification of turbulence and its influence on mass, momentum and energy transfer. The state-of-the art for multiphase flows is considerably more primitive in that the correct formulation of the governing equations is still subject to debate. For this reason, the study of multiphase flows represents a challenging and potentially fruitful area of endeavor for the scientist or engineer.

Gas-liquid flows	Bubbly flows
	Separated flows
	Gas-droplet flows
Gas-solid flows	Gas-particle flows
	Pneumatic transport
	Fluidized beds
Liquid-solid flows	Slurry flows
	Hydrotransport
	Sediment transport
Three-phase flows	Bubbles in a slurry flow
	Droplets/particles in gaseous flows

Table 1.2. Categories and examples of multiphase flows.

Multiphase flows can be subdivided into four categories; gas-liquid, gas-solid, liquid-solid and three-phase flows. Examples of these four categories are shown in Table 1.2. A gas-liquid flow can assume several different configurations. For example, the motion of bubbles in a liquid in which the liquid is the continuous phase is a gas-liquid flow. On the other hand, the motion of liquid droplets in a gas is also a gas-liquid flow. In this case the gas is the continuous phase. Also, a separated flow in which the liquid moves along the bottom of a pipe and the gas along the top is also a gas-liquid flow. In this situation both phases are continuous. The first two examples, bubbles in a liquid and droplets in a gas, are known as dispersed phase flows since one phase is dispersed and the other is continuous. By definition, one can pass from one point to another in the continuous phase while remaining in the same medium. One cannot pass from one droplet to another without going through the gas.

Gas-solid flows are usually considered to be a gas with suspended solid particles. This category includes pneumatic transport as well as fluidized beds. Another example of a gas-solid flow would be the motion of particles down a chute or inclined plane. These are known as granular flows. Particle-particle and particle-wall interactions are much more important than the forces due to the interstitial gas. If the particles become motionless the problem reduces to flow through a porous medium in which the viscous force on the particle surfaces is the primary mechanism affecting the gas flow. An example is a pebble-bed heat exchanger. It is not appropriate to refer to flow in a porous medium as a gas-solid flow since the solid phase is not in motion. Gas-solid flow is another example of a dispersed phase flow since the particles constitute the dispersed phase and the gas the continuous phase.

Liquid-solid flows consist of flows in which solid particles are carried by the liquid and are referred to as slurry flows. Slurry flows cover a wide spectrum of applications from the transport of coals and ores to the flow of mud. These flows can also be classified as dispersed phase flows and are the focus of considerable interest in engineering research. Once again it is not appropriate to refer to the motion of liquid through a porous medium as a liquid-solid flow since the solid phase is not in motion.

Three-phase flows are also encountered in engineering problems. For example, bubbles in a slurry flow gives rise to the presence of three phases flowing together. There is little work reported in the literature on three-phase flows.

The subject of this book is the flow of particles or droplets in a fluid, specifically the flow of particles and/or droplets in a conveying gas as well as particles in a conveying liquid. The other area of dispersed phase flows, namely, bubbly flows, will not be addressed here.

The flow of particles and droplets in fluids has a wide application in industrial processes. The removal of particulate material from exhaust gases is essential to the control of pollutants generated by power plants fired by fossil fuels. The efficient combustion of droplets and coal particles in a furnace depends on the interaction of particles or droplets with air. The generation of many food products depends on the drying of liquid droplets to powders in high temperature gas streams. The transport of powders in pipes is common to many chemical and processing industries.

For many years, the design of systems with particle/droplet flows was based primarily on empiricism. However, more sophisticated measurement techniques have lead to improved process control and quantification of fundamental parameters. Increased computational capability has enabled the development of numerical models that can be used to complement engineering system design. The improved understanding of this is a rapidly growing field of technology which will have far-reaching benefits in upgrading the operation and efficiency of current processes and in supporting the development of new and innovative approaches.

1.1   Industrial applications

The objective of this book is to provide a background in this important area of fluid mechanics to assist those new to the field and to provide a resource to those actively involved in the design and development of multiphase systems. In this chapter, examples of multiphase flows in industrial and energy conversion processes are outlined to illustrate the wide application of this technology.

1.1.1   Spray drying

Many products such as foods, detergents and pharmaceuticals are produced through spray drying (Masters, 1972). This is a process in which a liquid material is atomized, subjected to hot gases and dried into the form of a powder. The general configuration of a counter current flow spray dryer is shown in Figure 1.1. A slurry or concentrated mixture is introduced at the top of the dryer and atomized into droplets. Hot gases are fed into the bottom with a swirl component and move upward through the dryer. The droplets are dried as they fall through the hot rising gases to form a powder which is collected at the bottom and removed as the final product.

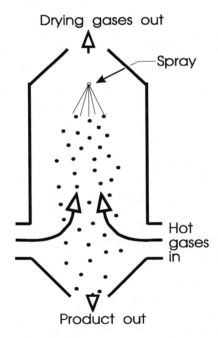

Figure 1.1: Counter-current flow spray dryer.

Accumulation of the dried product on the wall is to be avoided because of uncontrolled drying and the possibility of fire. Also, in the case of food production, the product cannot become too hot to avoid altering the taste.

The gas-droplet (particle) flow within the dryer is very complex. The swirling motion of the gases transports the particles toward the wall which may lead to impingement and accumulation. The temperature distribution in the dryer will depend on the local concentration of the droplets as they fall through the dryer. High local concentrations will depress the local gas temperature and lead to less effective drying. The result may be a nonuniformly dried product reducing product quality.

Even though spray drying technology has been continuously improved through the years, it is still difficult to scale up models to prototype operation. It is also difficult to determine, without actual testing, how modifying the design of a conventional dryer will affect performance. There currently is a need to develop numerical and analytic tools that adequately simulate the gas-droplet flow field in the dryer. Such models or analyses could be effectively used to improve the efficiency of current designs, predict off-design performance and serve as a tool for scale-up of promising bench-scale designs to prototype operation.

1.1.2 Pollution control

The removal of particles and droplets from industrial effluents is a very important application of gas-particle and droplet flows (Jorgensen & Johnson, 1981). Several devices are used to separate particles or droplets from gases. If the particles are sufficiently large (greater than 50 microns), a settling chamber can be used in which the condensed phase simply drops out of the flowing gas and is collected. For smaller particles ($\sim$ 5 microns), the cyclone separator shown in Figure 1.2 is used. The gas-particle flow enters the device in a tangential direction as shown. The resulting vortex motion in the separator causes the particles to migrate toward the wall due to centrifugal acceleration and then fall toward the bottom where they are removed. The gases converge toward the center and form a vortex flow which exits through the top. The performance of the cyclone is quantified by the "cut size" which is the particle diameter above which all the particles are collected. Years of experience in cyclone design have resulted in "standard" designs that, under normal operating conditions, have predictable performance. However, there is little information to design cyclones for special applications such as hot-gas clean up.

The particles issuing from power plants operating with fossil fuels are on the order of a micron in diameter. In these applications, the electrostatic precipitator is generally used. The top view of a conventional electrostatic precipitator is shown in Figure 1.3. The high voltage applied to the wires creates a corona with charged ions. These ions travel along the electric lines of force to the particles and accumulate on the particles. The resulting charged particles are moved toward the wall by Coulomb forces and deposited on the wall. Periodically the plates are vibrated (rapped) and the particles fall into a collection bin. The fluid mechanics of the electrostatic precipitator is quite complex. The particle-fluid interaction obviously influences the particle concentration and the charge den-

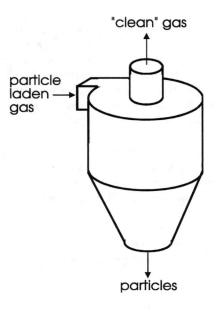

Figure 1.2: Cyclone separator.

sity. These, in turn, affect the electric field. Flow turbulence is also introduced by the structural ribs in the system. Electrostatic precipitators are still designed using empirical formulas because of the complexity of the fluid-particle-electrical field interactions.

Another pollution control device is the wet gas scrubber which is designed to remove particulate as well as gaseous pollutants. Scrubbers come in many configurations but the venturi scrubber shown in Figure 1.4 represents a simple design. Droplets are introduced upstream of the venturi and the particles are collected on the droplets. The droplets, being much larger than the particles, can be more easily separated from the flow. Sulfur dioxide can also be removed by using droplets mixed with lime. The sulfur dioxide is absorbed on the surface of the droplets. These droplets are collected, the sulfur products are removed and the droplets are reused in the scrubber.

1.1.3   Transport systems

Materials can be transported by either gases or liquids, depending on the specific application. The transport of materials by air is known as pneumatic transport. The movement of materials by liquids (usually water) is slurry transport.

Pneumatic transport

Pneumatic transport is used widely in industry for the transport of cement, grains, metal powders, ores, coal, etc. (Klinzing, 1981; Cheremisinoff & Cherem-

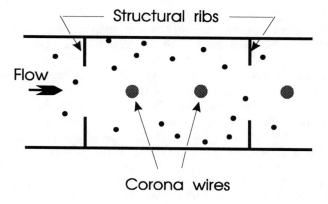

Figure 1.3: Electrostatic precipitator.

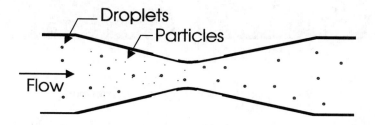

Figure 1.4: Venturi scrubber.

isinoff, 1984). The major advantage over a conveyer belt is continuous operation, flexibility of location and the possibility to tap into the pipe at arbitrary locations. Pneumatic transport has been used in industry for many years for the transport of solid materials. It has been particularly useful in layouts where obstacles prohibit straight-line transport or systems which require tapping the line at arbitrary locations.

Flow patterns will depend on many factors, such as solids loading, Reynolds number and particle properties. Vertical pneumatic transport corresponds to gas flow velocities exceeding the fast fluidization velocity. The following regimes, illustrated in Figure 1.5, have been identified for horizontal, gas-particle flows. In homogeneous flow, the gas velocity is sufficiently high that the particles are well mixed and maintained in a nearly homogeneous state by turbulent mixing as illustrated in Figure 1.5a. As the gas velocity is reduced the particles begin to settle out and collect on the bottom of the pipe as shown in Figure 1.5b. The velocity at which deposition begins to occur in the pipe is called the saltation velocity. After a layer builds up, ripples begin to form due to the gas flow. These ripples resemble "dunes". As powder continues to fill the pipe, there are alternate regions where particles have settled and where they are still in suspension as shown in Figure 1.5c. This is called slug flow. Finally, at even lower gas velocities, the powder completely fills the pipe and the flow of gas represents flow through a packed bed depicted by Figure 1.5d. At this point pneumatic transport ceases to exist.

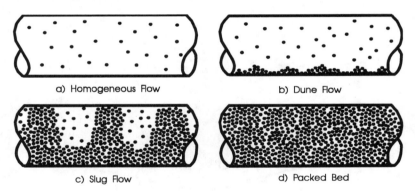

Figure 1.5: Horizontal pneumatic transport.

Pneumatic conveying systems are generally designated as dilute, or dense-phase transport. Dilute-phase transport is represented by Figure 1.5a. These systems normally operate on low pressure differences with low solids loading and high velocity (higher than the saltation velocity). Dense-phase transport is represented by Figure 1.5c in which the pressure drop and solids loading are higher. The lower velocity leads to less material degradation and line erosion.

Many studies on pressure drop in pneumatic transport have been reported. There are considerable discrepancies in the data. Extensive experience with the design, installation and operation of pneumatic transport systems has given rise to design criteria which ensure a functional system. In dense-phase transport the pressure drop is proportional to the square of the length of the slug, so various schemes have been devised to achieve this end. Still, there are situations where extensive experience is insufficient. One such case is the transport of wet particles. Little information is available on the tendency of these particles to accumulate on the wall or in bends, and plug the pipe. Even more fundamental, there is essentially no information on the head loss associated with the conveying of wet solids.

Slurry transport

The transport of particles in liquids is identified as slurry flow (Shook & Roco, 1993). The term "hydrotransport" is sometimes used for the transport of large particles like rock or chunks of coal. The flow of mud is regarded as a slurry flow. Considerable effort has also be devoted to the development of coal-water slurries which could be substituted for fuel oils.

Slurries are classified as homogeneous, heterogeneous, moving bed or stationary bed. Homogeneous slurries normally consist of small particles which are kept in suspension by the turbulence of the carrier fluid. On the other hand, heterogeneous slurries are generally composed of coarse particles which tend to settle on the bottom of the pipe. The velocity at which the particles settle out is the deposition velocity which is equivalent to the saltation velocity in pneumatic transport. Of course no slurry flow will be completely homogeneous. The rule of thumb is that the slurry is homogeneous if the variation in particle concentration from the top to bottom of the pipe is less than 20 percent. There are no well-established rules which predict whether a slurry will be homogeneous or not. The moving bed regime occurs when the particles settle on the bottom of the pipe and move along as a bed. In this case, the flow rate is considerably reduced because the bed moves more slowly than the fluid above the bed. Finally, when the particles fill the duct and no further motion is possible define the stationary bed. The flow is now analogous to the flow through a porous medium.

The fluid mechanics of the liquid-solid flows is complex because of the particle-particle and fluid-particle interaction. Usually the homogeneous slurry is treated as a single-phase fluid with modified properties which depend on solids loading. The various correlations for head loss which have been developed for slurry flows can only be used with confidence for slurries with properties identical to those for which the correlations have been obtained. Extrapolation of the correlations to other slurries may lead to significant errors in pressure drop predictions.

1.1.4   Fluidized beds

The fluidized bed is also another example of an important industrial operation involving multiphase flows. A fluidized bed consists of a vertical cylinder containing particles where gas is introduced through holes (distributor) in the bottom of the cylinder as shown in Figure 1.6. The gas rising through the bed suspends the particles. At a given flow rate "bubbles", which are regions of low particle density, appear and rise through the bed which intensifies mixing within the bed. Fluidized beds are used for many chemical processes such as coal gasification, combustion, liquefaction as well as the disposal of organic, biological and toxic wastes.

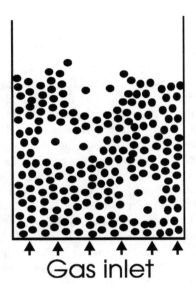

Figure 1.6: Fluidized bed.

Currently the design and operation of a fluidized bed is the result of many years of experience in building, modifying and testing to achieve the best performance possible. The mechanics of the multiphase flow in a fluidized bed has been, and continues to be, a challenge to the scientist and practicing engineer. Some have chosen to treat this flow as a fluid with a modified viscosity and thermal conductivity. Others have approached the problem using the discrete particle approach in which the motion of each particle is considered. The former approach depends on developing relationships for the transport properties of the particulate phase. The latter approach requires extensive computer capability to include sufficient particles to simulate the system.

Even though the numerical models for multiphase flow in a fluidized bed appear promising, there are still many issues that have to be included such as the cohesiveness of particles, the sticking probability of a wet particle as well as particle-wall interaction.

1.1.5 Manufacturing and material processing

The flow of droplets and particles in gases is important to many manufacturing and material processing techniques, a few of which are described below.

Spray forming

A new area in the manufacturing industry is spray forming or spray casting. This is a process in which molten metals are atomized into fine droplets, transported by a carrier gas and deposited on a substrate as shown in Figure 1.7. This casting technique has several advantages. The rapid solidification of the small metal droplets gives rise to a fine grain structure and improved material properties of the deposited material. Also, by moving the substrate it is possible to produce a shape close to the final product, which minimizes material waste.

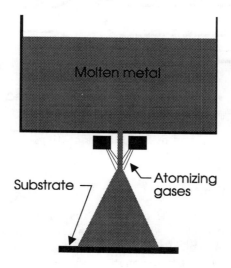

Figure 1.7: Spray forming process.

The importance of gas-particle flow is evident in spray forming. The state of the droplet upon impact with the substrate is important. A completely solidified droplet will have to be remelted to form a homogeneous deposit. A liquid droplet may splatter complicating the disposition pattern. Also, the energy associated with the latent heat of the droplet will have to be conducted through the substrate which may slow the cooling of the material on the surface. The cooling rate of the droplets depends on the droplet size and the local temperature in the spray. Droplets on the edge of the spray will cool faster than those near the core. The understanding and ability to predict the thermal behavior of the droplets in the spray is important to the continued development of the spray casting process.

Plasma spray coating

Another important area of gas-particle flows in manufacturing is plasma coating. The typical plasma torch consists of a chamber where a plasma is produced by an rf or dc source. Particles, introduced into the plasma flow, are melted and convected toward the substrate as shown in Figure 1.8. The heat transfer and drag on a particle in a plasma is fundamental to the operation of the torch. At present the plasma torch is designed primarily by experience. The improvement of efficiency, operation in off-design conditions and scale-up will be dependent on improved multiphase flow models for particles in a plasma.

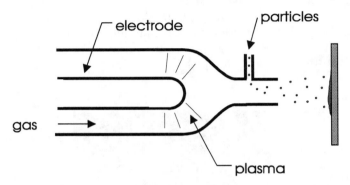

Figure 1.8: Plasma coating.

Abrasive water-jet cutting

The rapid and accurate cutting of various materials through the use of high velocity water jets with entrained abrasive materials is another new application of multiphase flows. A typical abrasive water jet is shown in Figure 1.9. Water issues through a small orifice from a high pressure source. Some systems operate at 60,000 psi with jet velocities of 3000 ft/sec. The material is moved to provide the desired shape. The inclusion of an abrasive material (usually garnet) in the jet enables cutting of hard materials such as concrete or glass. The effectiveness of the abrasive water jet depends on the speed with which the abrasive material impacts the surface. The abrasive material is accelerated by the drag of the fluid which is an intriguing multiphase flow problem.

Synthesis of nanophase materials

A very new and emerging area of multiphase flow applications in materials processing is the generation of nanophase materials. These materials have grain sizes of the order of 5 to 50 nanometers and can be produced by the compaction of noncrystalline material. One of the approaches currently being investigated to produce nanoclusters is gas phase synthesis. One technique is the injection

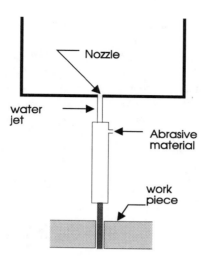

Figure 1.9: Abrasive water-jet cutting.

of precursors into diffusion or premixed flames. The resulting particles are from 1 to 500 nanometers in diameter. The wide range of sizes results from the lack of control of the steep thermal gradients. Another approach is with thermal reactors in which precursors are introduced into the furnace in the form of an aerosol. Chemical reactions occur as the multiphase mixture passes through the furnace. The control of particle size is highly dependent on the regulation of temperature and flow velocity. This represents an important application of multiphase flows in which thermal coupling between the gaseous and particulate phases is important. Spray pyrolysis is also being used in connection with furnace reactors in which a precursor is atomized and convected through the furnace. The solvent evaporates and reaction occurs between the particles to form the material. This approach is promising because of the possibility of making multicomponent materials. It is important in this approach to control the temperature as well as the time of the particles in the furnace. This is a challenging problem in multiphase flow technology.

1.2 Energy conversion and propulsion

There are many examples of droplet/particle flows in energy conversion and propulsion systems ranging from coal-fired or oil-fired furnaces to rocket propulsion.

1.2.1   Pulverized-coal-fired furnaces

Furnaces fired by pulverized coal operate by blowing a coal particle-air mixture from the corner of a furnace as shown in Figure 1.10. The corner-fired furnace produces a swirling flow in the furnace which enhances mixing. When the particles enter the furnace, the radiative heat transfer heats the particles, and the volatiles (mostly methane) are released. These gases serve as the primary fuel for combustion. Ultimately, the remaining char burns but at a lower rate. Obviously the effective mixing of the volatiles and the gas is important for efficient combustion with minimum pollutant production. The gas-particle flow in the furnace is very complex because of the interaction of heat transfer, combustion and particle dynamics.

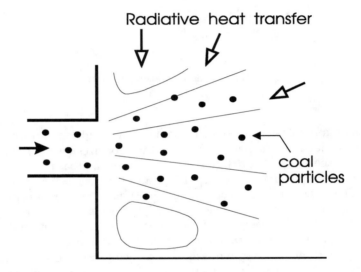

Figure 1.10: Corner-fired furnace with pulverized coal.

1.2.2   Solid propellant rocket

An example of a gas-particle flow in a propulsion system is the solid propellant rocket shown in Figure 1.11. The fuel of solid propellant rocket can be aluminum powder. When the aluminum burns, small alumina droplets about a micron in diameter are produced and convected out the nozzle in the exhaust gases. The presence of these particles lowers the specific impulse of the rocket. The principles of gas-particle flows are used to design nozzles to achieve the best specific impulse possible within the design constraints of the system.

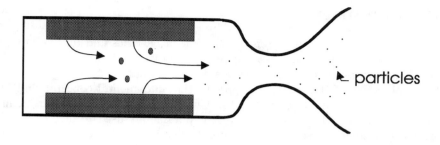

Figure 1.11: Solid propellant rocket.

1.3   Fire suppression and control

Fire suppression systems in buildings usually consist of nozzles located in ceilings
that are activated in the event of a fire as shown in Figure 1.12. Usually the
high temperatures produced by the fire melt wax in the nozzle which allows
the water to flow. The suppression systems are designed to deliver the amount
of water flux per unit area (required delivery density) to extinguish the fire.
This design criterion is generally established by experiment. The phenomena
associated with the spray is very complex. As the droplets are projected toward
the fire, they are evaporated by the hot gases and may not penetrate to the
location of the fire. The evaporating droplets will cool the gases and reduce the
radiative feed back to the flame. Modeling fire suppression by droplets is an
active area of research.

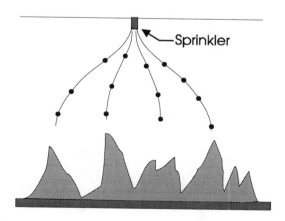

Figure 1.12: Ceiling sprinkler fire suppression system.

1.4   Summary

There are many other applications of the flow of particles and droplets in fluids which have not been addressed here. Those discussed above illustrate the technological significance of this important area of fluid mechanics.

The objective of this book is to present the fundamental concepts and approaches to address fluid particle flows. Chapter 2 provides important definitions to characterize the flow. The various parameters used to define particle and droplet size are presented in Chapter 3. The interaction between the fluid and the particle and droplet phase are addressed in Chapter 4. The equations describing particle-particle interaction are discussed in Chapter 5. The continuous phase equations are presented in Chapter 6 and those for the dispersed phase in Chapter 7. The fundamentals for numerical modeling are introduced in Chapter 8 and the experimental techniques in Chapter 9. Finally, Chapter 10 provides some examples of measurements and numerical modeling for the applications discussed in Chapter 1.

Chapter 2

Properties of Dispersed Phase Flows

Dispersed phase flows are flows in which one phase, the dispersed phase, is not materially connected. These include gas-droplet, gas-particle and liquid-particle flows in which the particles and droplets constitute the dispersed phase. Bubbles in a bubbly flow also represent the dispersed phase. The objective of this chapter is to introduce definitions of dispersed phase flows which apply specifically to particle or droplet flows.

2.1 Density and volume fraction

For a continuum one can define density at a point as

$$\rho = \lim_{\delta V \to 0} \frac{\delta M}{\delta V} \qquad (2.1)$$

where δM is the mass associated with volume δV. However real materials are not a continuum so, strictly speaking, one cannot take the limiting volume as zero. The volume must be large enough to contain sufficient molecules to yield a stationary average; that is, if the volume was increased or decreased slightly the average would remain unchanged. A mole of gas at standard conditions contains 10^{23} molecules and occupies a volume of approximately 22 liters. In order to have a stationary average, we need the volume to contain about 10^4 molecules. The volume containing 10^4 molecules is

$$\delta V \sim 22 \cdot 10^{-3} \frac{10^4}{10^{23}} m^3$$

which corresponds to a cube approximately 10^{-7} meters or 0.1 microns along an edge. This volume can be considered a point if this dimension is much smaller than the dimensions of the flow system. For example, the flow system

17

dimension for flow in a pipe would be pipe diameter. This is the situation for most problems, so the use of differential equations which describe conditions at a point is justified. On the other hand, if one were addressing flows with small dimensions, such as flow in a 10 micron diameter duct, then the continuum assumption may no longer be valid. This is called a rarefied flow and has to be treated on a molecular basis. The concepts of volume averaging and other averages will be discussed in more detail later.

The same concept extends to dispersed phase flows. Consider the mixture of dispersed phase elements in a volume, δV shown in Figure 2.1. The *number density* is defined as the number of particles per unit volume and is

$$n = \lim_{\delta V \to \delta V^o} \frac{\delta N}{\delta V}$$

where δN is the number of elements in the volume and δV^o is the limiting volume that ensures a stationary average. Of course, the limiting volume will be much larger than the limiting volume for a gas at standard conditions.

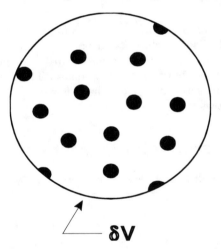

Figure 2.1: Volume with dispersed phase elements.

The *volume fraction* of the dispersed phase is defined as

$$\alpha_d = \lim_{\delta V \to \delta V^o} \frac{\delta V_d}{\delta V} \tag{2.2}$$

where δV_d is the volume of the dispersed phase in the volume. Equivalently, the volume fraction of the continuous phase is

$$\alpha_c = \lim_{\delta V \to \delta V^o} \frac{\delta V_c}{\delta V} \tag{2.3}$$

where δV_c is the volume of the continuous phase in the volume. This volume fraction is sometimes referred to as the *void fraction*. By definition, the sum of the volume fractions must be unity.

$$\alpha_d + \alpha_c = 1 \qquad (2.4)$$

The *bulk density* (or *apparent density*) of the dispersed phase is the mass of the dispersed phase per unit volume of mixture or, in terms of a limit, is defined as

$$\bar{\rho}_d = \lim_{\delta V \to \delta V^\circ} \frac{\delta M_d}{\delta V} \qquad (2.5)$$

where δM_d is the mass of the dispersed phase. If all the particles in a volume have the same mass, m, the bulk density is related to the number density by

$$\bar{\rho}_d = nm$$

The corresponding definition for the bulk density of the continuous phase is obvious.

The sum of the bulk densities for the dispersed and continuous phases is the mixture density,

$$\bar{\rho}_d + \bar{\rho}_c = \rho_m. \qquad (2.6)$$

The bulk density can be related to the volume fraction and material density. The mass associated with the dispersed phase can be written as

$$\delta M_d = \rho_d \delta V_d \qquad (2.7)$$

where ρ_d is the material, or actual, density of the dispersed phase material. For example, if the dispersed phase were water droplets, this would be the density of the water. Substituting Equation 2.7 into Equation 2.5 gives

$$\bar{\rho}_d = \lim_{\delta V \to \delta V^\circ} \frac{\rho_d \delta V_d}{\delta V} = \rho_d \lim_{\delta V \to \delta V^\circ} \frac{\delta V_d}{\delta V} = \rho_d \alpha_d \qquad (2.8)$$

Correspondingly, the bulk density of the continuous phase is

$$\bar{\rho}_c = \rho_c \alpha_c \qquad (2.9)$$

so an alternate expression for the mixture density is

$$\rho_m = \alpha_c \rho_c + \alpha_d \rho_d \qquad (2.10)$$

Another parameter important to the definition of dispersed phase flows is the dispersed phase *mass concentration* or

$$C = \frac{\bar{\rho}_d}{\bar{\rho}_c} \qquad (2.11)$$

which is the ratio of the mass of the dispersed phase to that of the continuous phase in a mixture. This parameter will sometimes be referred to as the particle or droplet mass ratio. The ratio of mass flux of the dispersed phase to that of the continuous phase is the *loading*,

$$z = \frac{\dot{m}_d}{\dot{m}_c} = \frac{\bar{\rho}_d v}{\bar{\rho}_c u} \tag{2.12}$$

where v and u are the velocities of the dispersed and continuous phases respectively. This is the "local" loading or the mass flux ratio at a local region in the flow. The total loading is the ratio of the overall mass flow rate of the dispersed phase flow to the overall mass flow rate of the continuous phase,

$$Z = \frac{\dot{M}_d}{\dot{M}_c} \tag{2.13}$$

For example, if droplets were sprayed into a gas flow, then z would be the local mass flux ratio which would vary throughout the spray and would be zero outside the spray envelope. The overall loading, Z, would be the ratio of total liquid mass flow rate supplied to the atomizer to the total gas flow rate. Obviously the total loading can be indicative of the magnitude of the local loading in the flow field.

2.2 Particle or droplet spacing

The mechanics of a dispersed phase flow depends significantly on the average distance between the dispersed phase elements. This information is important to determine if a particle or droplet can be treated as an isolated element.

Consider the elements in Figure 2.2 that are enclosed in cubes with side L which is the distance between element centers. The volume fraction of the dispersed phase is

$$\alpha_d = \frac{\pi D^3}{6L^3} \tag{2.14}$$

If the particles where in contact in this configuration the volume fraction of the dispersed phase would be $\pi/6$[1].

The particle or droplet spacing is related to the volume fraction by

$$\frac{L}{D} = (\frac{\pi}{6\alpha_d})^{\frac{1}{3}} \tag{2.15}$$

For a dispersed volume fraction of 10%, the spacing is 1.7 which would suggest that the dispersed phase elements are too close to be treated as isolated. That is, the mass, momentum and heat transfer for each element are influenced by the neighboring elements.

The volume fraction of the dispersed phase can be also expressed in terms of the dispersed phase mass concentration and material density ratio. The

---

[1] This lattice arrangement does not represent the maximum volume fraction for uniform-size spheres.

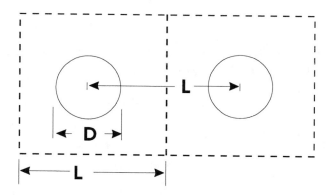

Figure 2.2: Interparticle spacing.

dispersed phase volume fraction is given by

$$\alpha_d = \frac{\bar{\rho}_d}{\rho_d} = \frac{\bar{\rho}_d}{\bar{\rho}_c} \frac{\bar{\rho}_c}{\rho_c} \frac{\rho_c}{\rho_d}$$

This equation can be written as

$$\alpha_d = C\alpha_c \frac{\rho_c}{\rho_d} \tag{2.16}$$

Using the relationship $\alpha_c = 1 - \alpha_d$ in Equation 2.16 yields

$$\alpha_d = \frac{\kappa}{1 + \kappa} \tag{2.17}$$

where $\kappa = C\rho_c/\rho_d$. Thus the distance between elements in a dispersed phase flow can be expressed as

$$\frac{L}{D} = (\frac{\pi}{6} \frac{1 + \kappa}{\kappa})^{\frac{1}{3}} \tag{2.18}$$

For most gas-particle and gas-droplet flows the ratio of material densities, ρ_c/ρ_d, is of the order of 10^{-3} so the interparticle spacing for flows with a mass ratio of unity is

$$\frac{L}{D} \sim 10$$

In this case, individual particles or droplets could be treated as isolated droplets with little influence of the neighboring elements on the drag or heat transfer rate. However, in fluidized beds, the mass ratio is large and the particles may be located less than 3 diameters apart so the particles cannot be treated as isolated droplets. For slurry flow, the material density ratio can be the order of unity and the particle spacing is insufficient to treat the particles as isolated elements.

Example: A sand-water slurry has a concentration of 0.3. Find the particle volume fraction and the interparticle spacing.

Solution: The density ratio of water to sand is $2500/1000 = 2.5$. The value of κ is $0.3 \times 2.5 = 0.75$. Thus, the particle volume fraction is

$$\alpha_d = \frac{\kappa}{1 + \kappa} = 0.43$$

and the spacing ratio is

$$\frac{L}{D} = (\frac{\pi}{6}\frac{1 + \kappa}{\kappa})^{\frac{1}{3}} = 1.07$$

This value must be regarded as an estimate since the particle positions will not form a lattice configuration, which is assumed to derive Equation 2.18.

It is also of interest to estimate the size of the limiting volume, δV^o, which would be needed to have a stationary average. If 10,000 particles were needed to form a stationary average, then the volume would be

$$\delta V^o = 10^4 (\frac{L}{D})^3 D^3 \tag{2.19}$$

where $\delta V^o = L^3$. The size of the cube representing the volume would be

$$L \sim 20(\frac{L}{D})D \tag{2.20}$$

Thus if the average spacing was 10 diameters and the element size was 100 microns, the size of the limiting volume would be a 2 cm. cube. If this dimension is comparable to the system dimensions, then the dispersed phase cannot be regarded as a continuum in this flow system.

2.3   Response times

The response time of a particle or droplet to changes in flow velocity or temperature are important in establishing nondimensional parameters to characterize the flow. The momentum response time relates to the time required for a particle or droplet to respond to a change in velocity. The equation of motion for a particle in a gas is given by

$$m\frac{dv}{dt} = \frac{1}{2}C_D\frac{\pi D^2}{4}\rho_c(u - v)\left|u - v\right| \tag{2.21}$$

where v is the particle velocity and u is the gas velocity. Defining the dispersed phase Reynolds number as

$$Re_r = \frac{\rho_c D\left|u - v\right|}{\mu_c} \tag{2.22}$$

and dividing through by the particle mass gives

$$\frac{dv}{dt} = \frac{18\mu}{\rho_d D^2} \frac{C_D Re_r}{24}(u - v) \tag{2.23}$$

where μ_c is the viscosity of the gas (continuous phase). For the limits of low Reynolds numbers (Stokes flow), the factor $C_D Re_r/24$ approaches unity. The other factor has dimensions of reciprocal time and defines the *momentum (velocity) response* time,

$$\tau_V = \frac{\rho_d D^2}{18\mu} \tag{2.24}$$

so the equation can be rewritten as

$$\frac{dv}{dt} = \frac{1}{\tau_V}(u - v) \tag{2.25}$$

The solution to this equation for constant u and an initial particle velocity of zero is

$$v = u(1 - e^{-t/\tau_V}) \tag{2.26}$$

Thus the momentum response time is the time required for a particle released from rest to achieve 63% ($\frac{e-1}{e}$) of the free stream velocity. The response time for a 100 micron water droplet in air at standard conditions is 30 ms. As one can see, the momentum response time is most sensitive to the particle size.

The *thermal response time* relates to the responsiveness of a particle or droplet to changes in temperature in the carrier fluid. The equation for particle temperature, assuming the temperature is uniform throughout the particle and radiative effects are unimportant, is

$$mc_d \frac{dT_d}{dt} = Nu\pi k_c D(T_c - T_d) \tag{2.27}$$

where Nu is the Nusselt number, c_d is the specific heat of the particle material and k_c is the thermal conductivity of the continuous phase. Dividing through by the particle mass and specific heat gives

$$\frac{dT_d}{dt} = \frac{Nu}{2} \frac{12k_c}{\rho_p c_d D^2}(T_c - T_d) \tag{2.28}$$

For low Reynolds numbers the ratio $Nu/2$ approaches unity. The other factor is the thermal response time defined as

$$\tau_T = \frac{\rho_d c_d D^2}{12k_c} \tag{2.29}$$

Thus the thermal equation for the particle becomes

$$\frac{dT_d}{dt} = \frac{1}{\tau_T}(T_c - T_d) \tag{2.30}$$

Once again, this is the time required for a particle to achieve 63% of a step change in the temperature of the carrier phase. The thermal response time for a 100 micron water droplet in air at standard conditions is 145 ms which is somewhat larger than the velocity response time.

The momentum and thermal response times are related through the properties of the fluid and the particles. Dividing the momentum response time by the thermal response time gives

$$\frac{\tau_V}{\tau_T} = \frac{\rho_d D^2}{18\mu} \frac{12k_c}{\rho_d c_d D^2} = \frac{2}{3} \frac{k_c}{\mu c_d} = \frac{2}{3} \frac{k}{\mu c_c} \frac{c_c}{c_d}$$

where c_c is the specific heat of the fluid (if a gas, specific heat at constant pressure). Thus the relationship between response times becomes

$$\frac{\tau_V}{\tau_T} = \frac{2}{3} \frac{c_c}{c_d} \frac{1}{\text{Pr}}$$

where Pr is the Prandtl number. For gases, the Prandtl number is on the order of unity so the response times are of the same order of magnitude. For a liquid, the Prandtl number can be the order of 10^2 which means that velocity equilibrium is achieved much more rapidly than thermal equilibrium in a liquid.

Even though the above relations for the ratio of response times have been derived for low Reynolds number (Stokes flow), the ratio changes little for higher Reynolds numbers.

2.4   Stokes number

The *Stokes number* is a very important parameter in fluid-particle flows. The Stokes number related to the particle velocity is defined as

$$St_V = \frac{\tau_V}{\tau_F} \tag{2.31}$$

where τ_F is some time characteristic of the flow field. For example, the characteristic time for the flow through a venturi may be D_T/U where D_T is the throat diameter and U is the flow velocity. The Stokes number then becomes

$$St_V = \frac{\tau_V U}{D_T} \tag{2.32}$$

If $St_V << 1$, the response time of the particles is much less than the characteristic time associated with the flow field. Thus the particles will have ample time to respond to changes in flow velocity. Thus the particle and fluid velocities will be nearly equal (velocity equilibrium). On the other hand, if $St_V >> 1$, then the particle will have essentially no time to respond to the fluid velocity changes and the particle velocity will be little affected during its passage through the venturi.

An approximate relationship for the particle/fluid velocity ratio as a function of the Stokes number can be obtained from the "constant lag" solution. The velocity ratio is expressed as $\phi = v/u$ and is assumed to vary slowly with time. Substituting this variable into the particle motion equation, Equation 2.25, one has

$$\phi\frac{du}{dt} = \frac{u}{\tau_V}(1 - \phi) \qquad (2.33)$$

The carrier phase acceleration can be approximated by

$$\frac{du}{dt} \sim \frac{u}{\tau_F} \qquad (2.34)$$

which, when substituted into Equation 2.33 yields

$$\phi St_V \sim (1 - \phi). \qquad (2.35)$$

Finally solving for ϕ gives

$$\phi = \frac{u_d}{u_c} \sim \frac{1}{1 + St_V} \qquad (2.36)$$

One notes that as $St_V \to 0$, the particle velocity approaches the carrier phase velocity and as $St_V \to \infty$, the particle velocity approaches zero. This means that the particle velocity is unaffected by the fluid.

2.5 Dilute versus dense flows

A dilute dispersed phase flow is one in which the particle motion is controlled by the fluid forces (drag and lift). A dense flow, on the other hand, is one in which the particle motion is controlled by collisions. A qualitative estimate of the dilute or dense nature of the flow can be made by comparing the ratio of momentum response time of a particle to the time between collisions. Thus the flow can be considered *dilute* if

$$\frac{\tau_V}{\tau_C} < 1 \qquad (2.37)$$

where τ_C is the average time between particle-particle collisions because the particles have sufficient time to respond to the local fluid dynamic forces before the next collision. On the other hand, if

$$\frac{\tau_V}{\tau_C} > 1 \qquad (2.38)$$

then the particle has no time to respond to the fluid dynamic forces before the next collision and the flow is *dense*.

The time between collisions can be estimated from the classic equations for collision frequency. Consider a group of particles with uniform diameter D as shown in Figure 2.3 through which one particle is traveling with a relative velocity v_r with respect to the other particles. In a time δt, the one particle will

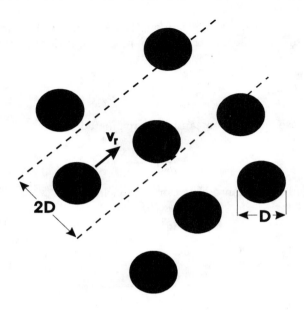

Figure 2.3: Particle-particle collisions.

intercept all the particles in tube with radius $2D$ and length $v_r\delta t$. The number of particles in this tube is

$$\delta N = n\pi D^2 v_r \delta t \tag{2.39}$$

where n is the number density of particles. Thus the collision frequency is

$$f_c = n\pi D^2 v_r \tag{2.40}$$

and the time between collisions is

$$\tau_C = \frac{1}{f_c} = \frac{1}{n\pi D^2 v_r} \tag{2.41}$$

Thus the time ratio τ_V/τ_C can be expressed as

$$\frac{\tau_V}{\tau_C} = \frac{n\pi \rho_d D^4 v_r}{18\mu} \tag{2.42}$$

However, the product of the number density and the mass of an individual particle is equal to the bulk density of the dispersed phase so the above equation can be written as

$$\frac{\tau_V}{\tau_C} = \frac{\bar{\rho}_d D v_r}{3\mu} \tag{2.43}$$

Thus the flow would be considered dilute if

$$\frac{\bar{\rho}_d D v_r}{3\mu} < 1 \tag{2.44}$$

The diameter of the particle corresponding to dilute flow would be

$$D < \frac{3\mu}{\bar{\rho}_d v_r} \tag{2.45}$$

The bulk density of the dispersed phase can be approximated by the product of the loading and the density of the carrier phase so the above expression can be rewritten as

$$D < \frac{3\mu}{Z\rho_c v_r} \tag{2.46}$$

The conditions responsible for a relative velocity between the particles could be the fluctuation of the particle motion due to turbulence of the carrier fluid. Thus the relative velocity should be related to the standard deviation of the particle fluctuation velocity, σ. A more detailed analysis by Sommerfeld (1994) based on a normal distribution of the particle number density with particle velocity shows

$$D < \frac{1.33\mu}{Z\rho_c \sigma} \tag{2.47}$$

The standard deviation of the particle fluctuation velocity will be the order of the root mean square of the turbulence fluctuations of the carrier phase.

The variation of the particle diameter with loading for dilute flow in air at standard conditions is shown in Figure 2.4. One notes that the maximum particle size for a dilute flow at a loading of unity would be about 200 microns.

2.6 Phase coupling

An important concept in the analysis of multiphase flows is coupling. If the flow of one phase affects the other while there is no reverse effect, the flow is said to be *one-way coupled*. If there is a mutual effect between the flows of both phases, then the flow is *two-way coupled*.

A schematic diagram of coupling is shown in Figure 2.5. The carrier phase is described by the density, temperature, pressure and velocity field. Of course, it may be important to include the concentrations of the gaseous species in a gas phase. The particle or droplet phase is described by concentration, size, temperature and velocity field. Coupling can take place through mass, momentum and energy transfer between phases. Mass coupling is the addition of mass through evaporation or the removal of mass from the carrier stream by condensation. Momentum coupling is the result of the drag force on the dispersed and continuous phase. Momentum coupling can also occur with momentum addition or depletion due to mass transfer. Energy coupling occurs through heat transfer between phases. Thermal and kinetic energy can also be transferred between phases owing to mass transfer. Obviously, analyses based on one-way coupling are straight forward. One must estimate through experience or parameter magnitude if two-way coupling is necessary.

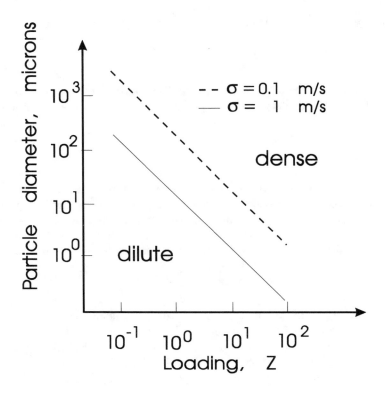

Figure 2.4: Dilute-dense flow regions.

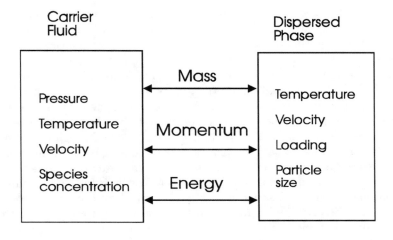

Figure 2.5: Schematic diagram of coupling effects.

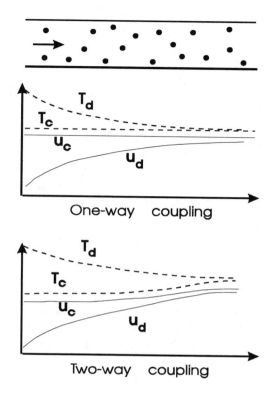

Figure 2.6: Droplets evaporating in a duct.

The best way to illustrate coupling is through an example. Assume that hot particles were injected into a cool gas flowing in a pipe as shown in Figure 2.6. Also assume that the void fraction is near unity. One can calculate the trajectory and the corresponding thermal history of the particles by using the local velocity and temperature of the gas. Thus particle temperature will decrease toward the gas temperature while the particle velocity will increase toward the gas velocity. One-way coupling implies that the presence of the particles does not affect the gas flow field while the gas flow field is responsible for the change in particle temperature and velocity.

If the effect of the particles on the gas were included (two-way coupling), the temperature of the gas would increase and the gas density would decrease. This, in turn, would lead to an increased gas velocity to satisfy mass conservation. Thus the particle cooling rate would be reduced (smaller temperature difference) and the particles would be accelerated to a higher velocity. The acceleration of the particles together with the increased gas velocity would lead to a more negative pressure gradient. The coupling parameters provide a measure of the importance of coupling effects (Crowe, 1991).

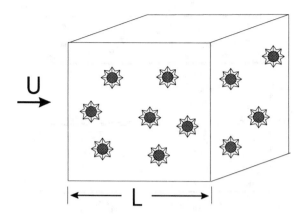

Figure 2.7: Droplets evaporating in a control volume.

2.6.1   Mass coupling

Suppose there are n droplets per unit volume in a box with side L as shown in Figure 2.7. Each droplet is evaporating at a rate $\dot{m}$. Thus the mass generated by the dispersed phase per unit time due to evaporation is

$$\dot{\mathsf{M}}_d = nL^3\dot{m} \tag{2.48}$$

The mass flux of the continuous phase through this volume is

$$\dot{M}_c \sim \bar{\rho}_c uL^2 \tag{2.49}$$

A *mass coupling parameter* is defined as[2]

$$\Pi_{mass} = \frac{\dot{\mathsf{M}}_d}{\dot{M}_c} \tag{2.50}$$

If $\Pi_{mass} \ll 1$, then the effect of mass addition to the continuous phase would be insignificant and mass coupling could be treated as one-way coupling. This ratio can be expressed as

$$\frac{\dot{\mathsf{M}}_d}{\dot{M}_c} \sim \frac{\bar{\rho}_d}{\bar{\rho}_c}\frac{L\dot{m}}{um} \tag{2.51}$$

The ratio $\frac{\dot{m}}{m}$ scales as the reciprocal of a characteristic evaporation, burning or condensation time, τ_m, so the ratio in Equation 2.51 can be rewritten as

$$\Pi_{mass} \sim C\frac{L}{u\tau_m} \tag{2.52}$$

---

[2]Note that $\dot{\mathsf{M}}_d$ is the mass generated by the dispersed phase and not $\dot{M}_d$ which is the mass flow of the dispersed phase.

Taking L as some characteristic dimension of the system, this ratio can be used to assess the importance of mass coupling on the continuous phase flow. Notice that the ratio $\tau_m u/L$ can be regarded as the ratio of a time associated with mass transfer to a time characteristic of the flow. Thus the ratio can be thought of as the Stokes number associated with mass transfer

$$St_{mass} = \frac{\tau_m u}{L} \qquad (2.53)$$

so the mass transfer parameter can be expressed as

$$\Pi_{mass} = \frac{C}{St_{mass}} \qquad (2.54)$$

If the velocity of both phases is comparable, then the loading is a measure of the concentration and the mass coupling parameter can be evaluated as

$$\Pi_{mass} = \frac{Z}{St_{mass}}$$

Thus if the dispersed phase mass concentration (or loading) and the mass exchange rate of the dispersed phase are low (large τ_m and St_{mass}), then mass coupling can probably be neglected.

2.6.2 Momentum coupling

The importance of momentum coupling can be assessed by comparing the drag force due to the dispersed phase with the momentum flux of the continuous phase. A *momentum coupling parameter* can be defined as

$$\Pi_{mom} = \frac{D_d}{Mom_c} \qquad (2.55)$$

where D_d is the drag force due to the particles in the volume and Mom_c is the momentum flux through the volume. The drag associated with droplets in volume with side L is

$$D_d = nL^3 3\pi\mu D(u - v) \qquad (2.56)$$

based on Stokes drag. The momentum flux of the continuous phase is given by

$$Mom_c = \bar{\rho}_c u^2 L^2 \qquad (2.57)$$

The momentum coupling parameter can be expressed as

$$\Pi_{mom} = \frac{nmL}{\bar{\rho}_c u \tau_V}\left(1 - \frac{v}{u}\right) \qquad (2.58)$$

where m is the mass of an individual element of the dispersed phase. However, the product nm is the bulk phase density of the dispersed phase, so the importance of momentum coupling can be assessed by the parameter

$$\Pi_{mom} \sim C\frac{L}{u\tau_V}(1-\frac{v}{u}) \qquad (2.59)$$

The velocity factor is not included since it depends on the dynamics of the flow field. As with mass transfer, the ratio $\tau_V u/L$ is the ratio of the time associated with momentum transfer to a time characteristic of the flow or the Stokes number for momentum transfer.

$$St_{mom} = \frac{\tau_V u}{L} \qquad (2.60)$$

The momentum coupling parameter can be expressed as

$$\Pi_{mom} = \frac{C}{St_{mom}}(1-\frac{v}{u}) \qquad (2.61)$$

As the Stokes number approaches zero, the velocity of the dispersed phase approaches that of the continuous phase so the above equation develops an indeterminacy $(0/0)$. However, using Equation 2.36 for the velocity ratio, the momentum coupling parameter becomes

$$\Pi_{mom} = \frac{C}{1+St_{mom}} \qquad (2.62)$$

which shows the correct limit as the Stokes number approaches zero. Momentum coupling effects become less important for small concentrations (loadings) and large Stokes numbers.

Example: Coal particles, 100 μm in diameter and with a material density of 1200 kg/m³, flow in an air stream into a venturi section with a 2 cm diameter throat. The throat velocity is 10 m/s, the loading is 1 and the air viscosity is 1.8×10^{-5} Ns/m². Evaluate the magnitude of the momentum coupling parameter. Assume that the loading is a measure of the concentration.

Solution: The velocity response time of the coal particles is

$$\tau_V = \frac{\rho_d d^2}{18\mu} = 0.037\,\text{sec}$$

The Stokes number is

$$St_{mom} = \frac{\tau_V U}{L} = 18.5$$

The momentum coupling parameter is

$$\Pi_{mom} = \frac{Z}{St_{mom}+1} = 0.051$$

One would estimate that the momentum coupling effects are unimportant.

2.6.3 Energy coupling

Energy coupling follows the same model as used for mass and momentum coupling. The significance of energy coupling can be assessed by comparing the heat transfer to (or from) the dispersed phase and the energy flux of the continuous phase. The *energy coupling parameter* is defined as

$$\Pi_{ener} = \frac{\dot{Q}_d}{\dot{E}_c} \tag{2.63}$$

Obviously if $\Pi_{ener} \ll 1$, then energy coupling is unimportant.

The heat transfer associated with the dispersed phase elements in volume L^3 is

$$\dot{H}_d = nL^3\pi Nuk_cD(T_d - T_c) \tag{2.64}$$

where Nu is the Nusselt number and k_c is the thermal conductivity of the continuous phase. The energy flux of the continuous phase is

$$\dot{E}_c = \bar{\rho}_c uc_pT_cL^2 \tag{2.65}$$

where c_p is the specific heat of the continuous phase. The thermal coupling parameter can be expressed as

$$\Pi_{ener} \sim \frac{C}{St_T}(1-\frac{T_c}{T_d}) \tag{2.66}$$

where St_T is the Stokes number based on the thermal response time ($St_T = u\tau_T/L$). As with the momentum coupling parameter, there is an indeterminacy as the Stokes number approaches zero. Using the same arguments as before, the energy coupling parameter can be expressed as[3]

$$\Pi_{ener} = \frac{C}{St_T + 1} \tag{2.67}$$

For gaseous flows, the momentum and thermal response times are of the same order so

$$\Pi_{mom} \sim \Pi_{ener}$$

and justification for one-way coupling for momentum usually justifies one-way coupling for energy transfer.

If most of the energy transfer in a system is associated with latent heat in the dispersed phase, there is another form of the energy coupling parameter which may be appropriate. The energy associated with change of phase in a volume L^3 is

$$\dot{E}_d = nL^3\dot{m}h_L \tag{2.68}$$

---

[3] Another factor should be included in the energy coupling parameter; namely, the ratio of the specific heat of the particle to the specific heat of the continuous phase, c_d/c_c. In general this ratio is of the order unity.

where h_L is the latent heat. Comparing this energy with the energy flux of the continuous phase results in the following energy coupling parameter

$$\Pi_L \sim C \frac{L}{u\tau_m} \frac{h_L}{c_p T_c} \tag{2.69}$$

which is the mass coupling parameter with an additional factor, $h_L/c_p T_c$. This factor can be large causing energy transfer due to phase change to be important even though mass transfer itself is unimportant. In many spray problems, the energy coupling due to phase change may be the only two-way coupling which must be included in developing an analysis for a gas-droplet flow.

2.7   Properties of an equilibrium mixture

A limiting case in the flow of a dispersed phase mixture is when the two phases are in dynamic and thermal equilibrium. This is the situation when the Stokes number approaches zero. In this situation the mixture can be regarded as the flow of a single phase with modified properties.

When the flow is in dynamic equilibrium, the loading and mass ratio have the same value.

$$z = \frac{\bar{\rho}_d v}{\bar{\rho}_c u} = C \tag{2.70}$$

The mixture density can then be written as

$$\rho_m = \bar{\rho}_d + \bar{\rho}_c = \bar{\rho}_c(1 + z) = \rho_c \alpha_c(1 + z) \tag{2.71}$$

If the void fraction (α_c) were near unity, the mixture could be treated as a gas with a modified gas constant.

The enthalpy of the equilibrium mixture would be

$$h_m = (\bar{\rho}_c c_c + \bar{\rho}_d c_d)T = \bar{\rho}_c c_c(1 + z\frac{c_d}{c_c})T \tag{2.72}$$

where c_c is the specific heat of the carrier phase (specific heat at constant pressure for a gas) and c_d is the specific heat of the dispersed phase material. The specific heat of the mixture would be

$$c_m = c_c\left(\frac{1 + z\frac{c_d}{c_c}}{1 + z}\right) \tag{2.73}$$

These relationships are useful in establishing the limiting cases of dispersed phase flows as the Stokes number approaches zero.

2.8   Summary

The discrete nature of a dispersed phase flow requires the use of volume average properties such as bulk density. In addition to the parameters that characterize a single-phase flow, two additional nondimensional parameters are important in fluid-particle flows: Stokes number and concentration (or loading). Dilute and dense refer to the importance of particle-particle interaction. Coupling refers to the property exchange between phases where a one-way coupled flow describes the condition where the carrier phase affects the dispersed phase but the dispersed phase has no influence on the continuous phase. The importance of coupling is quantified through the coupling parameters. A flow with a Stokes number approaching zero is characterized by velocity and thermal equilibrium and can be regarded as a single-phase flow with modified properties.

Exercises

2.1. A sand-water slurry has a solids volume fraction of 40%. Find the average ratio of the average interparticle spacing to the particle diameter.

2.2. The bulk density of water droplets in air at standard conditions (25°C and 1 atm) is 0.5 kg/m^3. The droplets are monodispersed (uniform size) with a diameter of 100 μm. Find the number density, void fraction and the concentration.

2.3. Particles of uniform size are stacked in a packed bed such that each particle sits in the "pocket" of the four particles adjacent to it. This is a "body-centered" cubic configuration and represents the maximum void fraction. Find the solids volume fraction. Also evaluate the volume fraction for the "lattice" configuration.

2.4. Droplets are released in a hot air stream and evaporate as they are convected with the flow as shown. The mass coupling parameter is small while the latent heat coupling parameter is large. Sketch the variation of the gas temperature, gas velocity, gas density and droplet velocity for both one-way and two-way coupling in the duct.

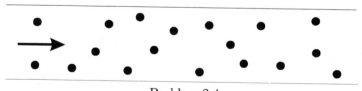

Problem 2.4

2.5. A pneumatic flow consists of particles conveyed by an air stream in a circular duct with diameter D_p. Ten percent of the particle mass impinges on the wall per diameter of length. The rebound velocity is 1/2 the impingement velocity. The gas velocity is u and the average particle velocity is v. The loading is Z. The skin friction coefficient is C_f. Find an expression for the Darcy-Weisbach friction factor as a function of C_f, Z, v and u. Assume a void fraction of unity.

2.6. A gas-particle flow accelerates through a venturi as shown. Sketch the pressure distribution through the venturi for a small momentum and large momentum coupling parameter and explain the rationale for your choice.

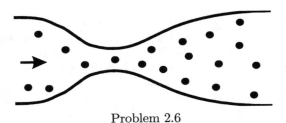

Problem 2.6

2.7. Water droplets are evaporating in an air flow at standard pressure and temperature. The air flow rate is 1 kg/s and the water flow rate is 0.01 kg/s. The droplet velocity is 30 m/s and the droplet diameter is 100 microns. The pipe diameter is 5 cm. The evaporation time for a water droplet is given by

$$\tau_e = \frac{D^2}{\lambda}$$

where D is the droplet diameter and λ is the evaporation constant equal to 0.02 cm^2/s for water droplets. Evaluate the mass coupling parameter and the energy coupling due to mass transfer (latent heat). Is mass coupling or energy coupling important?

2.8. Ice particles flow in a duct with air at 60°F. The ice particles are not completely melted at the end of the duct. The ice particles and air are in dynamic equilibrium. Sketch how the air temperature, particle temperature, gas velocity and pressure would change along the duct assuming one-way and two-way coupling. Provide a physical argument explaining your results.

Chapter 3

Size Distribution

Droplet or particle size can be a very important parameter governing the flow of a dispersed two-phase mixture. For this reason, it is useful to have a basic knowledge of the statistical parameters relating to particle size distributions. For spherical particles or droplets, a measure of the size is the diameter. For nonspherical particles, an equivalent diameter must be selected to quantify the size. (See Chapter 9)

The most general definition of the spread of the particle size distribution is monodisperse or polydisperse. A monodisperse distribution is one in which the particles are close to a single size whereas polydisperse suggests a wide range of particle sizes. An approximate definition of a monodisperse distribution is one for which the standard deviation is less than 10% of the mean particle diameter.

3.1 Discrete size distributions

Let us begin by assuming that one has the task of determining the size distribution of a sample of particles. Assume that the sizes of many particles in the sample have been measured by some technique, such as photography. One would choose size intervals, ΔD, which would be large enough to contain many particles yet small enough to obtain sufficient detail. The representative size for the interval could be the diameter corresponding to the mid-point of the interval. The number of particles in each size interval would be counted, recorded, and divided by the total number of particles in the sample. The results would be plotted in the form of a histogram (bar chart) shown in Figure 3.1. This is identified as the discrete number frequency distribution for the particle size.[1] The ordinate corresponding to each size interval is known as the *number frequency*, $\tilde{f}_n(D_i)$. The sum of each bar is unity since the number in each size category has been divided by the total number in the sample. The distribution

---

[1] This frequency distribution is often referred to as the probability density function or "pdf".

has been normalized; that is,

$$\sum_{i=1}^{N} \tilde{f}_n(D_i) = 1 \tag{3.1}$$

where N is the total number of intervals.

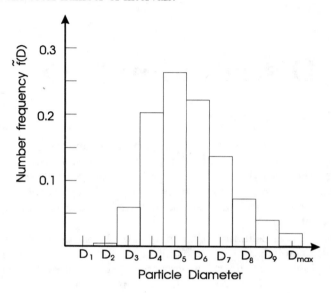

Figure 3.1: Discrete number frequency distribution.

The *number-average* particle diameter in the distribution is obtained from

$$\overline{D}_n = \sum_{i=1}^{N} D_i \tilde{f}_n(D_i) \tag{3.2}$$

and the *number variance* is defined as [2]

$$\sigma_n^2 = \sum_{i=1}^{N} (D_i - \overline{D_n})^2 \tilde{f}_n(D_i) \tag{3.3}$$

The variance is a measure of the spread of the distribution; a large variance implying a wide distribution of sizes. An alternate expression for the number variance is

$$\sigma_n^2 = \sum_{i=1}^{N} D_i^2 \tilde{f}_n(D_i) - \overline{D}_n^2 \tag{3.4}$$

---

[2] The sum should be multiplied by the factor $N/(N-1)$ to account for the one degree of freedom removed by using the average diameter. However for $N >> 1$ we have $N/(N-1) \sim 1$.

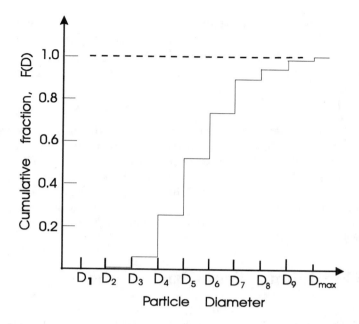

Figure 3.2: Discrete cumulative number distribution.

which is sometimes more convenient for calculations. The *standard deviation* is the square root of the variance.

$$\sigma_n = \sqrt{\sigma_n^2}.$$

Another approach to describing size distribution is to use the particle or droplet mass (or volume) in lieu of the number as the dependent variable. Thus the mass of each particle would be obtained, or inferred, from measurement and the fraction of mass associated with each size interval would be used to construct the distribution. This is known as the discrete mass frequency distribution and identified as $\tilde{f}_m(D_i)$. With this distribution one can calculate the *mass-average* particle diameter and mass variance,

$$\overline{D}_m = \sum_{i=1}^{N} D_i \tilde{f}_m(D_i) \tag{3.5}$$

$$\sigma_m^2 = \sum_{i=1}^{N}(D_i - \overline{D})^2 \tilde{f}_m(D_i) = \sum_{i=1}^{N} D_i^2 \tilde{f}_m(D_i) - \bar{D}_m^2 \tag{3.6}$$

A very large number of particles or droplets have to be counted to achieve a reasonably smooth frequency distribution function. This is feasible with modern experimental techniques.

Another commonly used method to quantify particle size is the cumulative distribution which is the sum of the frequency distribution. The cumulative number distribution associated with size D_k is

$$\tilde{F}_n(D_k) = \sum_{i=1}^{D_k} \tilde{f}_n(D_i) \tag{3.7}$$

The cumulative number distribution corresponding to the number frequency distribution shown in Figure 3.1 is illustrated in Figure 3.2. The value of $\tilde{F}$ is the fraction of particles with sizes less than or equal to D_k. Thus $\tilde{F}(100\mu) = 0.4$ means that 40% of the distribution has particles of 100 microns or less. Also, by definition, the value of cumulative distribution is unity for the largest particle size, provided the frequency distribution has been normalized. Both cumulative number and mass distributions can be generated from the corresponding frequency distributions.

3.2 Continuous size distributions

If the size intervals were made progressively smaller then, in the limit, as ΔD approaches zero, the continuous frequency function would be obtained.

$$f_n(D) = \lim_{\Delta D \to 0} \frac{\tilde{f}_n(D)}{\Delta D} \tag{3.8}$$

The number fraction of particles with diameters between D and $D+dD$ is given by the differential quantity $f_n(D)dD$. The variation of the frequency distribution with particle size is a continuous function as shown in Figure 3.3. Similarly, the differential quantity $f_m(D)dD$ is the fraction of particle mass associated with sizes between D and $D + dD$.

If the distribution has been normalized, then the area under the frequency distribution curve is unity,

$$\int_0^{D_{max}} f(D)dD = 1 \tag{3.9}$$

where D_{max} is the maximum particle size.

The continuous cumulative distribution is obtained from the integral of the continuous frequency distribution,

$$F_n(D) = \int_0^D f_n(\lambda)d\lambda \tag{3.10}$$

and illustrated as the S-shaped curve in Figure 3.4. By definition, the cumulative distribution approaches unity as the particle size approaches the maximum size.

The equivalent continuous frequency and cumulative distributions for mass (or volume) fraction are obvious.

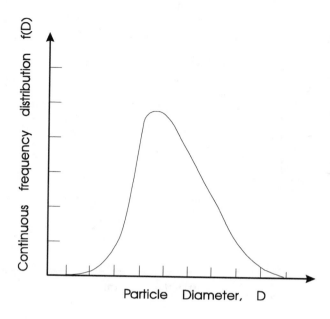

Figure 3.3: Continuous frequency size distribution.

In general, size data are not available as a continuous distribution. It is common practice to consider the data obtained for a discrete distributions as values on a curve for a continuous distribution and to proceed accordingly to evaluate the appropriate statistical parameters.

3.3 Statistical parameters

There are several parameters used to quantify distribution functions. Those used most commonly for dispersed phase flows are presented below.

3.3.1 Mode

The mode corresponds to the point where the frequency function is a maximum as shown in Figure 3.3. The modes of the number and mass frequency distributions for a given sample will not be the same. A distribution which has two local maxima is referred to as a "bimodal" distribution.

3.3.2 Mean

The mean of a continuous distribution is analogous to the average of a discrete distribution. The mean is calculated from the frequency distribution by

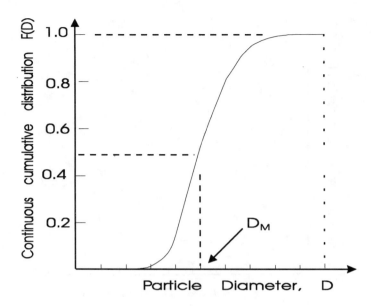

Figure 3.4: Continuous cumulative size distribution.

evaluating the integral,

$$\mu = \int_0^{D_{max}} D f(D) dD \tag{3.11}$$

Of course there is a number mean (μ_n) or mass mean (μ_m) size depending on the frequency function used.

3.3.3   Variance

The variance of the distribution is calculated from

$$\sigma^2 = \int_0^{D_{max}} (D - \mu)^2 f(D) dD \tag{3.12}$$

or by the equivalent expression,

$$\sigma^2 = \int_0^{D_{max}} D^2 f(D) dD - \mu^2 \tag{3.13}$$

Once again the variance can be based on the number or mass distribution for particle size. As previously mentioned, the variance is a measure of the spread of the distribution. The standard deviation is the square root of the variance and has dimensions of diameter.

A distribution can be classified as monodisperse if

$$\frac{\sigma}{\mu} < 0.1.$$

3.3.4 Median

The median diameter (D_M) corresponds to the diameter for which the cumulative distribution is 0.5. The number median diameter (D_{nM}) is shown in Figure 3.4. The corresponding mass median diameter (D_{mM}) is determined from the cumulative mass distribution function.

3.3.5 Sauter mean diameter

The Sauter mean diameter (SMD) is encountered frequently in the spray and atomization literature. It is defined as

$$D_{32} = \frac{\int_0^{D_{max}} D^3 f_n(D)dD}{\int_0^{D_{max}} D^2 f_n(D)dD} \tag{3.14}$$

The Sauter mean diameter can be thought of as the ratio of the particle volume to surface area in a distribution which may have physical significance in some applications.

3.4 Frequently used size distributions

There are two size distribution functions that are frequently used to correlate particle or droplet size measurements. The characteristics of these distributions are discussed below.

3.4.1 Log-normal distribution

The log-normal distribution is frequently used to represent the size of solid particles. The log-normal distribution derives from the normal or Gaussian distribution which is defined as

$$f(x) = \frac{1}{\sqrt{2\pi}\sigma} \exp\left[-\frac{1}{2}\left(\frac{x-\mu}{\sigma}\right)^2\right] \tag{3.15}$$

where σ is the standard deviation and μ is the mean. This distribution is the well-known "bell-shaped" curve, shown in Figure 3.5a, which is symmetric about the mean and trails off to plus and minus infinity. The distribution has been normalized so the area under the curve is unity; that is,

$$\int_0^{\infty} f(x)dx = 1. \tag{3.16}$$

The log-normal distribution is obtained from the normal distribution by replacing the independent variable with the logarithm of the particle diameter. The number of particles with diameters between D and $D+dD$ is

$$f(D)dD = \frac{1}{\sqrt{2\pi}\sigma_o} \exp\left[-\frac{1}{2}\left(\frac{\ln D - \mu_o}{\sigma_o}\right)^2\right]\frac{dD}{D} \tag{3.17}$$

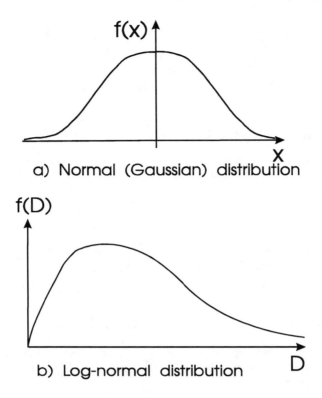

Figure 3.5: Normal (Gaussian) and log-normal distributions.

where, now, σ_o and μ_o are the standard deviation and the mean of the log normal distribution. The shape of the log-normal distribution is shown in Figure 3.5b. The limits of the log-normal distribution range from zero to infinity.[3]

The corresponding cumulative number distribution is given by

$$F(D) = \frac{1}{\sqrt{2\pi}\sigma_o} \int_0^D \exp\left[-\frac{1}{2}\left(\frac{\ln D - \mu_o}{\sigma_o}\right)^2\right]\frac{dD}{D} \qquad (3.18)$$

This integral can be expressed in terms of the error function

$$F(D) = \frac{1}{2}\left[1 + erf\left(\frac{\ln D - \mu_o}{\sigma_o}\right)\right] \qquad (3.19)$$

where the error function is defined as

$$erf(t) = \frac{2}{\sqrt{\pi}} \int_0^t e^{-\frac{\lambda^2}{2}} d\lambda. \qquad (3.20)$$

---

[3] The independent variable $\ln D$ approaches $-\infty$ as D approaches zero.

The number median diameter is defined as the particle diameter for which the cumulative distribution is 0.5 which corresponds to an error function of zero. The value of the error function is zero when its argument is zero. Thus, the mean of the log-normal distribution is the logarithm of the number median diameter, $\ln D_{nM}$, so the log-normal number frequency distribution can be conveniently expressed as

$$f_n(D) = \frac{1}{\sqrt{2\pi}\sigma_o} \exp\left[-\frac{1}{2}\left(\frac{\ln D - \ln D_{nM}}{\sigma_o}\right)^2\right]\frac{1}{D} \tag{3.21}$$

The corresponding mean for the log-normal mass frequency distribution is the logarithm of the mass median diameter, $\ln D_{mM}$, so

$$f_m(D) = \frac{1}{\sqrt{2\pi}\sigma_o} \exp\left[-\frac{1}{2}\left(\frac{\ln D - \ln D_{mM}}{\sigma_o}\right)^2\right]\frac{1}{D}$$

The variance of the log-normal distribution can be found by plotting the cumulative distribution on log-probability paper ($\ln D$ vs. probability) as shown in Figure 3.6. The distribution plots as a straight line on these coordinates. One can also obtain σ_o from the relation

$$\sigma_o = \ln\frac{D_{84\%}}{D_M} \tag{3.22}$$

where $D_{84\%}$ is the diameter corresponding to the 84th percentile on the log-probability curve. This relation derives from the fact that the value of the cumulative distribution (Equation 3.19) is 0.84 when the argument of the error function is unity. The value for σ_o is the same for both the number and mass distributions.

A very useful relationship for manipulation of the log-normal distribution is

$$D_M^k e^{\sigma_o^2 k^2/2} = \frac{1}{\sqrt{2\pi}\sigma_o} \int_0^\infty D^k \exp\left[-\frac{1}{2}\left(\frac{\ln D - \ln D_M}{\sigma_o}\right)^2\right]\frac{dD}{D} \tag{3.23}$$

or

$$D_M^k e^{\sigma_o^2 k^2/2} = \int_0^\infty D^k f(D) dD \tag{3.24}$$

An example illustrating the use of this relationship is calculating the number mean of the distribution which is defined as

$$\mu_n = \int_0^\infty D f_n(D) dD \tag{3.25}$$

For this example, k in equation 3.23 is unity so the number mean diameter is

$$\mu_n = D_{nM} e^{\sigma_o^2/2} \tag{3.26}$$

The number median diameter is related to the mass median diameter through the relation

$$D_{mM} = D_{nM} e^{3\sigma_o^2} \tag{3.27}$$

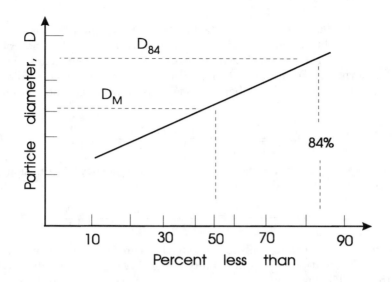

Figure 3.6: Log-normal distribution on log-probability coordinates.

With information on the median diameter and σ_o, one can easily evaluate all the statistical parameters corresponding to the log-normal distribution by applying Equations (3.23) and (3.27).

Example: The mass median diameter of a log-normal distribution with $\sigma_o = 0.3$ is 200 microns. Find the standard deviation of the mass distribution.

Solution: From Equation 3.13, the variance is given by

$$\sigma^2 = \int D^2 f(D)dD - \mu^2$$

The value of the mass mean is

$$\mu_m = D_{mM}e^{\sigma_0^2/2}$$

The mass variance is

$$\sigma_m^2 = D_{nM}^2(e^{2\sigma_0^2} - e^{\sigma_0^2})$$

so the standard deviation is

$$\sigma_m = 200\sqrt{1.197 - 1.094} = 64.2\mu m$$

An off-shoot of the log-normal distribution is the log-normal upper limit which is designed to set a maximum particle diameter as the upper limit of the distribution. This is accomplished by replacing $\ln D$ with the independent variable

$$\ln\left[\frac{aD}{D_{max} - D}\right]$$

where D_{max} is the maximum diameter and a is a constant to be selected which normalizes the distribution. Obviously, the modified independent variable approaches $-\infty$ as D approaches zero and $+\infty$ as D approaches D_{max}.

3.4.2   Rosin-Rammler distribution

The Rosin-Rammler distribution (Mugele & Evans, 1951) is frequently used for representing droplet size distributions in sprays. It is expressed in terms of the cumulative mass distribution in the form

$$F_m(D) = 1 - \exp\left[-\left(\frac{D}{\delta}\right)^n\right] \qquad (3.28)$$

where δ and n are two empirical constants. One notes that $F_M(0) = 0$ and $F_M(\infty) = 1$.

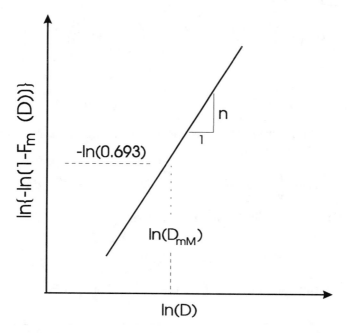

Figure 3.7: Rosin-Rammler distribution.

The empirical constants can be determined by plotting the cumulative distribution on log-log coordinates. Taking the logarithm of Equation 3.28 twice gives

$$\ln[-\ln(1 - F_m(D))] = n \ln D - n \ln \delta \qquad (3.29)$$

Thus the slope of the line obtained by plotting $-\ln[1 - F_m(D)]$ versus diameter on log-log paper as shown in Figure 3.7 provides n. The parameter δ can be obtained using n and the mass median diameter with

$$\delta = \frac{D_{mM}}{0.693^{\frac{1}{n}}} \tag{3.30}$$

The mass frequency distribution is obtained by taking the derivative of the cumulative distribution,

$$f_m(D) = \frac{dF_m}{dD} = e^{-(\frac{D}{\delta})^n}\frac{n}{\delta}\left(\frac{D}{\delta}\right)^{n-1} \tag{3.31}$$

A very useful function for evaluating the statistical parameters for the Rosin-Rammler distribution is the gamma function which is defined as

$$\Gamma(t) = \int_0^\infty e^{-\lambda}\lambda^{t-1}d\lambda \tag{3.32}$$

It can be shown that

$$\int_0^\infty nD^\alpha\left(\frac{D}{\delta}\right)^{n-1}\exp\left[-\left(\frac{D}{\delta}\right)^n\right]d\left(\frac{D}{\delta}\right) = \delta^\alpha\Gamma\left(\frac{\alpha}{n}+1\right) \tag{3.33}$$

so

$$\delta^\alpha\Gamma\left(\frac{\alpha}{n}+1\right) = \int_0^\infty D^\alpha f_m(D)dD \tag{3.34}$$

An example application of this equation is evaluating the mass mean diameter which is given by

$$\mu_m = \int_0^\infty D f_m(D)dD = \int_0^\infty D e^{-(\frac{D}{\delta})^n}\frac{n}{\delta}\left(\frac{D}{\delta}\right)^{n-1}dD \tag{3.35}$$

Setting α equal to unity in Equation 3.33 gives

$$\mu_m = \delta\Gamma\left(\frac{1}{n}+1\right) \tag{3.36}$$

Having n, one can obtain the value for the gamma function from a mathematics table.

Example: A Rosin-Rammler distribution has a mass median diameter of 120 microns with a n-value of 2.0. Find the mass mean of the distribution.

Solution: First we must evaluate δ from

$$\delta = \frac{D_{mM}}{0.693^{\frac{1}{n}}} = \frac{120}{0.693^{\frac{1}{2}}} = 144\mu m$$

The mass mean is

$$\mu_m = \delta\Gamma(\frac{1}{n} + 1) = 144\Gamma(\frac{3}{2}) = 128\mu m$$

The Rosin-Rammler distribution is a special case of the more general Nukiyama-Tanasawa distributions, descriptions of which are available through other sources.

3.4.3 Log-hyperbolic distribution

In some studies with sprays it has been shown that neither the log-normal nor the Rosin-Rammler distributions fit the data sufficiently well. Two droplet size distributions fitted with a Rosin-Rammler distribution may have the same parameters but the actual distribution may be quite different.

One of the shortcomings of both the log-normal and Rosin-Rammler distributions is the representation of the tails of the distribution (values of the frequency function for large and small values of the independent variable). Another distribution which is a better fit for the tails of the distributions is the log-hyperbolic distribution proposed by Barndorff-Nielsen (1977). The form of the frequency function is

$$f = A\exp\left[-\alpha\sqrt{\delta^2 + (x - \mu)^2} + \beta(x - \mu)\right] \qquad (3.37)$$

where α, β, δ and μ are fitting parameters and x is the random variable which, in this case, is the logarithm of the particle diameter. The coefficient A is a normalization factor which is related to the parameters by

$$A = \frac{\sqrt{\alpha^2 - \beta^2}}{2\alpha\delta K_1(\delta\sqrt{\alpha^2 - \beta^2})}$$

where K_1 is the third-order Bessel function of the third kind.

The logarithm of the frequency function is

$$\ln f = \ln A - \alpha\sqrt{\delta^2 + (x - \mu)^2} + \beta(x - \mu) \qquad (3.38)$$

which is the equation of a hyperbola. For $(x - \mu)/\delta < 0$ the slope of the asymptote is $\alpha + \beta$, while for $(x - \mu)/\delta > 0$ the slope of the asymptote is $-\alpha + \beta$. Thus the logarithm of the frequency data can be plotted versus $\ln D$ as shown in Figure 3.8 and the slopes of the asymptotes can be measured to find α and β. The parameter μ is the mode of the distribution. The value for δ must be obtained from some fitting procedure.

One of the problems with the log-hyperbolic distribution is the stability of the parameters. If the tails of the distribution are not long enough to yield accurate values of slopes of the asymptotes, there can be an instability in the parameters; that is, the parameters are not unique so the same distribution can

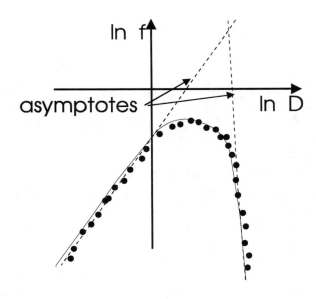

Figure 3.8: Log-hyperbolic distribution.

be matched by various combinations of the parameters. A method to circumvent this problem has been proposed by Xu et al. (1993). The data are fitted with a hyperbola which is rotated at an angle θ with respect to the y-axis as shown in Figure 3.9. The inclination angle θ, the slope of the asymptotes with respect the inclination angle (which will now be the same) and the mode are sufficient to define the distribution. This is referred to as the three-parameter log-hyperbolic distribution.

3.5   Summary

The distribution of particle or droplet sizes are usually formulated in terms of particle number or particle mass and the parameters used to describe the distribution are the mean, median, mode and variance. Example distribution functions are the traditional log-normal and Rosin-Rammler distributions and the more recently developed log-hyperbolic distribution. There are numerous other distributions which have been developed for specific purposes. The reader is referred to the abundant literature in this area.

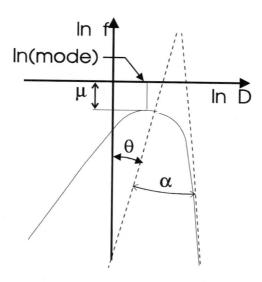

Figure 3.9: Log-hyperbolic distribution fitted with a rotated hyperbola.

Exercises

3.1. A measurement of a particle size distribution indicates that the distribution is log-normal with a mass median diameter of 60 μm and a slope (σ_0) of 0.3. Find the following parameters:

1. number median diameter

2. Sauter mean diameter

3. mass mean diameter

4. mode of mass distribution.

3.2. Data taken for a size distribution of droplets produced by an atomizer indicate that the data fit the Rosin-Rammler distribution with $n=1.8$ and a mass median diameter of 120 μm. Evaluate the following statistics:

1. the parameter δ for the distribution

2. the mass mean diameter and the mass variance for the distribution

3. the Sauter mean diameter.

3.3. The continuous number frequency distribution is uniform as shown. The maximum diameter is 100 μm.

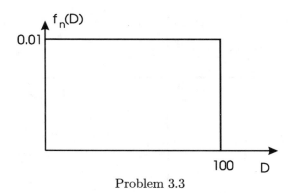

Problem 3.3

1. Find the mean, standard deviation and the Sauter mean diameter.

2. Derive an expression for the cumulative distribution and the number mean diameter.

3. Derive an expression for the mass frequency distribution and find the mode, mean and standard deviation.

4. Obtain an expression for the mass cumulative distribution and find the mass median diameter.

3.4. The following data were obtained for the cumulative mass distribution of a particle sample.

% less than	D (microns)
94	50
90	40
76	32
60	25.4
42	20
26	16
14	12.7
7	10
3	8

Carry out the following procedures.

1. Plot the data on log-probability paper and find D_{mM} and σ_0.

2. Calculate D_{nM} and SMD (Sauter mean diameter).

3. Find the standard deviation for both the mass and number distribution. Is the distribution monodisperse?

3.5. The data for a Rosin-Rammler distribution show that $n = 2$ and the mass median diameter is 20 microns.

1. Show that

$$\int_0^\infty nD^\alpha (\frac{D}{\delta})^{n-1} \exp(-\frac{D}{\delta})^n d(\frac{D}{\delta}) = \delta^\alpha \Gamma(\frac{\alpha}{n} + 1)$$

2. Using the expression generated above find an expression for the mass variance of the Rosin-Rammler distribution in the form $\sigma_m^2 = \delta^2 f[\Gamma(n)]$ and evaluate for the distribution parameters.

3. Given that

$$f_n(D) = \frac{\frac{f_m(D)}{D^3}}{\frac{\int_0^\infty f_m(D)dD}{D^3}}$$

show that

$$f_n(D) = \frac{n}{\delta}(\frac{D}{\delta})^{-3} \frac{\exp[-(\frac{D}{\delta})^n]}{\Gamma(1-\frac{3}{n})}$$

for the Roslin-Rammler distribution.

Also find an equation for the Sauter mean diameter in the form $SMD = \delta f[\Gamma(n)]$ and evaluate for the above parameters.

3.6. A bimodal distribution consists of two peaks as shown. Forty percent of the mass of the particles is associated with 60 microns and the remaining 60% of the mass is associated with 30 micron particles.

Find the number average and the mass average diameters and the corresponding standard deviations.

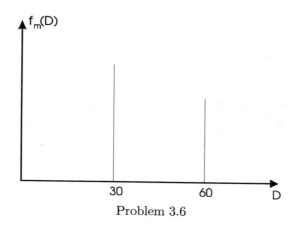

Problem 3.6

3.7. The continuous number frequency distribution for a particle sample is shown in the figure.

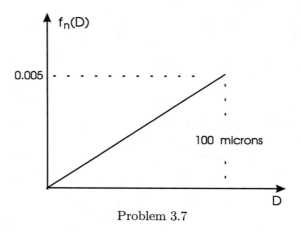

Problem 3.7

Find:

1. the number mean diameter, Sauter mean diameter and the number variance.

2. the mass mean and standard deviation, and

3. the number median and mass median diameter.

3.8. Show that

$$D_m^k \exp(\frac{\sigma_0^2 k^2}{2}) = \frac{1}{\sqrt{2\pi}\sigma_0} \int_0^\infty D^k \exp[-\frac{1}{2}(\frac{\ln D - \ln D_m}{\sigma_0})^2]$$

3.9. Using the definition

$$f_n(D) = \frac{D^3 f_n(D)}{\int_0^\infty D^3 f_n(D)dD}$$

find the relationship

$$\frac{D_{nm}}{D_{mm}} = f(\sigma_0)$$

3.10. The results of a digital image analysis of a particle sample yield the frequency distribution of the particle cross-sectional area. The following results for a sample were obtained

f(D)	A, m^2
0.2	0.286×10^{-8}
0.3	0.785×10^{-8}
0.5	1.13×10^{-8}

Assume each area is the projected area of a spherical particle. Find

1. the number averaged diameter

2. the mass averaged diameter

3. the Sauter mean diameter.

3.11. The following data were obtained with a Coulter Counter. The population refers to the number of particles of the given diameter.

Bin No.	Diameter (μm)	Population
1	1.75	0
2	2.21	3797
3	2.78	3119
4	3.51	2522
5	4.42	1878
6	5.75	1396
7	7.02	850
8	8.58	523
9	11.15	297
10	14.0	179
11	17.7	93
12	22.3	18
13	28.1	2

Form the cumulative number distribution and plot on log-probability paper. Find the

1. number median diameter,

2. mass median diameter, and

3. Sauter mean diameter.

3.12. The following data are obtained from the measurement of a particle size distribution. The particles are spherical.

D (microns)	Number
10	20
20	50
30	73
40	62
50	33
60	18
70	5

Find the mass and number averaged diameters. Estimate the number and mass median diameters. You may want to use a spread sheet or write a small computer program to address this problem.

3.13. Show for the log-hyperbolic distribution that the mode of the distribution is given by

$$x = \ln D = \mu + \frac{\beta\delta}{\sqrt{\alpha^2 - \beta^2}}$$

and the value of the frequency function at the mode is

$$f = A\exp(-\delta\sqrt{\alpha^2 - \beta^2})$$

3.14. Taking the following values for a number frequency log-parabolic distribution: $\alpha = 1.95$, $\beta = -0.768$, $\delta = 0.727$ and $\mu = 3.21$; find the mean, the mode and the variance. You will probably have to use a software package like MathCad to integrate the equations. Also generate some values for the distribution and plot the value on log-probability coordinates to give a best fit for a log-normal distribution. Find, also the mean, mode and variance and compare with those for the log-probability distribution.

Chapter 4

Particle-fluid Interaction

Particle-fluid interaction refers to the exchange of properties between phases and is responsible for coupling in dispersed phase flows. The conservation equations for single particles or droplets are introduced. The phenomena responsible for mass, momentum and energy transfer between phases are then presented. The purpose of this chapter is not to provide an extensive review of fluid-particle interaction but to present some of the basic ideas and observations. Reference to Clift et al. (1978) and other review papers in the literature provides more details.

4.1 Single particle equations

The equations for the mass, velocity and temperature of a given mass (system) are

$$M = const \tag{4.1}$$

$$\mathbf{F} = \frac{d}{dt}(M\mathbf{u}) \tag{4.2}$$

$$\mathbf{T} = \frac{d}{dt}\left[M(\mathbf{r} \times \mathbf{u})\right] \tag{4.3}$$

$$\frac{dE}{dt} = \dot{Q} - \dot{W} \tag{4.4}$$

where M is the system mass, $\mathbf{u}$ is the velocity with respect to an inertial reference frame, $\mathbf{F}$ is the force on the system, $\mathbf{T}$ is the torque acting on the system, E is the total energy (kinetic plus internal), $\dot{Q}$ is the heat transfer rate to the system and $\dot{W}$ is the work per unit time done by the system on the surroundings. Equation 4.1 is simply the definition of a system. Obviously, Equations 4.2 and 4.3 are Newton's second law and Equation 4.4 is the first law of thermodynamics for a closed system. These equations, referred to as the Lagrangian form of the equations, would suffice to calculate the properties of a system with a constant mass. However, they are not appropriate for a droplet which may be evaporating or condensing or a coal particle which is undergoing devolatization. In this case the Reynolds transport theorem is needed to account for the change in mass. A detailed derivation of the Reynolds transport equation applicable to particles with mass transfer (droplets) is provided in Appendix A.

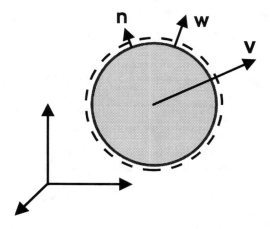

Figure 4.1: Droplet moving with respect to coordinate system.

Consider the droplet shown in Figure 4.1. The droplet has a mass m and the center of mass of the droplet is moving with a velocity $\mathbf{v}$ with respect to an inertial reference frame. A control surface surrounds the droplet and is adjacent to the surface of the droplet so assumes the instantaneous surface area of the droplet. The vector $\mathbf{n}$ is the unit outward normal vector from the control surface. The velocity of the gases through the control surface with respect to the control surface is $\mathbf{w}$.

4.1.1   Continuity equation

The droplet continuity equation simply states that the rate of change of droplet mass is the negative value of the mass efflux through the droplet surface,

$$\frac{dm}{dt} = -\rho_s w S \qquad (4.5)$$

where S is the area of the control (droplet) surface and $\rho_s w$ is the average mass flux over the surface due to evaporation or condensation.

4.1.2   Translational momentum equation

The formal statement of the momentum equation is that the rate of change of momentum within the control volume plus the net efflux of momentum through the control surface is equal to the forces acting on the system (both surface and body forces). The final result (see Appendix A) is

$$\mathbf{F} = m\frac{d\mathbf{v}}{dt} + \int_{cs} \rho(\dot{r}\mathbf{n} + \mathbf{w})\mathbf{w} \cdot \mathbf{n}dS \tag{4.6}$$

where $\dot{r}$ is the regression rate of the surface and the integral is taken over the control surface. The last term represents the thrust on the droplet due to a momentum mass flux from the droplet surface. If all the mass were to come off the droplet in one direction, the droplet would develop a thrust analogous to a rocket motor. However, if the momentum flux is uniform over the droplet surface, which is usually a good approximation, then

$$\dot{r}\mathbf{n} + \mathbf{w} = (\dot{r} + w)\mathbf{n} = w'\mathbf{n} \tag{4.7}$$

the thrust term becomes

$$\int_{cs} (\dot{r}\mathbf{n} + \mathbf{w})\rho\mathbf{w} \cdot \mathbf{n}dS = w'w\rho\int_{cs}\mathbf{n}dS = 0 \tag{4.8}$$

so the droplet momentum equation simplifies to

$$\mathbf{F} = m\frac{d\mathbf{v}}{dt} \tag{4.9}$$

where m is the instantaneous mass of the droplet.

The forces acting on the droplet are subdivided into body forces and surface forces. The body force is a force acting on the mass of the droplet, such as gravity, and the surface force is due to drag and lift which represent a momentum coupling between the two phases.

4.1.3 Angular momentum equation

The conservation equation for the angular momentum states that the rate of change of angular momentum within the control volume plus the net efflux of angular momentum through the control surfaces is equal to the torque acting on the particle. If the particle is spherical, the mass efflux from the surface contributes no torque so the equation for rotation (see Appendix A) becomes

$$\mathbf{T} = I\frac{d\boldsymbol{\omega}}{dt} \tag{4.10}$$

where I is the instantaneous momentum of inertia about an axis of symmetry.

The torque acting on a spherical droplet comes directly from the shear forces on the droplet surface.

4.1.4 Energy equation

The energy of a droplet consists of the external (kinetic energy) and the internal energy as well as the energy associated with surface tension. The heat transfer term includes both convective and radiative heat transfer. The work term includes the flow work due to the efflux of fluid through the control surface as well as the work associated with the forces on the droplet. The detailed derivation of the droplet energy equation is provided in Appendix A. The final result is

$$mc_d\frac{dT_d}{dt} = \dot{Q} + \dot{m}(h_{cs} - h_d + \frac{w'w'}{2}) \tag{4.11}$$

where $\dot{m}$ is the rate of mass change of the droplet, h_{cs} is the enthalpy of the vapor at the control surface and w' is the efflux velocity through the droplet surface with respect to the droplet center. The enthalpy difference $h_{cs} - h_d$ is the enthalpy associated with the change of phase at the droplet temperature or, simply, the latent heat of vaporization. The kinetic energy associated with the efflux velocity is typically very small compared to the enthalpy change so it can be neglected. The final form of the energy equation reduces to

$$mc_d\frac{dT_d}{dt} = \dot{Q} + \dot{m}h_L \tag{4.12}$$

where h_L is the latent heat and c_d is the specific heat of the droplet material. During a change in phase, when the droplet temperature remains constant at the saturation temperature, the heat transfer is just sufficient to balance the energy required for the change of phase.

4.2 Mass coupling

Mass coupling can occur through a variety of mechanisms such as evaporation, condensation or chemical reaction.

4.2.1 Evaporation or condensation

Evaporation or condensation is the transport of the droplet vapor to or from the droplet surface and the change of phase at the surface. The driving force for evaporation is the difference in concentration of the droplet vapor between the droplet surface and the free stream. In general it is assumed that the mixture is a binary mixture consisting of the carrier gas and the vapor of the droplet liquid. For example, a water droplet evaporating in a nitrogen stream would be a binary mixture of water vapor and nitrogen. It is common practice to consider air as a single species and to regard water vapor in air as a binary mixture.

By *Fick's law* the mass flux at the surface of an evaporating or condensing droplet in a binary mixture is

$$\rho_s w = -D_v\frac{\partial \rho_A}{\partial n} \tag{4.13}$$

where n is the coordinate normal to and away from the surface, ρ_A is the mass density of species A (one species of the binary mixture) and D_v is the diffusion coefficient. This equation can be expressed as

$$\rho_s w = -\rho_s D_v \frac{\partial(\rho_A/\rho_s)}{\partial n} = -\rho_s D_v \frac{\partial \omega_A}{\partial n} \tag{4.14}$$

where ω_A is the mass fraction of species A in the mixture.[1]

For a droplet of diameter D the gradient in mass fraction must be proportional to the mass fraction difference between the surface and the freestream and inversely proportional to the droplet diameter,

$$\rho_s w = -\rho_s D_v \frac{\partial \omega_A}{\partial n} \sim \rho_c D_v \frac{\omega_{A,s} - \omega_{A,\infty}}{D} \tag{4.15}$$

where $\omega_{A,s}$ is the mass fraction of species A (vapor corresponding to the droplet liquid) at the droplet surface and $\omega_{A,\infty}$ in the freestream. The density ρ_c is the representative density and may be the average density between the surface and the freestream. The average properties between the surface and the freestream are called the *film conditions*. Except for very high evaporation rates, the density change through the boundary layer is small so ρ_c could be considered the gas density in the freestream (continuous phase). The sign on the difference in mass fraction between the surface and freestream indicates evaporation or condensation. For evaporation $\omega_{A,s} > \omega_{A,\infty}$ so $\rho_s w > 0$.

From Equation 4.5 the mass rate of change of the droplet due to mass flux through the control surface is

$$\frac{dm}{dt} = -\rho_s w S$$

Thus the rate of change of droplet mass should be proportional to

$$\frac{dm}{dt} \sim \pi D^2 \rho_c D_v \frac{\omega_{A,\infty} - \omega_{A,s}}{D} \tag{4.16}$$

The constant of proportionality is the *Sherwood number* so the equation becomes

$$\frac{dm}{dt} = Sh\pi D \rho_c D_v (\omega_{A,\infty} - \omega_{A,s}) \tag{4.17}$$

The Sherwood number is 2 for a droplet which is evaporating with radial symmetry (no forced or free convection effects)

The mass fraction of the vapor at the droplet surface can be evaluated if the droplet temperature is known. The partial pressure of the vapor at the surface is the saturation pressure corresponding to the droplet temperature. The mole fraction of the droplet vapor at the surface is the ratio of the partial pressure to the local pressure

$$\frac{n_A}{n_M} = \frac{p_A}{p} \tag{4.18}$$

---

[1] The development here is based on the mass fraction of species A being much less than unity which is the situation for typical evaporation problems. More general information of mass transfer can be found in standard text books (Bird et al., 1960).

The corresponding vapor mass fraction is

$$\omega_{A,s} = \frac{\mathcal{M}_A}{\mathcal{M}_M} \frac{p_A}{p} \tag{4.19}$$

where $\mathcal{M}_A$ is the molecular weight of the species A and $\mathcal{M}_M$ is the molecular weight of the mixture. If the mole fraction of species A is small, then the molecular weight of the mixture is essentially the molecular weight of the carrier phase. For water vapor in air

$$\frac{\mathcal{M}_A}{\mathcal{M}_M} \sim \frac{18}{29}$$

The effect of a relative velocity between the droplet and the conveying gas is to increase the evaporation or condensation rate. This effect is usually represented by the *Ranz-Marshall* correlation[2]

$$Sh = 2 + 0.6 Re_r^{0.5} Sc^{0.33} \tag{4.20}$$

where Re_r is the *relative Reynolds number* based on the relative speed between the droplet and the carrier gas

$$Re_r = \frac{D|\mathbf{u} - \mathbf{v}|}{\nu}$$

where $\mathbf{u}$ and $\mathbf{v}$ are the velocity vectors of the continuous and dispersed phases, respectively, and Sc is the *Schmidt number* defined by

$$Sc = \frac{\nu}{D_v}$$

where ν is the kinematic viscosity of the mixture at the film conditions. One notes that the Sherwood number reduces to the stagnant flow case as the Reynolds number approaches zero.

The Ranz-Marshall correlation holds for Reynolds numbers from 2 to 200. Rowe et al. (1965) report that a better correlation for Sherwood number is provided if 0.69 is used in lieu of 0.6 in Equation 4.20 for Reynolds numbers from 30 to 2000.

The evaporation of a droplet is sometimes represented by the D^2 - *law* which states that the square of the droplet diameter varies linearly with time. This can be shown by application of the evaporation equation which can be rewritten as

$$\frac{dm}{dt} = \frac{d}{dt}(\rho_d \frac{\pi D^3}{6}) = \frac{\pi}{2}\rho_d D^2 \frac{dD}{dt} \tag{4.21}$$

Thus

$$\frac{\pi}{2}\rho_d D^2 \frac{dD}{dt} = Sh\pi D\rho_c D_v(\omega_{A,\infty} - \omega_{a,s}) \tag{4.22}$$

---

[2] This correlation was originally proposed by Frössling (1938) but commonly attributed to Ranz and Marshall (1952).

or

$$D\frac{dD}{dt} = -\frac{2Sh\rho_c D_v}{\rho_d}(\omega_{A,s} - \omega_{a,\infty}) \tag{4.23}$$

Taking the right side as constant and integrating gives

$$D^2 = D_o^2 - \lambda t \tag{4.24}$$

where λ is the *evaporation constant* and is equal to

$$\lambda = \frac{4Sh\rho_c D_v}{\rho_d}(\omega_{A,s} - \omega_{A,\infty}) \tag{4.25}$$

This form of the evaporation equation has been used extensively in the past. Data for droplet evaporation and combustion have been frequently reported as a value for λ. Obviously λ will not be a constant in a flow with changing freestream conditions but for many situations the approximation may be adequate.

The lifetime or evaporation time of a droplet is obtained by setting $D = 0$ resulting in

$$\tau_m = \frac{D_o^2}{\lambda} \tag{4.26}$$

A model has been developed for droplets evaporating in clusters (Bellan & Harstad, 1987). They find that in dense clusters, evaporation occurs primarily due to diffusional effects ($Sh \sim 2$), while convection plays the dominant role in very dilute clusters.

4.2.2 Mass transfer from slurry droplets

The mass transfer from a slurry droplet represents an important technological problem. For example, food products to be dried (such as powdered milk) consist of water and solids. These slurries are atomized and sprayed into a hot gas stream where the water is driven off and the dried products are collected. The droplet material can be thought of as a porous medium formed by the solids. As the drying proceeds, the size of the droplet may not change appreciably (it may actually increase slightly in diameter), but the mass decreases as the moisture is removed.

The drying process is generally regarded as happening in two stages; the *constant rate* and the *falling rate periods* (Masters, 1972). As the droplet dries the liquid is brought to the surface through capillary forces. During the constant rate period, the drying rate proceeds as if the slurry droplet were a liquid droplet. In this case a liquid layer forms on the slurry droplet so the rate of mass decrease is

$$\frac{dm}{dt} = \pi Sh D\rho D_v(\omega_{A,\infty} - \omega_{A,s}) \tag{4.27}$$

Because the droplet diameter will not change significantly with time, the mass removal rate is nearly constant so this is called the constant rate period.

The amount of moisture in the droplet is quantified by the moisture ratio or *wetness* defined as

$$x = \frac{m_w}{m_s}$$

where m_w is the mass of moisture and m_s is the mass of the solids when "bone dry".

When the wetness reaches the critical moisture ratio, the drying enters the falling rate period. In this period the primary resistance to mass transfer is the transfer of the liquid through the pores of the solid phase and the mass transfer rate may be considerably slower. The rate becomes progressively slower as the moisture content is reduced, hence is called the falling rate period.

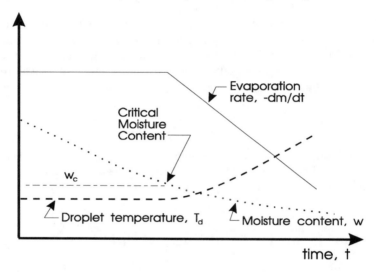

Figure 4.2: Property variation with time in a drying slurry droplet.

A typical variation of drying rate, moisture content and particle temperature with time is shown in Figure 4.2. During the initial period of the drying process, the drying rate and droplet temperature are constant while the moisture content is continuously reduced. After the critical moisture content is reached, the drying rate decreases and the temperature rises toward the temperature of the drying medium.

Few models or data are available in the open literature for the mass transfer rate during the falling rate period. A common approach is to assume that the rate of moisture removal is proportional to the moisture content

$$\frac{dm_w}{dt} = m_s \frac{dx}{dt} \sim m_s x \qquad (4.28)$$

Thus the moisture content would vary with time as

$$x \sim exp(-kt) \qquad (4.29)$$

The proper drying model will depend on the nature of the porous material in the droplet as well as the bound and free moisture and the thermal conductivity of the droplet.

4.2.3   Combustion

The combustion of a single droplet is modeled as a liquid fuel droplet surrounded by a flame as shown in Figure 4.3. The simplest model (Hedley et al., 1971) is based on a spherically symmetric flow of vapor from the droplet and oxidizer from the surroundings. A flame front occurs where the fuel vapor and oxidizer meet and react. The position of the flame front is established by the heat transfer necessary to evaporate the droplet and supply fuel to support the flame.

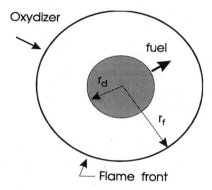

Figure 4.3: Burning fuel droplet.

The basic model equates the convective and conductive heat transfer

$$-\dot{m}c_p\frac{dT}{dr} = -k_c\frac{d}{dr}(r^2\frac{dT}{dr}) \tag{4.30}$$

where $\dot{m}$ is the evaporation rate of the droplet, k is the thermal conductivity of the gas and c_p is the specific heat of the gas at constant pressure. Integration of this equation between the flame radius r_f and the droplet radius r_d yields

$$\dot{m} = \frac{4\pi k_c \ln(1 - \frac{c_p\Delta T}{h_L})}{c_p(\frac{1}{r_d} - \frac{1}{r_f})} \tag{4.31}$$

Expressing the standoff distance between the flame and droplet surface as

$$r_f - r_d = \delta$$

allows one to write the above equation in the form

$$\dot{m} \cong \frac{2\pi k_c D \ln(1 - \frac{c_p\Delta T}{h_L})}{c_p\frac{\delta}{r_d}} \tag{4.32}$$

where D is the droplet diameter. The linear proportionality between burning rate and droplet diameter leads to a D^2 - law for burning rate where the *burning rate constant* is

$$\lambda = \frac{8k_c}{\delta c_p} \ln(1 - \frac{c_p \Delta T}{h_L})$$ (4.33)

The value for the standoff distance δ depends on the heat released by the flame. Typical values (Strehlow, 1984) for the burning rate constant are provided in Table 4.1.

Example: Calculate the burning time of a 200 micron diesel oil droplet.

Answer: The total burning time is given by Equation 4.26. From Table 4.1 the burning rate constant is $7.9 \times 10^{-7} m^2/s$ so the burning time is

$$\tau_b = (2 \times 10^{-4})^2/7.9 \times 10^{-7} = 0.051s$$

Fuel	$\lambda \times 10^7 m^2/s$
Benzene	9.9
Kerosene	9.6
Diesel oil	7.9
Iso-octane	11.4

Table 4.1. Burning rate constants for different fuels burning in air at 20°C and 100kPa.

The effect of a relative velocity between the droplet and the oxidizer is to increase the burning rate. One way to estimate this effect is to assume that the Ranz-Marshall correlation is valid in the form (Williams, 1990)

$$\lambda = \lambda_o(1 + 0.24 Re_r^{\frac{1}{2}} Sc^{\frac{1}{3}})$$ (4.34)

where λ_o is the burning rate with no convection effects.

Current conceptual models for droplet combustion in sprays (Chigier, 1995) suggest that combustion does not occur as flames around individual droplets but rather as flames around groups of droplets.

The combustion of a coal particle is an entirely different phenomenon (Smoot and Smith, 1985). A coal particle consists of four components: volatiles, moisture, char and ash. The volatiles are the combustible gases such as methane. The combustion takes place in two steps. First, the volatile and moisture are driven off. This occurs quickly and depends on the heating rate of the particle. The combustible volatiles then contribute to a gas-phase flame. There are no well-established models for the release rate of the volatiles but it appears to vary with temperature as

$$\dot{m} \sim exp(-\frac{E}{RT})$$ (4.35)

where E is an activation energy and T is the temperature.

After the combustibles are driven off, the char burns very slowly. Char burning is not well-quantified and depends strongly on the coal type and grade.

4.3 Momentum transfer

Momentum is transferred between phases through mass transfer and interphase drag and lift. Momentum transfer due to mass transfer was discussed in the derivation of the particle momentum equation. If the momentum efflux from the particle surface is uniform, then it does not contribute to the force on the particle.

4.3.1 Steady-state drag forces

The "steady-state" drag is the drag force which acts on the particle or droplet in a uniform pressure field when there is no acceleration of the relative velocity between the particle and the conveying fluid. The force is quantified by the drag coefficient through the equation

$$\mathbf{F}_D = \frac{1}{2}\rho_c C_D A |\mathbf{u} - \mathbf{v}|(\mathbf{u} - \mathbf{v}) \qquad (4.36)$$

where ρ_c is the density of the continuous (conveying) phase, C_D is the drag coefficient, A is the representative area of the droplet and $\mathbf{u}$ and $\mathbf{v}$ are the velocities of the continuous phase and the droplet or particle, respectively. Typically the area is the projected area of the particle or droplet in the direction of the relative velocity.

Reynolds number effects

In general, the drag coefficient will depend on the particle shape and orientation with respect to the flow as well as on the flow parameters such as Reynolds number, Mach number, turbulence level and so on. The most fundamental configuration is the sphere, which is addressed first.

The variation of the drag coefficient with Reynolds number for a nonrotating sphere is shown in the Figure 4.4.[3] At low Reynolds numbers the drag coefficient varies inversely with Reynolds number. This is referred to as the *Stokes flow* regime. With increasing Reynolds number the drag coefficient approaches a nearly constant value which is known as the *inertial range*.[4] For $750 < \text{Re} < 3.5 \times 10^5$ the drag coefficient varies by only 13% from $C_D = 0.445$. With increasing Reynolds number there is a sudden decrease in drag coefficient at the *critical Reynolds number*.

In the Stokes flow regime ($Re < 1$), the flow is regarded as a creeping flow in which the inertial terms in the Navier- Stokes equations are unimportant. Thus the governing equation is

$$\frac{\partial p}{\partial x_i} = \mu \frac{\partial^2 u_i}{\partial x_j \partial x_j} \qquad (4.37)$$

---

[3] This variation in drag coefficient with Reynolds number in steady flow is known as the *standard drag curve*.

[4] This regime is also known as "Newton's law" (Clift et al., 1978).

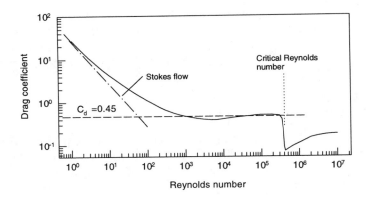

Figure 4.4: Variation of drag coefficient of a sphere with Reynolds number.

The application of this equation to the flow over a sphere was first addressed by
C.G. Stokes in 1851. He expressed the equations in spherical coordinates and
introduced a stream function (Stokes stream function for axisymmetric flows)
which reduced the equations to an ordinary fourth-order differential equation
for the stream function. The coordinate system used by Stokes is shown in
Figure 4.5.

Solution of Equation 4.37 yields the following velocity components

$$u_\theta = -U sin\theta [1 - \frac{3}{4}\frac{a}{r} - \frac{1}{4}(\frac{a}{r})^3] \tag{4.38}$$

$$u_r = U cos\theta [1 - \frac{3}{2}\frac{a}{r} + \frac{1}{2}(\frac{a}{r})^3] \tag{4.39}$$

where U is the freestream velocity and a is the sphere radius. One notes that the
velocity is zero at the surface $(r = a)$ and equal to the uniform flow condition
when $r \to \infty$.

Because of viscosity there is a pressure gradient along the surface of the
sphere; that is, a pressure gradient is needed to move the fluid adjacent to the
surface against the shear forces. This gives rise to a higher pressure at the
forward stagnation point (A) than at the rearward point (B). Thus there is a
form drag on the sphere which, when evaluated, is

$$F_p = \pi\mu DU \tag{4.40}$$

Similarly, the shear stress contribution to the drag force is

$$F_\tau = 2\pi\mu U D \tag{4.41}$$

The sum of both contributions gives a total drag force of

$$F_D = 3\pi\mu DU \tag{4.42}$$

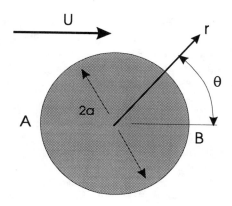

Figure 4.5: Coordinate system for Stokes flow over a sphere.

or, in terms of the velocity difference

$$\mathbf{F}_D = 3\pi\mu D(\mathbf{u} - \mathbf{v}) \tag{4.43}$$

Using this force to solve for the drag coefficient in Equation 4.36 results in

$$C_D = \frac{24}{Re_r} \tag{4.44}$$

where Re_r is the Reynolds number based on the relative velocity. This is the classic Stokes drag coefficient which is valid for $Re_r < 1$.

An extension of Stokes analysis is the *Hadamard-Rybczynski drag law* for a spherical fluid particle in which the shear stress on the surface induces an internal motion (see Clift et al., 1978). In this case the drag coefficient becomes

$$C_D = \frac{24}{Re_r} \left(\frac{1 + \frac{2}{3}\bar{\mu}}{1 + \bar{\mu}} \right) \tag{4.45}$$

where $\bar{\mu}$ is the ratio of the viscosity of the carrier phase to that of the fluid sphere. For a droplet in air, $\bar{\mu} \to 0$ and Stokes law is recovered. For a bubble in a liquid, $\bar{\mu} \to \infty$ so the drag coefficient becomes $16/\mathrm{Re}$

The Stokes drag force is based on a uniform free stream velocity. The Stokes drag has to be extended to account for the effect of a nonuniform flow field by the addition of the *Faxen force* (Happel and Brenner, 1973),

$$\mathbf{F}_D = 3\pi\mu D(\mathbf{u} - \mathbf{v}) + \mu\pi\frac{D^3}{8}\nabla^2\mathbf{u} \tag{4.46}$$

where $\nabla^2\mathbf{u}$ is evaluated at the position of the particle. For a uniform flow field, the Faxen force reduces to zero. The ratio of the Faxen force to Stokes drag varies as

$$\frac{F_{Faxen}}{F_{Stokes}} \sim \left(\frac{D}{\ell} \right)^2$$

where ℓ is a characteristic length associated with the carrier flow field velocity distribution, such as the radius of curvature of the velocity distribution.

For increasing Reynolds number the inertial forces become more important and the drag coefficient is higher than Stokes drag. In 1910 Oseen extended Stokes analysis to include first-order inertial effects and concluded

$$C_D = \frac{24}{Re_r}(1 + \frac{3}{16}Re_r) \tag{4.47}$$

which is valid up to a Reynolds number of 5.

With increasing Reynolds number (~ 100), the flow begins to separate and form vortices behind the sphere. With the formation of vortices the pressure in the wake is further reduced, increasing the form drag. Finally, as the flat portion of the C_D versus Re_r curve is approached, the drag is almost entirely due to form drag with the shear drag contributing little. In this region the drag coefficient can be approximated by a constant value of 0.42 and is referred to as Newton's drag law.

At the critical Reynolds number ($Re_r \sim 3 \times 10^5$) the boundary layer becomes turbulent and the separation point is moved rearward, sharply reducing the form drag and decreasing the drag coefficient. This phenomenon is entirely due to boundary layer effects. If the particle is rough, transition to turbulence occurs at a lower Reynolds number and the critical Reynolds number is reduced. Also the critical Reynolds number is less well-defined as the drop in C_D is less severe. The same trend is observed with increased free stream turbulence. If the particle has sharp edges, the separation is controlled by geometry (separation at the sharp edges) and the critical Reynolds number effect is not observed.

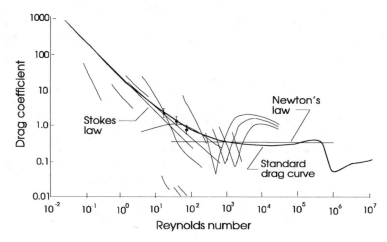

Figure 4.6: The spread in data obtained for the drag coefficient of a sphere.

There have been numerous experiments performed over the years to measure particle and droplet drag coefficients. Experiments have been performed in wind tunnels, ballistic ranges and fluidized beds. A summary of some of the data is shown in Figure 4.6. The many discrepancies between these data have never been resolved. A popular drag coefficient used for droplet motion in combustion studies is that of Ingebo (1956),

$$C_D = \frac{27}{Re^{0.84}}$$

This drag coefficient falls below the standard drag curve which Ingebo attributed to acceleration effects. Subsequent work suggested the observed trend was due to free stream turbulence effects (Crowe, 1961).

The equation of motion for a particle or droplet using the steady-state drag coefficient can be expressed as[5]

$$m\frac{d\mathbf{v}}{dt} = 3\pi\mu D f(\mathbf{u} - \mathbf{v}) + m\mathbf{g} \qquad (4.48)$$

where $\mathbf{g}$ is the acceleration due to gravity and f is the *drag factor* or the ratio of the drag coefficient to Stokes drag.[6]

$$f = \frac{C_D Re_r}{24} \qquad (4.49)$$

Obviously $f \to 1$ for Stokes flow. The variation of the drag factor with Reynolds number is shown in Figure 4.7.

Assuming a spherical droplet with material density of ρ_d, Equation 4.48 can be rewritten as

$$\frac{d\mathbf{v}}{dt} = \frac{f}{\tau_V}(\mathbf{u} - \mathbf{v}) + \mathbf{g} \qquad (4.50)$$

where τ_V is the velocity response time (see Chapter 2),

$$\tau_V = \frac{\rho_d D^2}{18\mu}$$

There are several correlations available in the literature for f as a function of Reynolds number. One correlation (Schiller and Naumann, 1933) that is reasonably good for Reynolds numbers up to 800 is

$$f = (1 + 0.15 Re_r^{0.687}) \qquad (4.51)$$

This correlation yields a drag coefficient with less than 5% deviation from the standard drag coefficient. A correlation suitable to higher Reynolds numbers has been proposed by Putnam (1961); namely,

$$f = 1 + \frac{Re_r^{2/3}}{6} \qquad Re_r < 1000$$

$$f = 0.0183 Re_r \qquad 1000 \le Re_r < 3 \times 10^5 \qquad (4.52)$$

---

[5] Bouyancy forces not included.

[6] The relative Reynolds number is based on the relative speed between the continuous and dispersed phase, $Re_r = D\,|\mathbf{u} - \mathbf{v}|\,/\nu$.

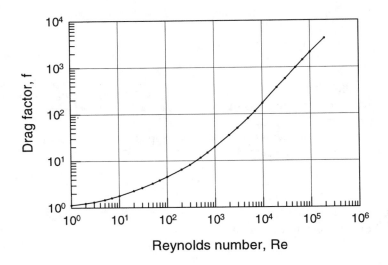

Figure 4.7: The variation of drag factor with Reynolds number.

The advantage of this correlation is that the equation for particle motion can be integrated analytically. A shortcoming is the discontinuity in the value for f at $Re_r = 1000$. A more accurate correlation over the entire subcritical Reynolds number range is that of Clift and Gauvin (1970) which is an extension of Schiller and Naumann's equation.

$$f = 1 + 0.15 Re_r^{0.687} + 0.0175(1 + 4.25 \times 10^4 Re_r^{-1.16})^{-1} \qquad (4.53)$$

This correlation provides a fit for f within $\pm 6\%$ of the experimental value over the entire subcritical Reynolds number range. A useful correlation is the product $f Re_r$ as a function of Re_r shown in Figure 4.8.

The terminal velocity of a particle is the ultimate velocity a particle achieves in free fall; that is, when the acceleration is zero. From Equation 4.50, for a particle falling in a quiescent environment ($\mathbf{u} = 0$), the terminal velocity is

$$\mathbf{v}_t = \frac{\mathbf{g}\tau_V}{f} \qquad (4.54)$$

In Stokes flow the terminal velocity is simply $\mathbf{g}\tau_V$. Otherwise the value for f has to be obtained iteratively.

Example: Calculate the terminal velocity of a 500-micron glass bead ($\rho_d = 2500$ kg/m^3) in air at standard conditions ($\nu = 1.51 \times 10^{-5}$ m^2/s).

Solution: Multiplying both sides of Equation 4.54 by D/ν and multiplying each side by f gives

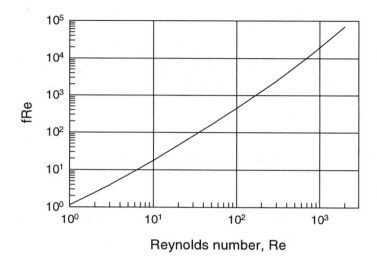

Figure 4.8: The variation of f Re with Reynolds number.

$$f\frac{v_t D}{\nu} = fRe_r = \frac{g\tau_V D}{\nu}$$

The velocity response time for the glass bead is 1.92 s. Evaluating the right side of this equation gives $fRe_r = 623$. From Figure 4.8 the value of the Reynolds number is 120 so the terminal velocity is

$$v_t = Re_r \frac{\nu}{D} = 120\frac{1.51 \times 10^{-5}}{5 \times 10^{-4}} = 3.62 \text{ m/s}$$

Compressibility and rarefaction effects

The typical variation of drag coefficient with the relative Mach number at high and low Reynolds number is shown in Figure 4.9. At a high Reynolds number, the drag coefficient shows an increase with Mach number reaching a maximum value for slightly supersonic flow. This increase is due to the formation of shock waves on the particle and the attendant wave drag (essentially form drag). Mach number effects become significant for a Mach number of 0.6 which is the *critical Mach number*; that is, when sonic flow first occurs on the sphere.

At a low Reynolds number, the drag coefficient uniformly decreases with increasing Mach number and does not display a maximum value near unity. This is due to the prevalence of rarefied flow.

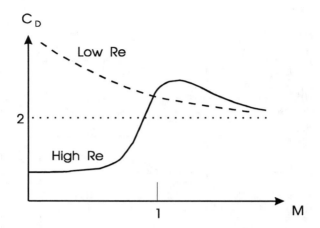

Figure 4.9: Dependence of drag coefficient on the Mach number for low and high Reynolds numbers.

The importance of rarefaction effects are assessed by the magnitude of the *Knudsen number* which is the ratio of the mean free path of the molecules to the particle diameter,

$$Kn = \frac{\lambda}{D} \tag{4.55}$$

where λ is the mean free path of the molecules. If the Knudsen number is large, the flow can not be regarded as a continuum. The wave drag due to a shock wave would no longer appear for particles with Knudsen numbers of the order of unity because the thickness of the shock wave would be comparable to the particle size so the particles would be engulfed by the wave.

The Knudsen number can be related directly to the Mach number and Reynolds number. The viscosity of a gas is proportional to

$$\mu \sim c\rho_c\lambda \tag{4.56}$$

where c is the speed of sound in the gas. Thus the Knudsen number can be written as

$$Kn = \frac{\lambda}{D} \sim \frac{\mu}{\rho_c cD} = \frac{M_r}{R_r} \tag{4.57}$$

Four regimes have been identified (Schaaf and Cambré, 1958), as shown in Table 4.2, to describe rarefaction effects. *Free molecule flow* occurs for Knudsen numbers greater than 10. In this regime, the flow is treated as the motion of individual molecules which impact and rebound from a surface with the approaching molecules unaffected by the rebounding molecules. *Transitional flow* occurs when approaching-rebounding molecular collisions become significant to the flow field. In the *slip flow* regime one assumes that there is a slip velocity and temperature jump between the fluid adjacent to the surface and the surface.

Finally *continuum flow* corresponds to a Knudsen number less than 10^{-3}. Studies (Crowe et al., 1969) show that a particle in a rocket nozzle passes through all these flow regimes.

Continuum	$Kn < 10^{-3}$	$M < 0.01\sqrt{\text{Re}}$
Slip flow	$10^{-3} < Kn < 0.25$	$0.01\sqrt{\text{Re}} < M < 0.1\sqrt{\text{Re}}$
Transitional flow	$0.25 < Kn < 10$	$0.1\sqrt{\text{Re}} < M < 3\,\text{Re}$
Free molecule flow	$Kn > 10$	$M > 3\,\text{Re}$

Table 4.2. Rarefied flow regimes.

There is no analytic nor numerical model available which provides the particle drag coefficient for particles over all the regimes of rarefied flows. The earlier methods to correct for rarefied flow effects were based on a correction to Stokes drag derived by Basset to account for velocity slip at the surface. In the case the drag coefficient can be exprssed as

$$\frac{C_D}{C_{D,Stokes}} = \frac{1 + 4C_m Kn}{1 + 6C_m Kn} \tag{4.58}$$

where C_m is the *momentum exchange coefficient*.[7] This equation is valid only in the limit of $Kn \to 0$.

The classic experiment for free molecule flow effects was performed by Millikan (1923) as part of his oil drop experiment. He found that the drag of the oil drop varied as

$$\frac{C_D}{C_{D,Stokes}} = \frac{1}{1 + Kn[2.49 + 0.84exp(-\frac{1.74}{Kn})]} \tag{4.59}$$

which can be regarded as an extension of the Basset correction. This equation has been used for many years as the correction for rarefied flow effects and is commonly referred to as the *Cunningham correction factor*. For large Mach numbers this equation reduces to

$$C_D \sim \frac{C_{D,Stokes}}{\frac{M}{Re_r}} \tag{4.60}$$

or

$$C_D \sim \frac{1}{M}$$

so as $M \to \infty$ the drag coefficient approaches zero. Analytic results available for free molecule flows (Schaaf and Cambré, 1958) show that the sphere drag coefficient approaches 2 as the Mach number approaches infinity. This is a shortcoming of the Cunningham correction factor but it is still useful for low Mach numbers.

Several authors (Crowe et al., 1973; Walsh, 1975; Bailey and Hiatt, 1972; Henderson, 1976) have proposed equations for the drag coefficient over the Mach

---

[7] From kinetic theory the best value for C_m appears to be 1.14 (Talbot, 1981).

number Reynolds number range for rarefied flows. An approximate surface contour of the drag coefficient of a sphere as a function of the Mach number and Reynolds number is shown in Figure 4.10. The steady-state drag curve corresponds to the plane for $M = 0$. For low Reynolds numbers, the drag coefficient decreases with an increasing Mach number as shown by Millikan's experiments. At high Reynolds numbers, the drag coefficient increases with an increasing Reynolds number. The contours for the drag coefficient beyond the critical Reynolds number are not included.

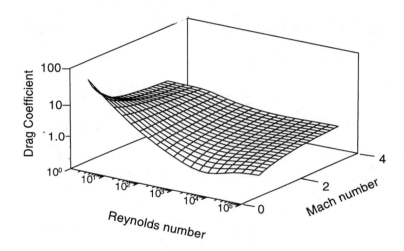

Figure 4.10: Variation of drag coefficient of a sphere with Mach number and Reynolds number.

The following empirical equation for drag coefficient proposed by Crowe et al. (1972) and simplified by Hermsen (1979) has been used extensively in the numerical analysis of the flow in solid propellant rocket nozzles.

$$C_D = 2 + (C_{Do} - 2)e^{-3.07\sqrt{k}g(\text{Re})M/\text{Re}} + \frac{h(M)}{\sqrt{k}M}e^{-\frac{Re}{2M}} \qquad (4.61)$$

where g and h are the two functions;

$$g(Re) = \frac{1 + Re(12.278 + 0.548Re)}{1 + 11.278Re}$$

and

$$h(M) = \frac{5.6}{1 + M} + 1.7\sqrt{\frac{T_d}{T_c}}$$

and where T_d is the particle temperature and T_c is the temperature of the gas. C_{Do} is the drag coefficient for a Mach number of zero, or the steady-state (standard) drag curve. Note that for large Knudsen numbers (large M/Re) the drag coefficient approaches two. For a Mach number approaching zero, the drag coefficient approaches the "standard" drag curve. The model is not valid beyond the Reynolds number where critical Reynolds number effects begin to appear (reduction in drag coefficient).

4.3.2 Pressure gradient and buoyancy force

The effect of the local pressure gradient gives rise to a force in the direction of the pressure gradient. The net pressure force acting on a particle is given by

$$\mathbf{F}_p = \int_{cs} -p\mathbf{n}dS \tag{4.62}$$

Applying the divergence theorem gives

$$\mathbf{F}_p = \int_{cv} -\nabla p dV \tag{4.63}$$

By assuming the pressure gradient is constant over the volume of the particle one has

$$\mathbf{F}_p = -\nabla p V_d \tag{4.64}$$

where V_d is the particle volume.

The pressure gradient produced by a hydrostatic pressure is

$$\nabla p = -\rho_c g \vec{e}_z \tag{4.65}$$

where z is in the direction opposed to gravity (upward). The corresponding pressure force is

$$F_p = \rho_c g V_d$$

which states that the force is equal to the weight of the fluid displaced. This is known as Archimedes principle.

The equation of motion of a particle including buoyancy effects is

$$m\frac{d\mathbf{v}}{dt} = 3\pi\mu D f(\mathbf{u} - \mathbf{v}) + m\mathbf{g} - \rho_c \mathbf{g} V_d \tag{4.66}$$

or

$$\frac{d\mathbf{v}}{dt} = \frac{f}{\tau_v}(\mathbf{u} - \mathbf{v}) + \mathbf{g}\left(1 - \frac{\rho_c}{\rho_d}\right) \tag{4.67}$$

The terminal velocity is affected by the buoyant force and is given by

$$\mathbf{v}_t = \frac{\mathbf{g}\tau_V}{f}\left(1 - \frac{\rho_c}{\rho_d}\right) = \frac{\mathbf{g}D^2(\rho_d - \rho_c)}{18\mu_c} \tag{4.68}$$

When $\rho_c = \rho_d$ the terminal velocity is zero and the particles are *neutrally buoyant*.

In a similar fashion, there is also a force on the particle due to the shear stress in the conveying fluid; namely,

$$\mathbf{F}_\tau = \int_{cs} \boldsymbol{\tau}_{ij} \cdot \mathbf{n} dS \qquad (4.69)$$

Using the divergence theorem, the force can be expressed as

$$\mathbf{F}_t = \int_{cv} \nabla \cdot \boldsymbol{\tau}_{ij} dV \qquad (4.70)$$

and if the shear stress gradient is constant over the particle, the force becomes

$$\mathbf{F}_t = \nabla \cdot \boldsymbol{\tau}_{ij} V_d \qquad (4.71)$$

The form of this equation is similar to, but is not, the Faxen force.

An estimate of the importance of the force due to the pressure gradient is obtained as follows. The magnitude of the pressure gradient in the continuous phase is the order of the flow acceleration.

$$\frac{\partial p}{\partial x} \sim \rho_c \frac{D u_c}{D t} \qquad (4.72)$$

Thus the ratio of the pressure force to the force to accelerate a particle is

$$\frac{\forall_d \frac{\partial p}{\partial x}}{m \frac{d u_d}{dt}} \sim \frac{\rho_c}{\rho_d} \frac{D u_c / D t}{d u_d / dt} \qquad (4.73)$$

If the accelerations of the two phases are of the same order, then the ratio of the pressure force to the acceleration is

$$\frac{V_d \frac{\partial p}{\partial x}}{m \frac{d u_d}{dt}} \sim \frac{\rho_c}{\rho_d} \qquad (4.74)$$

In a gas-particle flow the ratio of material densities is generally of the order of 10^{-3} so the pressure gradient force can be neglected. Of course this is not true for a slurry flow. In a bubbly flow the buoyant force is the most important force controlling bubble motion.

4.3.3   Particle clouds

There is little information available on the drag of particles in particle clouds. Analytical models are difficult because an adequate model must account for the surface of every particle. Experimental studies are hindered by the difficulties of measuring the force and the local flow field on an individual particle in a cloud. Most of the data for particle drag have been inferred from sedimentation and fluidization studies.

The classic study of pressure drop in a packed bed where the particles are motionless was performed by Ergun (1959). He found that the pressure drop through the bed varied as

$$\frac{\Delta p}{\Delta L} = 150\frac{\alpha_d^2\mu_c U}{\alpha_c^3 D^2} + 1.75\frac{\alpha_d\rho_c U^2}{\alpha_c^3 D} \tag{4.75}$$

where U is the superficial velocity. The phase velocity is related to the superficial velocity by $\alpha_c u = U$. The first term in this equation represents the pressure loss due to viscous drag on the particles while the second term adds the contribution of inertial drag. This correlation is valid up to minimum fluidization velocity; that is, just before the bed becomes fluidized (particles supported by hydrodynamic forces). The fluid phase volume fraction at which this occurs is about 0.42 for a bed of spherical particles.

By writing Equation 4.75 in terms of a velocity difference of the phase velocity, $U = \alpha_c(u - v)$, one finds[8]

$$\frac{\Delta p}{\Delta L} = 150\frac{\alpha_d^2\mu_c(u - v)}{\alpha_c^2 D^2} + 1.75\frac{\alpha_d\rho_c(u - v)\,|u - v|}{\alpha_c D} \tag{4.76}$$

The total force by the particles on the fluid in a volume $A\Delta L$ must equal the pressure drop times in the cross-sectional area of the flow[9]; that is,

$$\Delta pA = nA\Delta LF_D \tag{4.77}$$

or

$$\Delta pA = nA\Delta L\left[V_d\frac{\Delta p}{\Delta L} + 3\pi\mu_c Df(u - v)\right] \tag{4.78}$$

This equation can be reduced to

$$\alpha_c\frac{\Delta p}{\Delta L} = 3\pi n\mu_c Df(u - v) \tag{4.79}$$

Solving for f gives

$$f = 8.33\frac{\alpha_d}{\alpha_c} + 0.0972\mathrm{Re}_r \tag{4.80}$$

which should work reasonably well for very dense flows. Gidaspow (1994) suggests this equation is valid for $\alpha_c < 0.8$.

Richardson and Zaki (1954) carried out a series of experiments on sedimentation to determine how particle concentration affects the sedimentation velocity. They correlated the ratio of the settling velocity to the terminal velocity, Equation 4.68, with the volume fraction of the continuous phase. They also accounted for wall effects. For Reynolds numbers based on the terminal velocity between 200 and 500, the factor f works out to be

---

[8] The absolute value of the velocity difference is taken to ensure the correct sign for the pressure loss in the same way that it is used in the equation for particle motion.

[9] The pressure difference due to the weight of the carrier phase has been discounted.

$$f = \alpha_c^{-k} \qquad (4.81)$$

where $k = 4.45\,\mathrm{Re}^{-0.1}$ and $\mathrm{Re} = v_t D/\nu$.

Wen and Yu (1966) also conducted a series of fluidization experiments to infer the drag force on particles in dense mixtures. They were looking for a correction to the equation for drag force in the form

$$F_D = g(\alpha_c)3\pi\mu_c D f_o\,(u - v) \qquad (4.82)$$

where f_o is the drag factor for an isolated particle. Wen and Yu used the Schiller-Naumann correlation, Equation 4.51, in their analysis with the relative Reynolds number based on the superficial velocity. They were able to correlate their data and those of previous investigators (including Richardson and Zaki) by setting

$$g(\alpha_c) = \alpha_c^{-3.7} \qquad (4.83)$$

The contribution of their analysis is that they included Reynolds number effects on the terminal velocity and, in so doing, were able to develop an empirical correlation over the entire Reynolds number regime. The drag factor f now becomes

$$f = \alpha_c^{-3.7} f_0 \qquad (4.84)$$

As $\alpha_c \to 1$, $f \to f_0$.

Wen and Yu (1966) also claimed that their correlation provides the same results as Ergun's for volume fractions corresponding to minimum fluidization. This claim is somewhat dubious, however, since $g(\alpha_c)$ is so sensitive to α_c near minimum fluidization.

More recently, Di Felice (1994) found by analysis of various data available in the literature that

$$f = f_o \alpha_c^{-\beta} \qquad (4.85)$$

where β is a function of the relative Reynolds number. In the low Reynolds number regime the value of β approaches 3.65 based on the data of Richardson and Zaki (1954). At high Reynolds numbers β approaches 3.7 from the data of Wen and Yu (1966) and others. In the intermediate range of Reynolds numbers β goes through a minimum value of approximately 3 for Reynolds numbers in the range 20 to 80. Di Felice recommends the following empirical correlation for β,[10]

$$\beta = 3.7 - 0.65 \exp\left[-\frac{(1.5 - \log \mathrm{Re})^2}{2}\right] \qquad (4.86)$$

for relative Reynolds numbers from 10^{-2} to 10^4.

---

[10]The log Re is the log-based 10 Reynolds number.

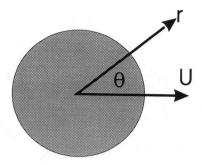

Figure 4.11: Coordinate system for sphere accelerating in a fluid.

The present state of knowledge on the drag of particles in a cloud is still very primitive. A better understanding may be forthcoming through the current development of numerical models for arrays of particles (Dasgupta et al., 1994; Feng et al., 1994; Hu, 1996). However, the correlation proposed by Di Felice (1994) is recommended for the present time.

4.3.4 Unsteady forces

The forces due to acceleration of the relative velocity can be divided into two parts: the *virtual mass effect* and the *Basset force*. The virtual mass effect relates to the force required to accelerate the surrounding fluid. The Basset term describes the force due to the lagging boundary layer development with changing relative velocity.

Virtual or apparent mass effect

When a body is accelerated through a fluid, there is a corresponding acceleration of the fluid which is at the expense of work done by the body. This additional work relates to the virtual mass effect.

Consider a sphere in a fluid as shown in Figure 4.11. The total kinetic energy of the fluid surrounding the sphere is

$$KE = \frac{1}{2}\rho_c \int_V u^2 dV \qquad (4.87)$$

where the integral is taken over all the fluid. It is assumed that the fluid is inviscid and incompressible (ideal fluid) so the velocity can be expressed as the derivative of a potential function

$$\mathbf{u} = \nabla\phi \qquad (4.88)$$

The continuity equation in terms of the potential function is

$$\nabla \cdot \mathbf{u} = \nabla^2\phi = 0 \qquad (4.89)$$

The kinetic energy of the fluid can be expressed in terms of the potential function as

$$KE = \frac{1}{2}\rho_c \int_V \nabla\phi \cdot \nabla\phi dV \tag{4.90}$$

However, because of the continuity equation, this integral can be written as

$$KE = \frac{1}{2}\rho_c \int_V [\nabla\phi \cdot \nabla\phi + \phi\nabla^2\phi]dV = \frac{1}{2}\rho_c \int_V \nabla \cdot (\phi\nabla\phi)dV \tag{4.91}$$

Using the divergence theorem, this volume integral can be expressed as a surface integral over the sphere (the sphere surface is the boundary "enclosing" the fluid)

$$KE = \frac{1}{2}\rho_c \int_V \nabla \cdot (\phi\nabla\phi)dV = \frac{1}{2}\rho_c \int_{cs} \phi\nabla\phi\cdot\mathbf{n}'dA \tag{4.92}$$

where $\mathbf{n}'$ is the unit outward normal vector from the fluid.

The potential function for a sphere moving with a relative velocity u through a fluid is

$$\phi = -\frac{ua^3}{2r^2}\cos\theta \tag{4.93}$$

where the angle θ is defined in the figure and a is the radius of the sphere. The radial component of velocity is

$$u_r = \frac{\partial\phi}{\partial r} = \frac{ua^3}{r^3}\cos\theta \tag{4.94}$$

which on the surface of the sphere reduces to

$$u_r = u\cos\theta \tag{4.95}$$

At $\theta = 0$ the velocity u_r is $-u$ which is the velocity at that point in the radial direction. The dot product in Equation 4.92 becomes

$$\nabla\phi \cdot \mathbf{n}' = \frac{\partial\phi}{\partial r}\vec{e}_r \cdot (-\vec{e}_r) = -\frac{\partial\phi}{\partial r} = -u\cos\theta \tag{4.96}$$

where $\vec{e}_r$ is the radial outward unit vector. Substituting the above expressions for ϕ and the gradient of ϕ into the equation for kinetic energy of the fluid gives

$$KE = \frac{1}{2}\rho_c \int_0^\pi \frac{ua}{2}\cos\theta U\cos\theta a^2 \sin\theta 2\pi d\theta \tag{4.97}$$

where $a^2\sin\theta 2\pi d\theta$ is the element of surface area on the sphere. This equation evaluates to

$$KE = \frac{\pi\rho_c a^3 u^2}{2} \int_0^\pi \cos^2\theta \sin\theta d\theta = \frac{\pi\rho_c a^3 u^2}{3} \tag{4.98}$$

The work rate required to change the kinetic energy is

$$uF_{vm} = \frac{dKE}{dt} \tag{4.99}$$

where F_{vm} is the "virtual mass" force. Thus

$$uF_{vm} = \frac{2\pi \rho_c a^3}{3} u \frac{du}{dt} \qquad (4.100)$$

so the force is equal to

$$F_{vm} = \frac{M_f}{2} \frac{du}{dt} \qquad (4.101)$$

where M_f is the mass of fluid displaced by the sphere. This force is the force of the particle on the fluid so the drag force is the opposite sense of this force. The relative acceleration of the fluid with respect to the particle acceleration is $\dot{\mathbf{u}} - \dot{\mathbf{v}}$. If the fluid was at rest, then the virtual mass force on the particle should be in the direction opposite the particle acceleration. Thus the virtual mass force acting on the particle is given by

$$\mathbf{F}_{vm} = \frac{\rho_c V_d}{2} (\dot{\mathbf{u}} - \dot{\mathbf{v}}) \qquad (4.102)$$

This force is sometimes called the *apparent mass* force because it is equivalent to adding a mass to the sphere. Analyses are available for shapes other than spheres for which the form of the equation is the same but the mass of fluid displaced is different.

Experiments for a sphere in simple harmonic motion (Odar & Hamilton, 1964) indicate that the virtual mass term also depends on the acceleration parameter which is defined as[11]

$$Ac = \frac{u_r^2}{D \frac{du_r}{dt}}$$

where u_r is the relative velocity. The acceleration parameter decreases as the relative velocity decreases or the relative acceleration increases. They proposed a coefficient to correct to the virtual mass term. Odar (1964) suggested the following empirical equation for the coefficient, C_{vm}, as a function of the acceleration parameter,[12]

$$C_{vm} = 2.1 - \frac{0..132}{0.12 + Ac^2} \qquad (4.103)$$

This correlation was developed from data using a sphere in simple harmonic motion. Subsequent work by Odar (1966) demonstrated the validity of the correlation for spheres dropping due to gravity in a tank of liquid. Further work by Schöneborn (1975) showed the utility of the correlation for predicting the fall velocity of particles in an tank of oscillating fluid.

---

[11] The accleration parameter is the reciprocal of the "acceleration modulus" defined by Clift et al. (1978).

[12] The expression Odar proposed approaches 0.5 as $Ac \rightarrow 0$ which replaces the 0.5 factor in the virtual mass equation. The expression given here approaches unity as $Ac \rightarrow 0$ so acts as a correction to the fundamental form of the virtual mass term.

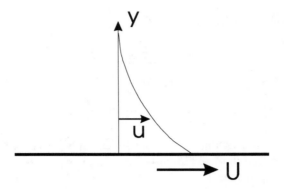

Figure 4.12: Impulsively accelerated flat plate.

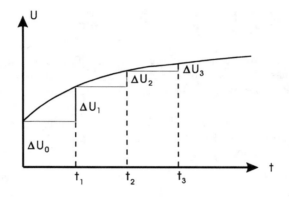

Figure 4.13: Stepwise impulsive acceleration of a flat plate.

Basset force

In that the virtual mass force accounts for the form drag due to acceleration, the Basset term accounts for the viscous effects. This term addresses the temporal delay in boundary layer development as the relative velocity changes with time. This term is sometimes called the "history" term.

The most direct approach to understanding the Basset force is to consider an impulsively accelerated infinite flat plate shown in Figure 4.12. The equation of motion for the fluid is

$$\frac{\partial u}{\partial t} = \nu \frac{\partial^2 u}{\partial y^2} \tag{4.104}$$

with the initial condition $u(0, y) = 0$ and the boundary conditions $u(t, 0) = u_0$ and $u(t, \infty) = 0$ where u_0 is the velocity of the plate. Thus the plate is started impulsively with a step change in velocity from 0 to u_0.

The solution to this equation is

$$u = u_0 \operatorname{erf}(\eta) \tag{4.105}$$

where $\eta = \frac{y}{2\sqrt{\nu t}}$ or

$$u = \frac{2u_0}{\sqrt{\pi}} \int_0^{\eta} e^{-\lambda^2} d\lambda \tag{4.106}$$

The local shear stress is

$$\tau = \mu \frac{\partial u}{\partial y}\Big|_{y=0} = \mu \frac{2u_0}{\sqrt{\pi}} \frac{\partial \eta}{\partial y} = \frac{\mu u_0}{\sqrt{\pi \nu t}} \tag{4.107}$$

or

$$\tau = \frac{\sqrt{\rho \mu} u_0}{\sqrt{\pi t}} \tag{4.108}$$

Now assume that a general temporal variation in plate velocity can be broken up into a series of step changes as shown in the Figure 4.13. At time 0 there is a change Δu_0, at time t_1 a change Δu_1 and so on. The cumulative effect on shear stress would be

$$\tau = \sqrt{\frac{\rho \mu}{\pi}} \left[\frac{\Delta u_o}{\sqrt{t}} + \frac{\Delta u_1}{\sqrt{t - t_1}} + \frac{\Delta u_2}{\sqrt{t - t_2}} ... \right] \tag{4.109}$$

For a time step $\Delta t'$ the change in velocity would be $\frac{du}{dt'} \Delta t'$ so the above sum can be expressed as

$$\tau = \sqrt{\frac{\rho \mu}{\pi}} \sum_{n=0}^{N} \frac{\frac{du}{dt'}}{\sqrt{t - n\Delta t'}} \Delta t' \tag{4.110}$$

where $N\Delta t'$ represents the time interval from the initiation of the acceleration to the present time; that is, from 0 to t. In the limit as $\Delta t'$ approaches zero and $N\Delta t' \to t'$ the equation becomes

$$\tau = \sqrt{\frac{\rho \mu}{\pi}} \int_0^t \frac{\frac{du}{dt'}}{\sqrt{t - t'}} dt' \tag{4.111}$$

Applying this same approach to the impulsive flow over a sphere at low Reynolds number, Basset found that the drag force was equal to

$$\mathbf{F}_{Basset} = \frac{3}{2} D^2 \sqrt{\pi \rho \mu} \int_0^t \frac{\dot{\mathbf{u}} - \dot{\mathbf{v}}}{\sqrt{t - t'}} dt' \tag{4.112}$$

The "historical" nature of this term is evident; the value of the Basset force depends on the acceleration history up to the present time. This term is often difficult to evaluate although important in many unsteady applications. According to the calculations of Hjemfelt and Mockros (1966), the Basset term and virtual mass term become insignificant for $\rho_c/\rho_d \sim 10^{-3}$ if $(\mu/\rho_c \omega D^2)^{1/2} > 6$. Thus the Basset term would not be important for a 10-μm particle in a stream oscillating at less than 700 Hz (Rudinger, 1980). Voir and Michaelides

(1994) have also shown that the Basset term is negligible for oscillatory velocity fields if $\rho_c/\rho_d < 0.002$ and $\omega\tau_V < 0.5$.

As with the virtual mass term, an empirical coefficient, C_B, has been proposed by Odar and Hamilton (1964) to account for the effect of acceleration on the Basset term. The coefficient as given by Odar (1966) is[13]

$$C_B = 0.48 + \frac{0.52}{(1 + Ac)^3} \qquad (4.113)$$

Reeks and McKee (1984) have shown that the Basset term has to be modified to include the case when there is an initial velocity. The term becomes

$$\mathbf{F}_{Basset} = \frac{3}{2}D^2\sqrt{\pi\rho\mu}\left[\int_0^t \frac{\dot{\mathbf{u}}-\dot{\mathbf{v}}}{\sqrt{t-t'}}dt' + \frac{(\mathbf{u}-\mathbf{v})_0}{\sqrt{t}}\right] \qquad (4.114)$$

where $(\mathbf{u}-\mathbf{v})_0$ is the initial velocity difference. Mei et al. (1991) developed a numerical model for stationary flow over a sphere with small free-stream velocity fluctuations. They found that the unsteady Stokes equation does not describe the character of unsteady drag at low frequencies and suggested that this effect may explain the observations of McKee and Reeks that the initial velocity difference has a finite contribution to the long-term particle diffusivity in a turbulent flow.

4.3.5   Basset-Boussinesq-Oseen equation

Equating the sum of the steady-state drag force, the pressure (buoyancy) force, virtual mass force, the Basset force and the body force to the mass times the acceleration of an isolated droplet or particle yields the Basset-Boussinesq-Oseen (BBO) equation for particle or droplet motion.[14]

$$m\frac{d\mathbf{v}}{dt} = 3\pi\mu D(\mathbf{u}-\mathbf{v}) + V_d(-\nabla p + \nabla\boldsymbol{\tau}) + \frac{\rho_c V_d}{2}(\dot{\mathbf{u}} - \dot{\mathbf{v}})$$

$$+ \frac{3}{2}D^2\sqrt{\pi\rho\mu}\left[\int_0^t \frac{\dot{\mathbf{u}}-\dot{\mathbf{v}}}{\sqrt{t-t'}}dt' + \frac{(\mathbf{u}-\mathbf{v})_0}{\sqrt{t}}\right] + m\mathbf{g} \qquad (4.115)$$

A rigorous derivation of the equation of motion of small particles in nonuniform flows has been performed by Maxey and Riley (1983). The equation they derive is essentially the same as the one above except for extra terms due to the nonuniformity of the velocity field and the additional term in the Basset force. If nonuniformity effects are included the Faxen force appears in the steady-state drag term (Equation 4.46). Also the virtual mass term becomes

$$\frac{\rho_c V_d}{2}(\dot{\mathbf{u}} - \dot{\mathbf{v}} - \frac{D^2}{40}\frac{d}{dt}\nabla^2\mathbf{u}) \qquad (4.116)$$

and an additional term appears in the Basset force. For the sake of simplicity, the terms due to flow field nonuniformity will not be included here.

---

[13] As with the correction term for the virtual mass, the correction coefficient C_B approaches unity as $Ac \to 0$.

[14] The Faxen force is not included here.

Dividing through Equation 4.115 by the droplet mass and rearranging the virtual mass term gives

$$\left(1 + \frac{1}{2}\frac{\rho_c}{\rho_d}\right)\frac{d\mathbf{v}}{dt} = \frac{1}{\tau_V}\left(\mathbf{u} - \mathbf{v}\right) + \frac{1}{\rho_d}(-\nabla p + \nabla\boldsymbol{\tau}_i) + \frac{1}{2}\frac{\rho_c}{\rho_d}\dot{\mathbf{u}}$$

$$+\sqrt{\frac{9}{2\pi}}\left(\frac{\rho_c}{\rho_d}\right)^{\frac{1}{2}}\frac{1}{\sqrt{\tau_V}}\left[\int_0^t \frac{\dot{\mathbf{u}}-\dot{\mathbf{v}}}{\sqrt{t-t'}}dt' + \frac{(\mathbf{u}-\mathbf{v})_0}{\sqrt{t}}\right] + \mathbf{g} \tag{4.117}$$

The pressure gradient and shear stress term can be related to the fluid acceleration and force due to gravity from the Navier-Stokes equation for the conveying fluid

$$-\nabla p + \nabla \cdot \tau_{ij} = \rho_c \frac{D\mathbf{u}}{Dt} - \rho_c \mathbf{g}$$

so the combination of the pressure gradient and shear stress term can be combined with the fluid acceleration in the virtual mass term to yield the following form of the BBO equation[15]

$$\left(1 + \frac{1}{2}\frac{\rho_c}{\rho_d}\right)\frac{d\mathbf{v}}{dt} = \frac{1}{\tau_V}\left(\mathbf{u} - \mathbf{v}\right) + \frac{3}{2}\frac{\rho_c}{\rho_d}\dot{\mathbf{u}}$$

$$+\sqrt{\frac{9}{2\pi}}\left(\frac{\rho_c}{\rho_d}\right)^{\frac{1}{2}}\frac{1}{\sqrt{\tau_V}}\left[\int_0^t \frac{\dot{\mathbf{u}}-\dot{\mathbf{v}}}{\sqrt{t-t'}}dt' + \frac{(\mathbf{u}-\mathbf{v})_0}{\sqrt{t}}\right] + \mathbf{g}\left(1 - \frac{\rho_c}{\rho_d}\right) \tag{4.118}$$

For flows, such as gas-particle flows, where the ratio of the continuous phase density to the droplet material density is very small ($\sim 10^{-3}$), the BBO equation can be justifiably simplified to

$$\frac{d\mathbf{v}}{dt} = \frac{1}{\tau_V}(\mathbf{u} - \mathbf{v}) + \mathbf{g} \tag{4.119}$$

However, the complete BBO equation must be used for liquid-solid flows for which the densities are comparable. In the dynamics of bubble motion, the virtual mass force is very important. Also, the complete form of the equation as derived by Maxey and Riley (1983) must be used if flow curvature effects are important.

Corrections factors to account for Reynolds number and acceleration effects are often incorporated into the BBO equation to yield the form

$$\left(1 + \frac{C_{vm}}{2}\frac{\rho_c}{\rho_d}\right)\frac{d\mathbf{v}}{dt} = \frac{f}{\tau_V}\left(\mathbf{u} - \mathbf{v}\right) + (1 + \frac{C_{vm}}{2})\frac{\rho_c}{\rho_d}\dot{u}_i$$

$$+C_B\sqrt{\frac{9}{2\pi}}\left(\frac{\rho_c}{\rho_d}\right)^{\frac{1}{2}}\frac{1}{\sqrt{\tau_V}}\left[\int_0^t \frac{\dot{\mathbf{u}}-\dot{\mathbf{v}}}{\sqrt{t-t'}}dt' + \frac{(\mathbf{u}-\mathbf{v})_0}{\sqrt{t}}\right] + \mathbf{g}\left(1 - \frac{\rho_c}{\rho_d}\right) \tag{4.120}$$

The challenge of integrating the BBO equation has occupied the interest of scientists and engineers for many years. Tchen (1949) was probably one of the first to try to use the BBO equation to study the motion of particles in

---

[15] Using the Navier-Stokes equation in this manner limits the BBO equation to the equation of motion for an isolated particle.

a turbulent flow. Michaelides and Feng (1996) provide an excellent review of approaches to solving the BBO equation.

Unfortunately the BBO equation does not account for many of the effects important to engineering calculations such as Reynolds number effects, the influence of neighboring particles, turbulence effects, particle asphericity, mass blowing from the surface and so on. Still, it is an important fundamental equation which continues to remind one of the inherent complexities of predicting particle motion in fluid flows. Some of the effects not included in the BBO equation are addressed below.

4.3.6   Turbulence effects

Two parameters are used to quantify the effect of carrier phase turbulence on the particle drag: the relative turbulence intensity and the turbulence length scale-particle diameter ratio. The *relative turbulence intensity* is defined as

$$I_r = \frac{\sqrt{\overline{u'^2}}}{|\mathbf{u} - \mathbf{v}|}$$

where $\sqrt{\overline{u'^2}}$ is the root mean square of the carrier fluid turbulence fluctuations. Obviously the relative turbulence intensity is augmented as the relative velocity between the carrier and dispersed phases is reduced.

There is a spectrum of length scales in a turbulent flow. The smallest length scale is the Komolgorov length scale. For particles smaller than the Komolgorov length scale, the primary effects are probably those due to unsteady flow and flow field curvature. The length scale ratio which has been the subject of experimental investigation (Neve and Shansonga, 1989) is L_x/D where L_x is the macroscale of turbulence. Although the results do not always show a consistent trend, the general tendency is for the drag coefficient to increase with an increasing length scale ratio at a given relative turbulence intensity. Intuitively one would expect that the length scale of the turbulence should be less than the particle size to affect the boundary layer, while length scales larger than the particle size would produce effects attributable to unsteady flow and velocity profile nonuniformity. Obviously if the particle is much smaller than large-scale turbulent structures, such as produced in a wake flow, the primary effect of the turbulence is to transport the particle in the nonstationary flow field.

It is well-known (Horner, 1965) that free stream turbulence reduces the critical Reynolds number for the drag coefficient of a sphere. This trend is the result of a premature transition to turbulence of the sphere boundary layer and the attendant decrease in form drag. Torobin and Gauvin (1961) report a series of experiments on the effect of relative turbulence intensity on the drag coefficient of a spherical particle. They found that increasing the relative turbulence intensity decreases the critical Reynolds number. In fact, the critical Reynolds number is reduced to the order of 500 with a relative turbulence intensity of 40%. Clamen and Gauvin (1969) found that, after the precipitous

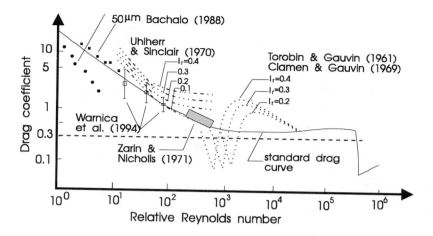

Figure 4.14: A summary of data for turbulence effects on particle drag coefficient.

decrease in drag coefficient at the critical Reynolds number, the drag coefficient increases again with Reynolds number and passes through a maximum.

The composite results from Torobin and Gauvin and Clamen and Gauvin, together with other data, are shown in Figure 4.14. Torobin and Gauvin's experiments were carried out in a ballistic wind tunnel using a radioactive tracer velocity measuring technique. The turbulence scale (integral scale) to particle diameter ratio ranged from 0.4 to 5 but was not used to correlate the data.

Torobin and Gauvin (1961) found that the relative turbulence intensity and critical Reynolds number correlated as

$$I_r Re_c = 45 \tag{4.121}$$

In order to provide a better correlation at the higher turbulence intensities, Clift and Gauvin (1970) suggested

$$
\begin{aligned}
\log_{10} Re_c &= 5.477 - 15.8 I_r \quad (I_r \le 0.15) \\
&= 3.371 - 1.75 I_r \quad (I_r > 0.15)
\end{aligned}
\tag{4.122}
$$

The data for the drag coefficient in the critical Reynolds number regime were found to vary as

$$C_D = 0.3 \left(\frac{Re}{Re_c} \right)^{-3} \quad \text{for } 0.9 Re_c < Re < Re_m \tag{4.123}$$

where Re_m is the Reynolds number at which the drag coefficient is a minimum. Clamen and Gauvin (1969) chose to define the Reynolds number at which the drag coefficient in the supercritical regime increased through 0.3 as the *meta-critical* Reynolds number. This Reynolds number was also found to correlate with the relative turbulence intensity as

$$
\begin{aligned}
\log_{10} Re_M &= 6.878 - 23.2.8 I_r \quad (I_r \le 0.15) \\
&= 3.633 - 1.8 I_r \quad (I_r > 0.15)
\end{aligned}
\tag{4.124}
$$

The particle drag coefficient in the supercritical Reynolds number regime was fitted with the following empirical equations.

$$
C_D = 0.3 \left(\frac{Re}{Re_M} \right)^{(0.45 + 20 I_r)} \qquad Re_m < Re < Re_M
$$

$$
C_D = 3990 Re^{-6.10} - 4.47 \times 10^5 I_r^{-0.97} Re^{-1.8} \quad Re_M < Re < 3 \times 10^4, I_r > 0.7
\tag{4.125}
$$

Neve and Jaafar (1982) conducted experiments on the drag coefficients of spheres in turbulent jets. Their data showed a connection between the supercritical data of Clamen and Gauvin (1969) and the limiting drag coefficient of 0.2 measured by Achenbach (1972) at a very high Reynolds number. Neve and Jaafar also propose a curve for the variation of drag coefficient of a sphere in turbulent flow over the entire Reynolds number range and suggest how the curve would change with a varying relative turbulence intensity. The need for more data was emphasized.

A very limited amount of data have been taken for the affect of turbulence on drag coefficient in subcritical turbulent flows. Uhlerr and Sinclair (1970) found that the drag coefficient was increased with turbulence intensity and suggested the following correlation

$$
\begin{aligned}
C_D &= 162 I_r^{0.33} Re^{-1} \qquad\qquad\quad Re < 50, \qquad 0.05 < I_r < 0.5 \\
C_D &= 0.133(1 + \tfrac{150}{Re})^{1.565} + 4 I_r \quad 50 < Re < 700, \qquad 0.07 < I_r < 0.5
\end{aligned}
\tag{4.126}
$$

These equations have to be used with extreme caution and not be extrapolated to other conditions. There is reason to doubt their validity in that they predict a drag coefficient of 0.48 for a Reynolds number of 500 and relative turbulence intensity of .07. This is less than the standard drag coefficient for a sphere at this Reynolds number. Zarin and Nichols (1971) found, on the other hand, that the drag coefficient in this Reynolds number range increased uniformly with relative turbulence intensity from the standard value. They also found that the turbulence scale-particle size ratio had an effect on the results in that the drag coefficient increased with smaller particles which is opposite to the general trend observed by Neve and Shansonga (1989).

Rudoff and Bachalo (1988) report the results of an experimental study to measure droplet drag in a spray using a PDA system. The Reynolds number range was from 0.1 to 10. Both the droplet velocities and carrier phase velocities were measured. No attempt was made to correlate the data with relative turbulence intensity. In general the drag coefficients were below the standard values. The data also showed an increase in drag coefficient with droplet size which parallels Zarin and Nichols' findings.

Recent studies on the drag coefficient of spherical liquid droplets in turbulent fields for Reynolds numbers from 20 to 100 have been reported by Warinca et al. (1994). In this study it was concluded that there is no experimental evidence to suggest that the drag of droplets is significantly different than standard drag coefficient.

Obviously there is considerable discrepancy in the data for drag coefficient dependence on turbulence. It seems most reasonable, at this point, to use the standard drag curve for particles in subcritical turbulent flow. The correlations developed by Torobin and Gauvin (1961) and Clamen and Gauvin (1969) seem most appropriate for critical and supercritical Reynolds numbers. More detailed experiments are needed to better understand the affect of turbulence on particle drag coefficients.

4.3.7 Drag on rotating particles in velocity gradients

Very little information is available on the drag coefficient of spinning particles primarily because of the difficulties in measuring drag coefficients and quantifying spin rate. The nondimensional parameter used for spin rate is the ratio of the equatorial velocity to the particle-fluid relative velocity,

$$\Omega = \frac{\omega D}{2 |\mathbf{u} - \mathbf{v}|} \qquad (4.127)$$

Davis (1949), who was interested in the dynamics of golf balls, measured drag and lift on smooth and roughened balls up to $\Omega = 0.6$ at a Reynolds number of 9×10^4. He found that the drag coefficient for a spinning smooth ball did not change appreciably with the spin parameter, but a dimpled ball showed a continual increase in drag coefficient with spin. Barkla and Auchterlonie (1971) measured drag coefficients at spin parameters from 1.5 to 12 at a Reynolds number of 2×10^3. They also found no significant trend in drag coefficient with spin parameter. Their data did not reduce to the standard drag coefficient at zero spin which they attributed to their experimental approach. A similar insensitivity of drag coefficient to spin parameter was found in the early work of Maccoll (1928).

The theoretical work of Rubinow and Keller (1961) also shows that spin has no effect on drag for Reynolds numbers the order of unity.

The general finding is that particle spin does not affect the drag coefficient of smooth balls. The data, however, are limited to Reynolds numbers exceeding 2×10^3, which is beyond the range of many fluid-particle flows.

Very little information is available on the drag coefficient of particles in velocity gradients. One study of note is that of Patnaik et al. (1992) who measured the drag coefficients of spheres suspended in a turbulent boundary layer. The Reynolds number range was 10^4 and the sphere diameter/boundary layer thickness varied from 0.29 to 1.0. They found that the drag coefficient based on the velocity of the undisturbed flow at the sphere center correlated well with the standard drag coefficient for a sphere.

4.3.8   Nonspherical particles

The degree of nonsphericity of particles is quantified by the *shape factor*. One shape factor commonly used is that proposed by Wadell (1933); namely,

$$\Psi = \frac{A_s}{A} \tag{4.128}$$

where A_s is the surface area of the sphere of the same volume (smallest possible area per unit volume) and A is the actual surface area. The surface area of the equivalent sphere is

$$A_s = \pi^{\frac{1}{3}}(6V)^{\frac{2}{3}} \tag{4.129}$$

where V is the particle volume. The corresponding diameter of the equivalent sphere is

$$D_s = (\frac{6V}{\pi})^{\frac{1}{3}} \tag{4.130}$$

The product of the drag coefficient and the projected area of the nonspherical particle to the same product for the spherical particle is represented as

$$(C_D A)_a = \frac{1}{K^2}(C_D A)_s \tag{4.131}$$

where K^2 depends on the shape factor and the relative Reynolds number. The equation of motion for the nonspherical particle now becomes

$$\frac{d\mathbf{v}}{dt} = \frac{f}{\tau_V}\frac{1}{K^2}(\mathbf{u} - \mathbf{v}) + \mathbf{g} \tag{4.132}$$

where τ_V is based on the equivalent sphere and both K and f are functions of the relative Reynolds number based on the equivalent sphere. The drag factor and the factor K can be combined to form an effective drag factor; that is, a drag factor that accounts for nonspherical effects.

$$\frac{d\mathbf{v}}{dt} = \frac{f}{K^2}\frac{1}{\tau_V}(\mathbf{u} - \mathbf{v}) + \mathbf{g} = \frac{f_e}{\tau_V}(\mathbf{u} - \mathbf{v}) + \mathbf{g} \tag{4.133}$$

A plot of the effective drag factor as a function of the relative Reynolds number based on an equivalent sphere and the shape factor is provided in Figure 4.15.

Example: Find the drag force on a 1-mm cubical particle in an airstream at standard conditions with a relative velocity of 1 m/s.

Solution: The area and diameter of the equivalent sphere is

$$A_s = \pi^{\frac{1}{3}}(6 \times 10^{-9})^{\frac{2}{3}} = 4.82 \times 10^{-6} \text{ m}$$
$$D_s = (6 \times 10^{-9}/\pi)^{\frac{1}{3}} = 1.24 \times 10^{-3} \text{ m}$$

The shape factor is

$$\Psi = \frac{4.82 \times 10^{-6} \text{ m}}{6 \times 10^{-6} \text{ m}} = 0.803$$

The Reynolds number of the equivalent sphere is

$$\text{Re} = UD/\nu_c = 1 \times 0.00124/1.51 \times 10^{-5} = 821$$

The effective drag factor from Figure 4.15 is 60, so the drag force is

$$F_D = 3\pi\mu_c f DU = 3\pi \times 1.81 \times 10^{-5} \times 60 \times .00124 \times 1 = 1.27 \times 10^{-5} \text{ N}$$

Waddel's correlation is widely accepted. However, there is evidence that it does not work well for oblate spheroids and cylinders (Clift et al., 1978). Obviously the equivalent sphere methods would not be adequate to correlate the drag on sliver- and flake-like shapes.

4.3.9 Blowing effects

The blowing from the surface of a burning or evaporating droplet tends to reduce the drag coefficient. The correlation generally used is (Eisenklam, 1967)

$$C_D = \frac{C_{D,0}}{1 + B} \qquad (4.134)$$

where B is the transfer number and $C_{D,0}$ is the drag coefficient with no blowing. For evaporating droplets the transfer number is

$$B = \frac{c_p \Delta T}{h_L} \qquad (4.135)$$

where h_L is the latent heat of vaporization and ΔT is the temperature difference between the droplet and the surroundings. For a burning droplet, the transfer number is expressed as

$$B = \frac{c_p \Delta T + x_{O_2} H/s}{h_L} \qquad (4.136)$$

where x_{O_2} is the oxygen concentration in gas, H is the heat of combustion and s is the stochiometric rate for oxygen.

Yuen and Chen (1976) suggest that blowing effects can be accounted for by adjusting the Reynolds number to account for the viscosity change in the film around the droplet. The standard drag coefficient based on this Reynolds

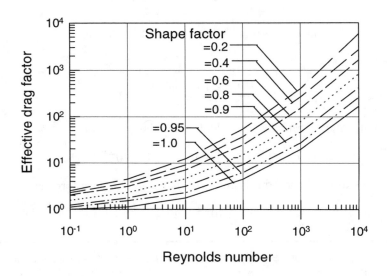

Figure 4.15: Variation of the effective drag factor for nonspherical particles as a function of the shape factor and the relative Reynolds number based on the equivalent sphere.

number is then used for the droplet drag coefficient. They found that application of the 1/3 rule for the film properties provides a good fit with the data. In this case the film temperature is

$$T_f = T_d + \frac{1}{3}(T_c - T_d) \tag{4.137}$$

The same relationship is used for the vapor mass fraction. This temperature and mass fraction are then used to calculate a viscosity. The Reynolds number is based on this viscosity and the free stream gas density. Yuen and Chen's correlation showed a good fit of the data, including Eisenklam's data.

It is generally found that the drag of burning or evaporating droplets is only slightly modified from the non-burning case unless the droplets are burning in a pure oxidizing environment. A recent review of droplet dynamics and evaporation in sprays can be found in Sirignano (1993).

4.3.10   Particle drag in plasmas

The dynamics of particles in plasmas are important in plasma spraying processes. Temperatures are in the range of 7000-9000K and particles are the order of 10 μm. Under these conditions the Knudsen number is near unity so rarefaction effects are significant. Pfender and Lee (1985) review the various formulations used to correlate the drag coefficient in a plasma with the standard drag curve. They suggest that two correction factors are necessary; one to account for the variable property effects (property changes between the particle and the free stream) and the other for the rarefaction effects. A correction factor proposed for variable property effects is

$$C_{Dvp} = C_{Df} \left(\frac{\rho_c \mu_c}{\rho_d \mu_d} \right)^{-0.45} \tag{4.138}$$

where C_{Df} is the drag coefficient based on the film properties (average temperature between particle and plasma gas). The rarefaction effects can be incorporated using correlations presented earlier.

Some experimental results for particle drag in a plasma have been reported in the Russian literature by Asanaliev et al. (1990). They found that the drag coefficient in a plasma is about 30-40% less than the standard drag coefficient. They proposed the following correlation

$$C_D = 16.6 Re^{-.75} + 0.2 \tag{4.139}$$

They also correlated their data with the variable property correction, Equation 4.138, and found reasonable agreement. No correction for rarefaction effects were necessary. One of the problems in interpreting the experimental results is the difficulty in measuring particle temperature.

4.3.11   Lift forces

Lift forces on a particle are due to particle rotation. This rotation may be caused by a velocity gradient or may be imposed from some other source such as particle contact and rebound from a surface.

Saffman lift force

The Saffman lift force is due to the pressure distribution developed on a particle due to rotation induced by a velocity gradient as shown in Figure 4.16. The higher velocity on the top of the particle gives rise to a low pressure, and the high pressure on the low velocity side gives rise to a lift force. Saffman (1965, 1968) analyzed this force for low Reynolds numbers and found the magnitude of the force to be

$$F_{Saff} = 1.61 \mu D |\mathbf{u} - \mathbf{v}| \sqrt{Re_G} \tag{4.140}$$

where Re_G is the shear Reynolds number defined as

$$Re_G = \frac{D^2}{\nu} \frac{du}{dy} \tag{4.141}$$

This can be thought of as the Reynolds number based on the velocity difference between the bottom and top of the particle. The above equation can also be expressed as

$$\mathbf{F}_{Saff} = 1.61 D^2 (\mu \rho_c)^{\frac{1}{2}} |\boldsymbol{\omega}_c|^{-\frac{1}{2}} [(\mathbf{u} - \mathbf{v}) \times \boldsymbol{\omega}_c] \tag{4.142}$$

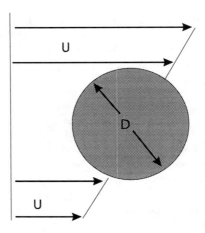

Figure 4.16: Particle in a shear flow.

where

$$\boldsymbol{\omega}_c = \nabla \times \mathbf{u}$$

One notes that if the relative velocity is positive, there is a lift force toward the higher velocity of the continuous phase. On the other hand, if the relative velocity is negative the lift force is toward the lower velocity of the continuous phase.

Saffman's analysis is based on the conditions that the Reynolds number based on the velocity difference is much less than the shear Reynolds number,

$$Re_r << \sqrt{Re_G}$$

and both Reynolds numbers are much less than unity. McLaughlin (1991) extended Saffman's analysis to allow the relative Reynolds number to exceed the shear Reynolds number and found that the lift force rapidly decreases. Mei (1992), using numerical results obtained by Dandy and Dwyer (1990), proposed a empirical fit based on Re_r and Re_G for a correction to the Saffman lift force in the form

$$
\begin{aligned}
F_L/F_{Saff} &= (1 - 0.3314\beta^{1/2})\exp(-\tfrac{Re_r}{10}) + 0.3314\beta^{1/2} & Re_r \le 40 \\
&= 0.0524(\beta\,Re_r)^{1/2} & Re_r > 40
\end{aligned}
$$

$$(4.143)$$

where

$$\beta = \frac{D}{2\,|\mathbf{u} - \mathbf{v}|}\,|\boldsymbol{\omega}_c|,\qquad 0.005 < \beta < 0.4$$

Mei (1992) showed that the above empirical equation provided a reasonable fit for Mclaughlin's results as well.

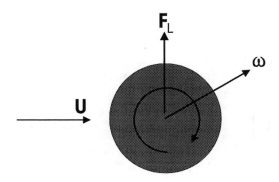

Figure 4.17: Magnus lift on particle rotating in a fluid.

Magnus force

The Magnus force is the lift developed due to rotation of the particle as shown in Figure 4.17. The lift is caused by a pressure differential between both sides of the particle resulting from the velocity differential due to rotation. The rotation may be caused by sources other than the velocity gradient.

The lift force derived by Rubinow and Keller (1961) for Reynolds numbers the order of unity is[16]

$$\mathbf{F}_{Mag} = \frac{\pi}{8} D^3 \rho_c \left[\left(\frac{1}{2} \nabla \times \mathbf{u} - \boldsymbol{\omega}_d \right) \times (\mathbf{u} - \mathbf{v}) \right] \tag{4.144}$$

where $\frac{1}{2} \nabla \times \mathbf{u}$ is the local fluid rotation and $\boldsymbol{\omega}_d$ is the particle rotation. One notes that the lift would be zero if the particle rotation is equal to the location rotation of the fluid.

The lift produced by the Magnus force can be quantified by a lift coefficient in the form

$$\mathbf{F}_{Mag} = \frac{1}{2} \rho_c |\mathbf{u} - \mathbf{v}| C_{LR} A \left(\frac{(\mathbf{u} - \mathbf{v}) \times \boldsymbol{\omega}_r}{|\boldsymbol{\omega}_d - \frac{1}{2} \nabla \times \mathbf{u}|} \right) \tag{4.145}$$

where A is the projected area of the particle, C_{LR} is the lift coefficient due to rotation and $\boldsymbol{\omega}_r$ is the relative spin of the particle with respect to the fluid

$$\boldsymbol{\omega}_r = \boldsymbol{\omega}_d - \frac{1}{2} \nabla \times \mathbf{u} \tag{4.146}$$

In this case, the lift coefficient corresponding to Equation 4.144 is

$$C_{LR} = \frac{D |\boldsymbol{\omega}_d|}{|\mathbf{u} - \mathbf{v}|} \tag{4.147}$$

which is related to the spin parameter defined in Equation 4.127 by

---

[16] Actually, Rubinow and Keller based their analysis on a particle rotating in a field with no fluid rotation. Their equation has been extended here to include flow rotation effects.

$$C_{LR} = 2\Omega \qquad (4.148)$$

This equation shows that the lift coefficient due to particle spin in Stokes flows should increase linearly with the spin parameter.

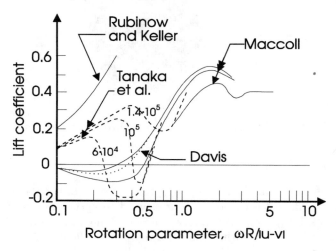

Figure 4.18: Variation of rotational lift coefficient with spin parameter.

There have been numerous experiments to measure C_{LR}. These experiments (Maccoll, 1928; Davis, 1929; Barkla & Auchterlonie, 1971; Tanaka et al., 1990) were conducted at Reynolds numbers exceeding 2×10^3. A summary of the results from Davis, Maccoll and Tanaka et al. is shown in Figure 4.18. At low spin ratios Maccoll and Davis both found the lift coefficient to be actually negative and then to become positive with increasing spin ratio. The negative lift coefficient is attributed to a premature transition to turbulence of the boundary layer of the side of the sphere with the largest relative velocity. This, in turn, causes the wake to deflect in the direction opposite to that expected due to rotation and a negative lift ensues.

Tanaka et al., on the other hand, found that the lift coefficient increases linearly with spin ratio at low spin ratios and then rapidly decreases with increasing spin ratio, actually becoming negative at the lower Reynolds number. These results suggest that the increasing spin gives rise to transition which changes the character of the wake and the pressure distribution, thereby affecting the lift. The observed trends still lack a definitive explanation. Thus the lift coefficient due to rotation in regions of interest to fluid-particle flows ($1 < Re_r < 10^3$) is still an open question.

Tanaka et al. (1990) suggest the following relationship for lift coefficient due to rotation

$$C_{LR} = \min(0.5, 0.5\Omega) \qquad (4.149)$$

which is a ramp function up to a lift coefficient of 0.5 and a constant value thereafter.

4.3.12 Torque

The torque applied to a particle in a fluid is due to the shear stress distribution on the particle surface. For a low Reynolds number flow (Stokes flow), the torque acting on a spherical particle is (Happel and Brenner, 1973),

$$\mathbf{T} = \pi \mu D^3 (\frac{1}{2} \nabla \times \mathbf{u} - \boldsymbol{\omega}_d) \qquad (4.150)$$

Dennis et al. (1980) performed an analytic study on the torque required to rotate a sphere in a viscous fluid which is at rest at large distances. A good representation of the results for Reynolds numbers from 20 to 1000 is

$$T = -2.01 \mu D^3 \omega_d (1 + 0.201 \, \overset{\frac{1}{2}}{\text{Re}})$$

where the Reynolds number is based on the rotational velocity of the sphere and defined as

$$\text{Re} = \frac{\rho_c \omega_d D^2}{4\mu}$$

There appears to be no data in the literature for the torque produced on particles at higher Reynolds numbers spinning in a moving fluid.

4.3.13 Body forces

The most common body force is gravity which is simply the product of the particle mass and the vector representing the acceleration due to gravity. There are other body forces which may be important depending on the application. For example, Coulomb forces are responsible for the operation of an electrostatic precipitator. Thermophoretic forces may be important to the motion of small particles in flows with high temperature gradients such as plasmas.

Coulomb forces

The Coulomb force is the product of the charge on the particle or droplet and the local electric field intensity.

$$\mathbf{F}_c = -q\mathbf{E} \qquad (4.151)$$

In an electrostatic precipitator ions are created by a corona around the charging wire. The charge on a particle can result from one of two sources: field charging or diffusion charging.

If a particle is in a uniform electric field, as shown in Figure 4.19, ions will travel along the electric field lines and deposit on the particle (White, 1963).

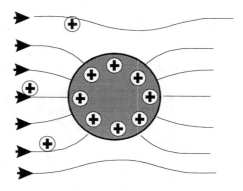

Figure 4.19: Field lines and charge accumulation on a sphere in an electric field.

This is *field charging*. Ultimately a charge is accumulated sufficient to repel additional ions and the *saturation* charge is attained. For a spherical particle of diameter D the saturation charge is

$$q_s = 3\pi D^2 \epsilon_0 E_0 \tag{4.152}$$

where ϵ_0 is the permittivity of free space and E_0 is the electric field strength.

Diffusion charging occurs because of collisions between ions and particles due to the random thermal motion of the ions. *Diffusion charging* is a much slower process than field charging but becomes the dominant charging mechanism for submicron particles. The rate of charge increase due to diffusion charging is given by

$$q(t) = \frac{DkT}{2e} \ln(1 + \frac{\pi D c N_0 e^2 t}{2kT}) \tag{4.153}$$

where k is Boltzmann's constant, e is the electronic charge, N_0 is the number density of the molecules, c is the rms velocity of the ions, t is time and T is the gas temperature. Note that the charge continues to accumulate with time but the accumulation rate decreases with increasing time.

Thermophoretic forces

The thermophoretic force arises due to a temperature gradient in the continuous phase. The higher molecular velocities on one side of a particle due to a higher temperature gives rise to more momentum exchange and a resulting force in the direction of decreasing temperature. Epstein (1929) reports one of the earliest efforts to develop an equation for the thermophoretic force. He proposed

$$\mathbf{F}_T = -\frac{9}{2}\mu_c D^2 Kn\sqrt{\frac{2\pi R}{T}} \frac{k_c/k_d}{1 + 2k_c/k_d} \mathbf{\nabla} T \tag{4.154}$$

where R is the gas constant for the carrier gas and k_c and k_d are the thermal conductivities of the carrier and dispersed phases, respectively. It was found that this equation is valid for Knudsen numbers and thermal conductivity ratios of the order of unity but greatly underpredicted the thermophoretic force on particles with high thermal conductivity.

An extensive review of thermophoresis by Talbot (1981) indicates that the equation developed by Brock (1962) provides the best fit with experimental data over a wide range of Knudsen numbers and thermal conductivity ratios. Brock's equation is

$$\mathbf{F}_T = -6\pi\mu_c\nu_c DC_s \frac{1}{1+6C_mKn}\frac{k_c/k_d+2C_tKn}{1+2k_c/k_d+4C_tKn}\frac{\nabla T}{T} \tag{4.155}$$

where C_s is the thermal slip coefficient and C_t is the thermal exchange coefficient.[17] The best values for the thermal slip and thermal exchange coefficients based on kinetic theory are $C_s = 1.17$ and $C_t = 2.18$.

4.4 Energy coupling

Energy coupling can be effected in three ways, convection, radiation and internal heating.

4.4.1 Radiative heat transfer

The radiative heat transfer is the net heat transfer due to absorption and emission of radiant energy

$$\dot{Q}_r = \pi D^2(\alpha J - \epsilon\sigma T_d^4) \tag{4.156}$$

where α is the absorptivity of the particle material, J is the radiosity, σ is the Stephan-Boltzmann constant and ϵ is the emissivity. If the particle is considered a *grey body* then the absorptivity is equal to the emissivity.. The radiosity depends on the surroundings including the presence of other particles. The equation for radiant heat transfer becomes more complex if the particle is transparent or smaller than the wave length of the light. A more extensive description of radiative heat transfer to an isolated particle can be found in Siegel and Howell (1981).

4.4.2 Convective heat transfer

The heat transfer at the surface of the particle or droplet is

$$\dot{Q}_c = -\int_{cs} \mathbf{q}_c \cdot \mathbf{n}dS \tag{4.157}$$

where $\mathbf{q}_c$ is the heat transfer per unit area per unit time to the particle and $\mathbf{n}$ is the unit outward normal vector. The overall heat transfer (excluding radiation) can be written as

$$\dot{Q}_c = -\bar{q}_cS \tag{4.158}$$

---

[17]The momentum exchange coefficient in introduced in Eqn. 4.58.

where $\bar{q}_c$ is the average heat transfer flux over the surface. From *Fourier's law* for heat transfer

$$\bar{q}_c \sim -k_c \frac{(T_c - T_s)}{D} \tag{4.159}$$

where k_c is the thermal conduction coefficient of the continuous phase, T_c is the temperature of the continuous phase and T_s is the temperature at the particle surface. Thus the heat transfer rate is

$$\dot{Q}_c \sim -\pi D^2 k_c \frac{(T_c - T_s)}{D} \tag{4.160}$$

Using the definition of *Nusselt number*, the equation becomes

$$\dot{Q}_c = Nu\pi D k_c (T_s - T_c) \tag{4.161}$$

The rate of heat transfer to a particle in a stagnant medium with no forced or free convection effects is

$$\dot{Q}_c = 2\pi D^2 k_c \frac{(T_c - T_s)}{D} = 2\pi D k_c (T_c - T_d) \tag{4.162}$$

in which case the Nusselt number is two.

Reynolds number effects

The rate of heat transfer increases with a relative flow with respect to the sphere in the same fashion as the Sherwood number increases. The *Ranz-Marshall correlation* (1952) for forced convection effects; namely,

$$Nu = 2 + 0.6 Re_r^{\frac{1}{2}} Pr^{\frac{1}{3}}, \tag{4.163}$$

where Re_r is the Reynolds number based on the relative velocity and Pr is the Prandtl number, $Pr = \frac{\mu c_p}{k_c}$, provides a good fit with the data up to a relative Reynolds number of approximately 5×10^4. One notes the similarity with the Sherwood number for mass transfer. Other values for the coefficient in Equation 4.163 are discussed in Rowe et al. (1965).

The energy equation for a particle based on convective heat transfer, assuming a *Biot number* less than 0.1, can written as

$$mc_d \frac{dT_d}{dt} = Nu\pi k_c D(T_c - T_d) \tag{4.164}$$

or

$$\frac{dT_d}{dt} = \frac{Nu}{2} \frac{12 k_c}{c_d \rho_d D^2}(T_c - T_d) \tag{4.165}$$

The parameter $c_d \rho_d D^2 / 12 k_c$ has the units of time and is thermal response time, τ_T. Thus the equation becomes

$$\frac{dT_d}{dt} = \frac{Nu}{2} \frac{1}{\tau_T}(T_c - T_d) \tag{4.166}$$

which is analogous to the particle momentum equation, Equation 4.50, with $Nu/2$ being the heat transfer factor.[18]

The energy equation for a droplet with mass transfer (Equation 4.12) can be expressed as

$$\frac{dT_d}{dt} = \frac{Nu}{2}\frac{1}{\tau_T}(T_c - T_d) + \frac{Sh\rho_c\pi DD_v(w_{A,\infty} - w_{A,s})h_L}{c_d\rho_d\pi D^2/6} \tag{4.167}$$

or

$$\frac{dT_d}{dt} = \frac{Nu}{2}\frac{1}{\tau_T}(T_c - T_d) + \frac{Sh}{2}\frac{1}{\tau_T}\frac{Pr}{Sc}\frac{h_L}{c_p}(w_{A,\infty} - w_{A,s}) \tag{4.168}$$

where c_p is the specific heat at constant pressure of the carrier gas. Because of evaporation the rate of increase in droplet temperature is reduced. When the droplet temperature reaches saturation conditions, the *wet-bulb* condition is achieved and the temperature remains constant until evaporation is complete.

Example: Find the wet-bulb temperature for a water droplet in air at $20°C$, at a pressure of 1 bar and with a relative humidity of 40%. Assume $Nu = Sh$, a Prandtl of 0.76 and a Schmidt number of 0.65. The specific heat of the air is 1.005 kJ/kg and the latent heat is 2454 kJ/kg.

Solution: Setting $dT_d/dt = 0$ in Equation 4.168, one has

$$\frac{Nu}{2}(T_c - T_d) = \frac{Sh}{2}\frac{Pr}{Sc}\frac{h_L}{c_p}(w_{A,s} - w_{A,\infty})$$

The saturation pressure of water vapor at 293 K is 0.02339 bar so the partial pressure is 0.02239. The mass fraction of water vapor in the free stream is

$$w_{A,\infty} = 0.02239 \times \frac{18}{29} \times 0.4 = 0.00556$$

Substituting in the above values, one has

$$293 - T_d - 2855(w_{A,s} - 0.00556) = F$$

where the value for T_d at which $F = 0$ is the solution. The solution must be obtained iteratively from the following table.

T_d	p_{sat}	$w_{a,s}$	F
280	0.009912	0.00615	11.31
285	0.01388	0.00862	-0.74
290	0.01919	0.01191	-15.13

Interpolating to find temperature for $F = 0$ gives $T_d = 284.7$ K or $11.7°C$.

---

[18] $Nu/2$ approaches unity as $Re_r \to 0$ which is the same limit for the drag fractor, f.

Nonuniform field effects

In the same fashion that the Faxen force relates to the nonuniformity of the velocity field over the particle, there is an equivalent correction to the particle heat transfer equation due to a spatially nonuniform temperature field. Michaelides and Feng (1994) indicate that the conductive heat transfer should be

$$\dot{Q}_c = 2\pi D k_c (T_s - T_c + \frac{D^2}{24}\nabla^2 T_c) \tag{4.169}$$

Rarefied flows

For rarefied flows the Nusselt number is reduced and a good correlation is (Kavanau, 1955)

$$Nu = \frac{Nu_0}{1 + 3.42 Nu_0 \frac{M}{Re_r Pr}} \tag{4.170}$$

where Nu_0 is the Nusselt number for incompressible flows. This is valid over all flow regimes and yields the correct limit for free molecule flow.

Blowing effects

Mass transfer at the surface tends to reduce the Nusselt number since it lowers the temperature gradient at the surface. Renksizbulut and Yuen (1983) proposed the following empirical correlation based on experimental data,

$$Nu_f = \frac{2 + 0.584 Re_r^{1/2} Pr^{1/3}}{(1 + B_f)^{0.7}} \tag{4.171}$$

where the Nusselt number, transfer number and Prandtl number are based on the film properties (average temperature) and the Reynolds number is defined as

$$Re_r = \frac{\rho_c D |\mathbf{u} - \mathbf{v}|}{\mu_f}$$

where μ_f is the viscosity of the mixture at the film temperature. This correlation provides a good fit with the available data. One notes that it essentially reduces to the Ranz-Marshall correlation as the transfer number approaches zero.

Turbulence effects

Experimental studies on the heat transfer to spheres in turbulent flows generated by grids have been reported by Raithby and Eckert (1971) for Reynolds numbers from 3×10^3 to 5×10^4. These studies show that the Nusselt number at the lower Reynolds numbers increases rapidly with turbulence intensity for turbulence intensities up to 1% and then appears to vary linearly with turbulence intensity thereafter. At the higher Reynolds numbers the linear variation of Nu with turbulence intensity extrapolates linearly back to the values for the flows with no grids. Clift et al. (1978) propose

$$\frac{Nu}{Nu_0} = 1 + 4.8 \times 10^{-4} \frac{I_r}{I_{rc}} Re_r^{0.57} \qquad (4.172)$$

where I_r is the relative turbulence intensity and I_{rc} is the critical turbulence intensity; that is, the turbulence intensity required to have a critical Reynolds number at the relative Reynolds number of the particle (see Section 4.3.6).

Yearling and Gould (1995) conducted an experimental study on the effects of free stream turbulence on the Nusselt number of evaporating water, ethanol and methanol droplets. They found that the Nusselt correlated with

$$Nu_f = \frac{2 + 0.584 Re_r^{1/2} Pr_f^{1/3}}{(1 + B_f)^{0.7}} (1 + 3.4 I_r^{0.843}) \qquad (4.173)$$

where the definitions for the Reynolds number, Prandtl number and the transfer number are the same as those for Equation 4.171. The data were obtained for $50 < Re_r < 1500$. The authors report that the same correlation is valid for nonevaporating particles ($B_f = 0$).

4.4.3 Unsteady flow

Michaelides and Feng (1994) have shown that there are additional contributions to particle heat transfer in an unsteady flow equivalent to the Basset term in the particle drag equation. For a particle in a nonstationary temperature field with no spatial gradients, the heat transfer due to unsteady effects is

$$\dot{Q}_{unsteady} = \pi D^2 k_c \int_0^t \frac{\dot{T}_c - \dot{T}_d}{\sqrt{\pi \alpha_c (t - \tau)}} d\tau \qquad (4.174)$$

where α_c is the thermal diffusivity of the continuous phase and the time rate of change of the temperature field is evaluated at the position of the particle. This equation for unsteady heat transfer is valid only for small Peclét numbers.

4.4.4 Dielectric heating

Dielectric, or microwave, heating is used in many industrial processes such as plywood manufacturing, drying of textiles and food products and curing rubber. A dielectric is a nonconductor in which there are no free electrons. In an electric field, the positive and electric charges within an atom or molecule (dipoles) are displaced and aligned in a direction opposite to the field direction. If the direction of the voltage is changed, the dipoles will attempt to rotate. This rotation dissipates energy and heat is released to the medium.

The power dissipated or heat released per unit volume produced by the alternating electric field is given by

$$\frac{\dot{Q}}{V} = 0.556 \times 10^{-10} f \epsilon''_{eff} E_{rms}^2 \qquad (4.175)$$

where E_{rms} is the electric field intensity, f is the frequency in Hz and $\epsilon"_{eff}$ is the effective dielectric constant. If a material experiences magnetic losses as well, this equation has to be extended to include magnetic wall domain and electron spin losses.

4.5 Summary

There have been a large number of experiments and analyses performed for the transfer of properties between particles and droplets and the carrier phase. Most of the studies have been done for isolated particles in uniform flow fields. The effect of turbulence on particle drag, especially at low Reynolds numbers, is still open to question. The drag on a particle in a cloud of particles has been inferred from experiments from fluidized beds because of the difficulty in making detailed local measurements. There also seems to be considerable confusion concerning the data on the Magnus lift force as a function of spin. The heat transfer relationships for isolated particles and droplets appear to be reasonably well-founded. Like particle drag, there is a need to better quantify the heat transfer between the carrier fluid and particles in a particle cloud.

Exercises

4.1. Measurements indicate that the mass transfer rate on the windward side of an evaporating droplet is 10% higher than that on the leeward side. Assume that the local mass efflux rate can be subdivided into a uniform flow over the windward hemisphere and a uniform flow over the leeward hemisphere. Determine the nondimensional thrust on the droplet in the form

$$\frac{T}{\rho_f \left(\frac{\dot{m}}{\rho_c S}\right)^2 A_d} = f\left(\frac{\rho_d}{\rho_c}\right)$$

where $\dot{m}$ is the evaporation rate of the droplet, A_d is the projected area of the droplet, ρ_c is the density of the gas adjacent to the droplet surface, S is the surface area of the droplet and ρ_d is the density of the droplet material.

4.2. The second law of thermodynamics for a system is

$$\frac{dS}{dt} \geq \frac{\dot{Q}}{T}$$

Using Reynolds transport theorem, write out the second law for an evaporating droplet.

4.3. A spherical droplet is rotating with an angular velocity ω_i as it moves with a translational velocity of U_i at its center of mass. Assume that the droplet is evaporating uniformly. Is the expression

$$m\frac{dU_i}{dt} = F_i$$

where F_i is the force acting on the particle, still applicable? Determine by applying Reynolds transport theorem to the rotating, evaporating droplet.

4.4. A rigid spherical droplet is spinning with an angular velocity about its z-axis as shown. The magnitude of the velocity due to spin is

$$|v_i| = \omega r \sin \phi$$

The velocity with respect to an inertial reference frame is $U_i + v_i$ where U_i is the velocity of the droplet center with respect to an inertial reference frame. The specific kinetic energy is

$$e_k = \frac{|U_i + v_i|^2}{2}$$

1. Show that the kinetic energy of the droplet is

$$\int_{cv} \rho_d e_k dV = m\frac{U^2}{2} + AI\omega^2$$

where A is a constant and I is the moment of inertia of the droplet. Evaluate the constant A.

2. The velocity at the surface of a nonevaporating spherical particle is given by

$$v_{i,s} = v_i + v_t t_i$$

where v_t is the magnitude of the tangential velocity and t_i is the unit vector in the tangent plane of the particle surface. Show that the work term in the energy equation due to surface forces becomes

$$W = -v_i F_{i,s} - \int_{cs} v_t t_i \tau_{ij} n_j dA$$

4.5. Droplets can have an internal circulation as shown in the figure. This is called a "Hill Vortex". For a droplet with internal circulation, the velocity field with respect to an inertial reference frame is $V_i = v_i + g_i$ where v_i is the velocity of the center of mass and g_i is the velocity of the internal motion with respect to the center of mass. Using Reynolds transport theorem, write down the continuity and momentum equation for the droplet. What affect do you think the internal motion will have on the momentum equation for the droplet? Also what affect might it have on the pressure and shear force distribution; that is, the aerodynamic forces acting on the droplet. Provide a qualitative assessment.

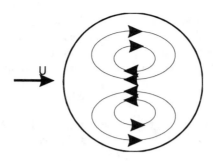

4.6. A coal particle is modeled as a cube as shown. Mass is ejected from the surface due to burning but the size does not change with time. The efflux is uniform and equal to w over the surface except for the leeward face where it is 2w. Find the equation of motion for the particle in the x-direction in the form

$$m\frac{dU_d}{dt} + A\dot{m}w = F_{d,x}$$

where $F_{d,x}$ is the drag force in the x-direction, m is the mass of the particle and $\dot{m}$ is the burning rate.

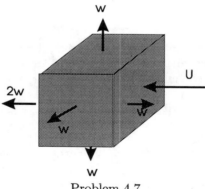

Problem 4.7

4.7. The following questions refer to the energy equation for a droplet (or particle).

1. How would the energy equation be altered if the dependence of surface tension on temperature was to be included?

2. What changes would occur to the energy equation if internal circulation of the fluid was included? Do not work the problem out, only indicate what changes you might expect and explain why.

3. How would the equation change if the particle was porous and the surface was a solid material with pores.

4.8. A droplet is injected with a velocity v_0 into a quiescent medium (no motion). The droplet evaporates according to the D^2-law and Stokes drag is applicable. The drag force is the only force acting on the droplet (no gravity) and the mass flux is uniform over the surface.

1. Derive an equation for the droplet velocity and distance as a function of v_0, t and τ_V/τ_m.

2. Setting $t = \tau_m$, evaluate the distance traveled for τ_V/τ_m much less than unity and much greater than unity.

4.9. A particle is in a harmonically oscillating gas flow field. The flow velocity is given by

$$u = u_0 e^{i\omega t}$$

Assume Stokes drag is applicable so the equation of motion is

$$\frac{dv}{dt} = \frac{u - v}{\tau_V}$$

1. Determine the amplitude and phase lag between the gas and the particle in terms of ω and τ_A. Provide a physical interpretation of the result.

2. Evaluate the amplitude ratio and phase lag of a 10-micron glass ($\rho_p = 2500$ kg/m^3) in air at standard conditions oscillating with a frequency of 100 Hz.

4.10. The energy equation for an evaporating droplet is

$$mc\frac{dT_d}{dt} = Nu\pi D_d k \left(T_g - T_d\right) + Sh\pi\rho D_v D_d \left(\omega_\infty - \omega_s\right) h_{fg}$$

1. Discuss the relative interaction of the terms ensuing as a cold droplet is injected into a hot stream.

2. Determine the wet-bulb temperature for a water droplet in air at 20°C and 40% relative humidity. Assume that Nu = Sh and Sc=0.65. You will need properties from a thermodynamics table for water to do this problem.

3. Write out an equation which represents a perturbation about the wet-bulb temperature; that is,

$$T_g = T_{g,0} + T'_g$$

and

$$T_d = T_{wb} + T'_d$$

yielding

$$\frac{dT'_p}{dt} = f\left(T'_p, T'_g\right)$$

4.11. During the falling rate period, the wetness of a spherical slurry droplet is assumed to vary as

$$\frac{dW}{dt} = -kW$$

where k is a rate constant. The wetness is defined as the ratio of the moisture to the dry solids in the droplet. Derive an equation for the velocity variation with time during this time assuming Stokes law is valid, the gas is quiescent and the droplet size remains constant. The result should be in the form

$$v = v_0 f\left(W_0, D, \mu, k, m_s, t\right)$$

where v_0 is the initial velocity, W_0 is the initial wetness, D is the droplet diameter, μ is the gas viscosity and m_s is the mass of dry solids.

4.12. A charged particle is injected into a channel with quiescent fluid and a uniform electric field strength. The initial velocity of the particle is U_0 and the distance from the injection point to the wall (collecting surface) is L. Find the axial distance the particle will travel before it impacts the wall in terms of electric field strength, E, the aerodynamic response time of the particle, the charge to mass ratio on the particle (q/m), the initial velocity and the distance to the wall. Neglect gravitational effects.

4.13. A cubical coal particle, with a side equal to 1 mm and a material density of 1400 kg/m^3 drops in air at standard conditions. Find the terminal velocity.

4.14. A spherical particle, 200 microns in diameter, drops in a duct with an upward moving flow of air at 10 m/s. The turbulence intensity is 20% and the particle density is 1500 kg/m^3. Will the terminal velocity be higher or lower than one dropping through the same flow with zero turbulence intensity?

4.15. Assume a 100 micron glass bead (density = 2500 kg/m^3) is injected into stagnant water at 5 m/s at 20°C. Solve the BBO equation numerically to find the velocity and distance as a function of time. Use the modified Euler method and any ingenious method to integrate the Basset term! Compare your result with those obtained if the Basset and virtual mass terms were neglected.

4.16. An evaporating droplet is accelerated from rest in a uniform flow with a velocity of v_0. The droplet evaporates according to the D^2- law and Stokes drag is applicable. Derive an expression for the droplet velocity as a function of aerodynamic response time based on initial droplet diameter, the fluid velocity and droplet evaporation time. How far does the droplet travel before completely evaporating? Neglect gravity.

4.17. An evaporating droplet is released from rest in a quiescent medium and drops due to gravity. Buoyancy effects are negligible. The D^2-law and Stokes law apply. Find an expression for the maximum velocity in terms of aerodynamic response time based on initial diameter, acceleration due to gravity and evaporation time.

4.18. An expression sometimes used for the drag coefficient of a particle is

$$C_D = \frac{1}{2} + \frac{24}{Re}$$

A 100 micron particle with a material density of 2000 kg/m³ is fired into still air at a velocity of 10 m/s. The air is at standard conditions. Using the above drag law, calculate how far it will travel before stopping. What will the velocity be after one aerodynamic response time? How does this velocity compare with the velocity calculated using Stokes drag? Explain the difference. Neglect gravitational effects.

4.19. The rate of mass decrease of a porous particle is modeled as

$$\frac{dm_w}{dt} = -k\frac{m_w}{m_s}$$

where m_w is the mass of the water in the particle and m_s is the mass of the solids. As the particle dries, its diameter remains constant but the mass decreases. Assuming Stokes law is valid, derive an expression for the penetration distance of a wet particle in terms of the initial velocity, initial wetness, particle diameter, gas viscosity and mass of the solid component. Neglect gravity.

4.20. A 0.1-micron particle with a material density of 800 kg/m³ falls in air at standard conditions (p = 101 kPa, T = 20°C). Find the terminal velocity assuming Cunningham correction is valid. What is the terminal velocity based on Stokes drag?

4.21. Using Waddel's correction for nonspherical particles, find the terminal velocity for a 100-micron cube in air at standard conditions. The material density is 1200 kg/m³.

4.22. How important do you think the Basset term is for a prismatic particle (with angular edges)? Explain.

4.23. Using the equation for the velocity field around a particle in Stokes flow and the Navier-Stokes equations in spherical coordinates, show that the form drag is $\pi\mu DU$.

4.24. Particles are injected into a cross flow as shown. The initial particle velocity is v_0 and the gas velocity has a linear profile from the wall in the form $u = ky$. The particles have diameter D, density ρ_p and the fluid viscosity is μ. Find an expression for the particle trajectory. Assume Stokes drag is valid. Upon what parameters does the maximum penetration from the wall depend?

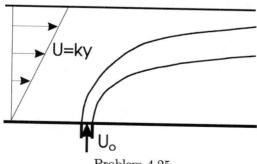

Problem 4.25

4.25. A glass particle with a material density of 2500 kg/m³ is located in the laminar sublayer of a turbulent boundary layer in water flow in a 5-cm duct.

The mean water velocity is 10 m/s and the temperature is 20°C. Assuming the particle has zero velocity, for what particle diameter will the Saffman lift force just balance the particle weight?

4.26. The equation of motion for a particle in a gas flow field is

$$\frac{dv}{dt} = \frac{f}{\tau_V}(u - v) + g$$

It is difficult to solve this equation for small τ_V because $u - v$ also becomes small. By using $(v - u)/\tau_V$ as the dependent variable, show that

$$v = u + \frac{\tau_V g}{f} - \frac{\tau_V}{f}\frac{du}{dt} + \left(\frac{\tau_V}{f}\right)^2 \frac{d^2u}{dt^2} + O(\tau_V^3)$$

Chapter 5

Particle-Particle Interaction

Particle-particle interaction controls the motion of particles in dense particle flows. Also particle-wall interaction is important in dense flows as well as wall-dominated dilute flows. This chapter first addresses particle-wall interaction and then particle-particle interaction. The application of this information in the development of numerical models is discussed in Chapter 8.

5.1 Particle-wall interaction

The problem of particle-wall interaction is encountered when analyzing gas-particle flows contained within walls such as pipe flows, channel flows and fluidized beds. The particle-wall interaction considered here falls into two categories: hydrodynamic forces due to the proximity of a wall and the purely mechanical interaction in the absence of a fluid. The Saffman lift force due to velocity gradient near the wall is one example of the hydrodynamic interaction. Another example is the fluid force acting on a particle approaching the wall in the normal direction. This interaction is explained in Section 5.2.4 where the case of two approaching particles is addressed. Assuming the diameter of one particle to be infinitely large, particle-particle interaction is reduced to particle-wall interaction. Hydrodynamic interaction of this type can prevent a particle from making contact with the wall. The post-collisional velocity of the particle is also affected by this interaction. This hydrodynamic interaction can be neglected if the particle inertia force is so large that collision takes place in a time small compared to the hydrodynamic relaxation time of the particle.

The treatment of the mechanical behavior associated with particle-wall interaction depends on the inertia of the particle. When a massive particle collides with a wall, it rebounds but loses kinetic energy due to friction and inelasticity effects. For a very small particle approaching a wall, molecular forces become dominant compared with the inertial force. As a result, the particle is captured by the wall due to cohesive forces, and neither rebounds from nor slides along the wall. This cohesive force is identified as the *van der Waals force*. Energy loss due to wall collision is considered in Section 5.1.1, where the analysis is based on a single spherical particle and a smooth wall. The energy loss at walls is readdressed in Section 5.1.2 where the case of an nonspherical particle impacting a rough wall is considered. The van der Waals force is explained in Section 5.1.4.

5.1.1   Momentum and energy exchange at walls

There are two models used to deal with particle-wall and particle-particle collisions, the hard sphere model and the soft sphere model. The *hard sphere model* is based on the integrated forms of the equations of motion, namely the impulse equations, and instantaneous deformation of the particle does not appear explicitly in the formulation. The relationship between the pre- and post-collisional velocities is obtained using the coefficient of restitution. In the *soft sphere model*, not only the relationship between the pre- and post-collisional velocities is obtained but the instantaneous motion during the whole collision process is obtained as well. The soft sphere model uses the concept of the mechanical spring and dash-pot instead of actual deformation. In this section, the hard sphere model is explained while the soft sphere model is dealt with in Section 5.2.2.

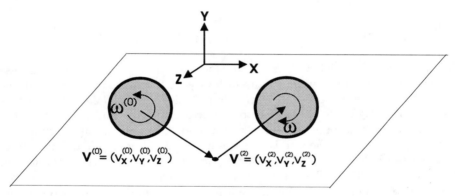

Figure 5.1: Three-dimensional particle-wall collision.

Consider a three-dimensional collision of a spherical particle with a smooth flat wall, as shown in Figure 5.1. The velocities $\mathbf{v}^{(0)}$ and $\boldsymbol{\omega}^{(0)}$ are the pre-collision translation and angular velocities, respectively. The corresponding post-collisional velocities are expressed by $\mathbf{v}^{(2)}$ and $\boldsymbol{\omega}^{(2)}$. The Y-axis is normal to the wall and the X and Z axes are in the plane of the wall. The subscripts X, Y and Z signify component directions. If the coefficient of restitution e and kinetic (sliding) friction f are known, the post-collision translation and angular velocities can be obtained by solving the impulse equations. Readers unfamiliar with the impulse equations should refer to text books on classical dynamics (Hibbeler, 1995). To solve the problem, the following assumptions are made:

1) particle deformation is neglected so, throughout the collision process, the distance between the particle center of mass and the contact point is constant and equal to the particle radius;

2) Coulomb's friction law applies to particles sliding along a wall; and

3) once a particle stops sliding, there is no further sliding.

In general, the process of collision is divided into two periods: one in which the material is compressed and the other in which the compressive force is released. These are often referred to as the compression and recovery periods. Depending on the period during which the particle slides along the wall, the formulation is separated into the following three cases:

Case I: the particle stops sliding in the compression period.

Case II: the particle stops sliding in the recovery period.

Case III: the particle continues to slide throughout the compression and recovery periods.

A detailed explanation for each case is given below.

Case I

The symbols used for the translation velocities, angular velocities and impulses for each moment and period in Case I are shown in Table 5.1. The *compression period* in Case I is subdivided into sliding and nonsliding periods. The superscripts of (0), (s), (1) and (2) on velocities relate to the values at the end of the period. The superscripts (s), (r) and (2) on impulses signify the values which act on the particle during the corresponding periods. The particle velocities change continuously from the pre-collisional to the post-collisional values, but the method based on the impulse equations considers only the momentum difference between periods. The difference in momentum is equal to the impulsive force acting on the particle. The *impulsive force* is defined by the integral of the force versus time over the interval in which the force acts on the particle. The only known variables in Table 5.1 are the initial velocities, $\mathbf{v}^{(0)}$ and $\boldsymbol{\omega}^{(0)}$. The other variables are unknown. The objective is to evaluate the post-collisional velocities, $\mathbf{v}^{(2)}$ and $\boldsymbol{\omega}^{(2)}$. Each variable in the table has three components corresponding to the X, Y and Z-axes and thus there are a total of 27 unknown variables.

The coordinate system used for the impulse equations is shown in Figure 5.2. The origin is at the point of contact between the sphere and the wall. The vector $\mathbf{r}$ is directed from the point of contact to the center of the particle. The cross product of the radius and impulse vector $\mathbf{J}$ is normal to the plane of the paper and the right hand rule is used for sign conventions. The impulse equations are written as

$$m(\mathbf{v}^{(s)} - \mathbf{v}^{(0)}) = \mathbf{J}^{(s)} \tag{5.1}$$

$$m(\mathbf{v}^{(1)} - \mathbf{v}^{(s)}) = \mathbf{J}^{(r)} \tag{5.2}$$

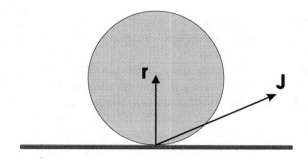

Figure 5.2: Coordinate system for analysis of particle-wall impact.

$$m(\mathbf{v}^{(2)} - \mathbf{v}^{(1)}) = \mathbf{J}^{(2)} \qquad (5.3)$$

$$I(\boldsymbol{\omega}^{(s)} - \boldsymbol{\omega}^{(0)}) = -\mathbf{r} \times \mathbf{J}^{(s)} \qquad (5.4)$$

$$I(\boldsymbol{\omega}^{(1)} - \boldsymbol{\omega}^{(s)}) = -\mathbf{r} \times \mathbf{J}^{(r)} \qquad (5.5)$$

$$I(\boldsymbol{\omega}^{(2)} - \boldsymbol{\omega}^{(1)}) = -\mathbf{r} \times \mathbf{J}^{(2)} \qquad (5.6)$$

Pre-collision	Compression period (1)		Recovery period (2)	Post-collision
	sliding period			
Trans. velocity	$\mathbf{v}^{(0)}$	$\mathbf{v}^{(s)}$	$\mathbf{v}^{(1)}$	$\mathbf{v}^{(2)}$
Angular velocity	$\boldsymbol{\omega}^{(0)}$	$\boldsymbol{\omega}^{(s)}$	$\boldsymbol{\omega}^{(1)}$	$\boldsymbol{\omega}^{(2)}$
Impulse	$\mathbf{J}^{(s)}$	$\mathbf{J}^{(r)}$	$\mathbf{J}^{(2)}$	
$\mathbf{J}^{(1)} = \mathbf{J}^{(s)} + \mathbf{J}^{(r)}$				

Table 5.1. Velocities and impulses for Case I.

The first three equations are for translational velocities and the last three are for angular velocities. The variables m and I are the mass and moment of inertia about the axis of the diameter, respectively; that is, $m = (3\pi/4)a^3\rho_p$ and $I = (2/5)ma^2$ where a is the sphere radius. Each impulse equation also has three components corresponding to the three axes. To obtain the post-collisional velocities, several auxiliary relations are necessary to close the system of equations for the unknown variables. Such auxiliary relations are deduced from boundary conditions, the definition of the coefficient of restitution and

Coulomb's friction law. The boundary conditions relate to the surface velocity of the particle. The surface velocity is given by

$$\mathbf{U} = \mathbf{v} + \mathbf{r} \times \boldsymbol{\omega} = (v_X + a\omega_Z)\mathbf{i} + v_Y\mathbf{j} + (v_Z - a\omega_X)\mathbf{k} \qquad (5.7)$$

where $\mathbf{i}$, $\mathbf{j}$ and $\mathbf{k}$ are unit vectors corresponding to the X, Y and Z-axes, respectively. The Y-component of $\mathbf{r} \times \boldsymbol{\omega}$ is zero because the direction of vector $\mathbf{r}$ coincides with the Y-axis. The condition that the particle stops sliding is obtained by setting the tangential component of the surface velocity equal to zero. Therefore, at the end of the sliding period, the following equation

$$(v_X^{(s)} + a\omega_Z^{(s)})\mathbf{i} + (v_Z^{(s)} - a\omega_X^{(s)})\mathbf{k} = 0 \qquad (5.8)$$

must be satisfied. Since all the components of the surface velocity are zero at the end of the compression period, one has

$$(v_X^{(1)} + a\omega_Z^{(1)})\mathbf{i} + v_Y^{(1)}\mathbf{j} + (v_Z^{(1)} - a\omega_X^{(1)})\mathbf{k} = 0 \qquad (5.9)$$

Also since there is no sliding in the *recovery period*, one has

$$(v_X^{(2)} + a\omega_Z^{(2)})\mathbf{i} + (v_Z^{(2)} - a\omega_X^{(2)})\mathbf{k} = 0 \qquad (5.10)$$

The above three Equations 5.8, 5.9 and 5.10 express the boundary conditions.

Up to this point, only the relationships derived from the impulse equations and boundary conditions have been established. From here, the relationships related to the coefficient of restitution and Coulomb's friction law are developed.

The definition of the *coefficient of restitution e* is somewhat arbitrary. In fact, there are few definitions available for it. The definition most widely used is the ratio of the post- and pre-collisional normal velocities . However, if this definition is adopted, the value of the coefficient for nonspherical particles depends on the location of the contact point between the particle and the wall. If the coefficient of restitution is regarded as a property of the material, its value should not depend on the location of the contact point. With this point of view, a reasonable definition for the coefficient of restitution is the ratio of the normal component of impulse in the compression period to that in the recovery period; that is,

$$e = \frac{J_Y^{(2)}}{J_Y^{(1)}} \qquad (5.11)$$

If the particle is a sphere, the impulse ratio is equal to the velocity ratio $v_Y^{(2)}/v_Y^{(1)}$, as shown later. In Case I, $J_Y^{(1)}$ is the sum of $J_Y^{(s)}$ and $J_Y^{(r)}$ as shown in Table 5.1. When the value of e is given, the Y−component of $\mathbf{J}^{(2)}$ is expressed by

$$J_Y^{(2)} = e(J_Y^{(s)} + J_Y^{(r)}) \qquad (5.12)$$

Coulomb's law states that the friction force is the product of the normal force and the coefficient of kinetic friction f. The friction force corresponds to the tangential component of impulse in the sliding period. Applying Coulomb's law to the sliding period in Case I, one has

$$J_X^{(s)}\mathbf{i} + J_Z^{(s)}\mathbf{k} = -\varepsilon_X f \cdot J_Y^{(s)}\mathbf{i} - \varepsilon_Z f \cdot J_Y^{(s)}\mathbf{k} \qquad (5.13)$$

where ε_X and ε_Z are factors indicating the proportion of the velocity in each component direction; that is, the direction cosines of the approach velocity in the $X - Z$ plane. They satisfy the relation

$$\varepsilon_X^2 + \varepsilon_Z^2 = 1 \qquad (5.14)$$

Since ε_X and ε_Z are new unknown variables, there are twenty-nine unknowns in total. There are also twenty-nine equations; eighteen from Equations 5.1 to 5.6, seven from Equations 5.8 to 5.10 and 5.12, and two from Equations 5.13 and 5.14. Whether Case I occurs or not is determined by the condition

$$J_Y^{(s)} > 0 \quad \text{and} \quad J_Y^{(r)} > 0 \qquad (5.15)$$

Case II

The procedure for Case II follows that for Case I except that the period of sliding falls in the recovery period. The appropriate variables are shown in Table 5.2.

Pre-collision	Compression period (1)		Recovery period (2)	Post-collision
	sliding period			
Trans. velocity	$\mathbf{v}^{(0)}$	$\mathbf{v}^{(1)}$	$\mathbf{v}^{(s)}$	$\mathbf{v}^{(2)}$
Angular velocity	$\boldsymbol{\omega}^{(0)}$	$\boldsymbol{\omega}^{(1)}$	$\boldsymbol{\omega}^{(s)}$	$\boldsymbol{\omega}^{(2)}$
Impulse	$\mathbf{J}^{(1)}$	$\mathbf{J}^{(s)}$	$\mathbf{J}^{(r)}$	
$\mathbf{J}^{(2)} = \mathbf{J}^{(s)} + \mathbf{J}^{(r)}$				

Table 5.2. Velocities and impulses for Case II.

The impulse equations are

$$m(\mathbf{v}^{(1)} - \mathbf{v}^{(0)}) = \mathbf{J}^{(1)} \qquad (5.16)$$

$$m(\mathbf{v}^{(s)} - \mathbf{v}^{(1)}) = \mathbf{J}^{(s)} \qquad (5.17)$$

$$m(\mathbf{v}^{(2)} - \mathbf{v}^{(s)}) = \mathbf{J}^{(r)} \qquad (5.18)$$

$$I(\boldsymbol{\omega}^{(1)} - \boldsymbol{\omega}^{(0)}) = -\mathbf{r} \times \mathbf{J}^{(1)} \qquad (5.19)$$

$$I(\boldsymbol{\omega}^{(s)} - \boldsymbol{\omega}^{(1)}) = -\mathbf{r} \times \mathbf{J}^{(s)} \tag{5.20}$$

$$I(\boldsymbol{\omega}^{(2)} - \boldsymbol{\omega}^{(s)}) = -\mathbf{r} \times \mathbf{J}^{(r)} \tag{5.21}$$

The boundary conditions are

$$v_Y^{(1)} = 0 \tag{5.22}$$

$$(v_X^{(s)} + a\omega_Z^{(s)})\mathbf{i} + (v_Z^{(s)} - a\omega_X^{(s)})\mathbf{k} = 0 \tag{5.23}$$

$$(v_X^{(2)} + a\omega_Z^{(2)})\mathbf{i} + (v_Z^{(2)} - a\omega_X^{(2)})\mathbf{k} = 0 \tag{5.24}$$

From the definition of the coefficient of restitution,

$$J_Y^{(s)} + J_Y^{(r)} = eJ_Y^{(1)} \tag{5.25}$$

and from Coulomb's friction law,

$$J_X^{(1)}\mathbf{i} + J_Z^{(1)}\mathbf{k} = -\varepsilon_X f \cdot J_Y^{(1)}\mathbf{i} - \varepsilon_Z f \cdot J_Y^{(1)}\mathbf{k} \tag{5.26}$$

$$J_X^{(s)}\mathbf{i} + J_Z^{(s)}\mathbf{k} = -\varepsilon_X^{(s)} f \cdot J_Y^{(s)}\mathbf{i} - \varepsilon_Z^{(s)} f \cdot J_Y^{(s)}\mathbf{k} \tag{5.27}$$

The proportion factors for the compression period, ε_X and ε_Z , are the same values as used in Case I. The proportion factors, $\varepsilon_X^{(s)}$ and $\varepsilon_Z^{(s)}$, for Case II are also constrained in the same way as expressed by Equation 5.14; namely,

$$\varepsilon_X^{(s)^2} + \varepsilon_Z^{(s)^2} = 1 \tag{5.28}$$

The condition for the occurrence of Case II is given by

$$J_Y^{(s)} > 0 \text{ and } J_Y^{(r)} > 0 \tag{5.29}$$

Case III

Case III is the simplest, because it is not necessary to distinguish between the period of sliding and nonsliding. The variables for Case III are shown in Table 5.3.

The impulse equations are

$$m(\mathbf{v}^{(1)} - \mathbf{v}^{(0)}) = \mathbf{J}^{(1)} \tag{5.30}$$

$$m(\mathbf{v}^{(2)} - \mathbf{v}^{(1)}) = \mathbf{J}^{(2)} \tag{5.31}$$

$$I(\boldsymbol{\omega}^{(1)} - \boldsymbol{\omega}^{(0)}) = -\mathbf{r} \times \mathbf{J}^{(1)} \tag{5.32}$$

$$I(\omega^{(2)} - \omega^{(1)}) = -\mathbf{r} \times \mathbf{J}^{(2)} \tag{5.33}$$

The boundary condition is

$$v_Y^{(1)} = 0 \tag{5.34}$$

From the definition of the coefficient of restitution,

$$J_Y^{(2)} = e J_Y^{(1)} \tag{5.35}$$

and from Coulomb's friction law, one has

$$J_X^{(1)}\mathbf{i} + J_Z^{(1)}\mathbf{k} = -\varepsilon_X f \cdot J_Y^{(1)}\mathbf{i} - \varepsilon_Z f \cdot J_Y^{(1)}\mathbf{k} \tag{5.36}$$

$$J_X^{(2)}\mathbf{i} + J_Z^{(2)}\mathbf{k} = -\varepsilon_X^{(2)} f \cdot J_Y^{(2)}\mathbf{i} - \varepsilon_Z^{(2)} f \cdot J_Y^{(2)}\mathbf{k} \tag{5.37}$$

Pre-collision	Compression period (1)		Recovery period (2)	Post-collision
Trans. velocity $\mathbf{v}^{(0)}$		$\mathbf{v}^{(1)}$		$\mathbf{v}^{(2)}$
Angular velocity $\boldsymbol{\omega}^{(0)}$		$\boldsymbol{\omega}^{(1)}$		$\boldsymbol{\omega}^{(2)}$
Impulse	$\mathbf{J}^{(1)}$		$\mathbf{J}^{(2)}$	

Table 5.3 Velocities and impulses in Case III.

In Case II, $\varepsilon_X^{(s)}$ and $\varepsilon_Z^{(s)}$ turn out to be equal to ε_X and ε_Z. The suffix (2) in Case III is practically the same as the suffix (s) in Case II. Therefore,

$$\varepsilon_X^{(2)} = \varepsilon_X \tag{5.38}$$

$$\varepsilon_X^{(2)} = \varepsilon_Z \tag{5.39}$$

The solutions for Cases I, II and III obtained using the above procedure are summarized in Table 5.4. There are two sets of solutions for three cases because the solutions for $v = v^{(2)}$ and $\omega = \omega^{(2)}$ turn out to be the same for Cases I and II.

Example: A 100-micron nylon particle bounces off a flat, steel wall. The particle speed is 10 m/s and the angle of incidence is 30°. The particle has no initial rotation and moves in the x-y plane. The coefficient of friction is 0.2 and the coefficient of restitution is 0.8. Find the rebound speed, the angle of rebound and the spin rate (with correct sign). In this problem, $\epsilon_X = 1.0$.

Solution: The incident velocities are $v_y = -5$ m/s, $v_x = 8.66$ m/s. Therefore $v_y/|v| = -0.5$. The value of $2/7f(e+1)$ is 0.79 so the second condition in Table 5.4 is satisfied. The rebound velocities are

$$v_x = 8.66 - 0.2 \times 1.8 \times 5 = 6.86 \text{ m/s}$$

$$v_y = -.8 \times (-5) = 4 \text{ m/s}$$

The rebound angle is $\arctan(v_y/v_x) = 30.2°$. The spin velocity on rebound will be

$$\omega_Z = 0 + \frac{5}{2(5 \times 10^{-5})} 0.2 \times 1.8 \times (-5) = -9 \times 10^4 \text{ rad/s}$$

Condition	$\dfrac{v_Y^{(0)}}{|v|} < -\dfrac{2}{7f(e+1)}$	$-\dfrac{2}{7f(e+1)} < \dfrac{v_Y^{(0)}}{|v|} < 0$
Translational velocity	$v_X = \left(\frac{5}{7}\right)\left(v_X^{(0)} - \frac{2a}{5}\omega_Z^{(0)}\right)$ $v_Y = -ev_Y^{(0)}$ $v_Z = \left(\frac{5}{7}\right)\left(v_Z^{(0)} + \frac{2a}{5}\omega_X^{(0)}\right)$	$v_X = v_X^{(0)} + \varepsilon_x f(e+1)v_Y^{(0)}$ $v_Y = -ev_Y^{(0)}$ $v_Z = v_Z^{(0)} + \varepsilon_Z f(e+1)v_Y^{(0)}$
Angular velocity	$\omega_x = \frac{v_Z}{a}$ $\omega_Y = \omega_Y^{(0)}$ $\omega_Z = -\frac{v_x}{a}$	$\omega_x = \omega_X^{(0)} - \frac{5}{2a}\varepsilon_Z f(e+1)v_Y^{(0)}$ $\omega_Y = \omega_Y^{(0)}$ $\omega_z = \omega_Z^{(0)} + \frac{5}{2a}\varepsilon_x f(e+1)v_Y^{(0)}$

Table 5.4. Relation between pre-and post-collisional velocities.

5.1.2 Nonspherical particles — rough walls

When the coefficients of restitution and dynamic friction are given as physical properties of the particles, the post-collisional velocities can be expressed in terms of the pre-collisional velocities. The relationship between these velocities for the case of a spherical particle and smooth wall is shown in Table 5.4. This relationship is useful for calculating particle trajectories in a field enclosed by walls. However, in some cases, the trajectory calculation in multiphase flows is not as straightforward as it might appear. The coefficient of restitution defined by Equation 5.11 must be less than unity. If trajectory calculations were made for a long horizontal pipe or duct, particles repeatedly colliding with the wall would ultimately lose their vertical velocity component and slide along the wall.

To avoid such an unrealistic result, the irregularity of collisions must be considered. If this irregularity is neglected, large particles cannot be suspended

in the carrier fluid. Fluid dynamic forces, such as Magnus or shear lift forces, and
fluid turbulence acting on the particles are sufficient to suspend small particles.
However, the mechanism responsible for suspending large particles in horizontal
pipes is the irregular bouncing of the particles against the wall. In general, the
irregularity is caused by the particle shape (nonspherical particles) and/or by
the roughness of the wall. In practice, both mechanisms are operative. Several
models in regard to irregular bouncing have been proposed. Those models
are classified into two types: one in which the irregularity is attributed to the
nonsphericity of the particle (Matsumoto & Saito, 1970a; Tsuji et al., 1975;
Tsuji et al., 1991) and the other to wall roughness (Matsumoto & Saito, 1970b;
Morikawa et al., 1987; Sommerfeld & Zivkovis, 1992; Frank et al., 1993).

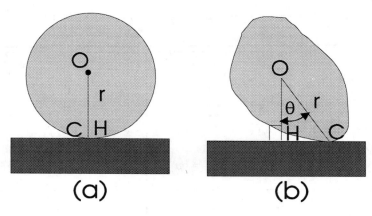

Figure 5.3: Particle-wall collision; a) spherical particle, b) nonspherical particle.

If the problem is two-dimensional, the collision of an arbitrarily shaped particle with the wall can be solved basically in the same way as shown in Section
5.1.1. A two-dimensional particle colliding with a wall is shown in Figure 5.3.
The difference between this case and Section 5.1.1 is that the vector $\mathbf{r}$ drawn
from the center of gravity to the contact point is not normal to the wall. When
the particle shape is spherical, the vector product $\mathbf{r} \times \boldsymbol{\omega}$ which occurs in the
expression of the surface velocity for the rotating particle has no component
normal to the wall. In the case of the nonspherical particle, the vector product has a velocity component normal to the wall. Taking this difference into
account, the calculational method outlined in Section 5.1.2 is modified for the
two-dimensional collision of a particle having arbitrary shape. While radius
$r = |\mathbf{r}|$ represents the particle shape for a spherical particle, a set of values for
r and θ is necessary to describe the shape of an nonspherical particle. It is also
possible to describe any particle shape by having a set of values for r and θ.

It is very difficult to deal, in a rigorous way, with the collision of an arbitrary nonspherical particle having three components of translational and angular velocity, so various assumptions to reduce the complexity of the problem are
necessary. Examples of such assumptions follow.

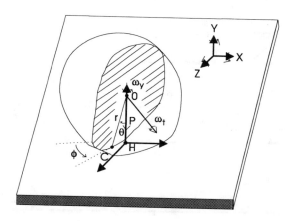

Figure 5.4: Collision of arbitrarily shaped particle with wall.

The collision of an nonspherical particle and a wall is shown in Figure 5.4. The X, Y and Z axes are the same as those in Figure 5.1; i.e., the X and Z axes are on the wall and the Y axis is normal to the wall. The point O is the center of gravity of the particle and the point C is the contact point at the wall. The line OH is drawn from the center of gravity O normal to the wall. If the particle is a sphere, the points C and H coincide.

The angular velocity can be divided into the two components, normal and tangential to the wall. It is assumed that the axes of rotation of both components pass through the center O. The shaded cross section in Figure 5.4 is normal to the axis of ω_t and includes the center of gravity O. It is assumed that the contacting point C is on the periphery of the shaded cross section. The nonspherical particle has various cross sections depending on the viewing direction but, for simplicity, it is further assumed that the shaded cross section in Figure 5.4 represents the shape of the particle. Therefore, viewing the cross section from the direction of the ω_t axis, the same particle figure as shown in Figure 5.3 is obtained. Using the above assumptions, the description of the three-dimensional collision of the nonspherical particle with a flat plate is manageable though the formulation is considerably more complicated than the two-dimensional case. The results of particle trajectory calculations in a two-dimensional channel are shown in Figure 5.5. Here the scale of the longitudinal distance is greatly reduced so that the motion of the particles in the whole channel can be seen. When the particle collides with the wall in the simulation, an attitude, which corresponds to one pair of r and θ values, is chosen by a random number generator before solving the impulse equations. As shown in Figure 5.5a, the particle used for the calculation is nearly spherical (major axis/minor axis = 1.043). This demonstrates that even a small deviation of shape from the sphere causes considerable irregular bouncing motion.

The wall roughness model for irregular bouncing is shown in Figure 5.6. The choice of wall roughness is arbitrary. Several configurations could be considered:

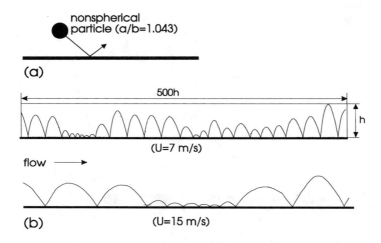

Figure 5.5: Motion of a near spherical particle in a horizontal channel. a) Particle impact with wall ($\bar{D} = 1.1\ mm$, $e = 0.93$ and $f = 0.28$). b) Trajectories at two flow velocities. (From Shen N. et al., $JSME$, 55, 2294, 1989. With permission.)

a wavy pattern, a combination of inclined plates and so on. The calculation for a spherical particle bouncing on the rough wall is simpler than that for an nonspherical particle on a flat wall, if the pattern of roughness is specified, because the results shown in Table 5.4 can be used with some modifications. In the rough wall model, the point at which the particle collides with the wall is assumed to be on an inclined plane. Each inclined plane has its own coordinate system. The relationship between the coordinates, with and without an inclination angle, is easily described using a matrix. Once particle velocities are transformed by the usual linear transformation technique into the coordinates corresponding to the inclined plane, the results shown in Table 5.4 can be used to obtain the post-collisional velocities. After that, the post-collisional velocities based on the coordinate on the inclined plate are re-transformed into the original coordinate system.

As with the nonspherical particle model, the rough wall models make it possible for the particle to move along with a bouncing motion in long, horizontal pipes or ducts.

Predicting the impact-rebound relations for particles with sharp edges, such as quartz particles, would be very difficult. Experiments (Tabakoff, 1982) with quartz particles impacting on an aluminum surface show that for a given impact angle, there is a distribution of rebound angles. In this case, it more pragmatic to use a Monte Carlo method with the measured distribution of rebound angles to model particle-surface impact.

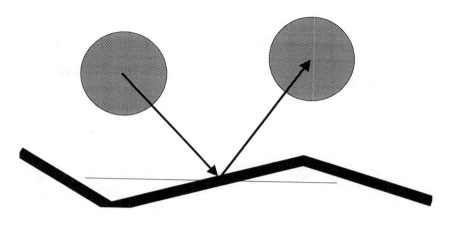

Figure 5.6: Particle collision with a rough wall.

5.1.3 Erosion

Particle-wall impact leads to erosion which is an important consideration in the design and operation of fluid-particle systems. There have been many studies reported on the erosion of materials by impinging particles. An important early work was that by Finnie (1972) who developed a model for the mass of surface material removed by the impact of a single particle. The model is based on the cutting and displacement process common to metal cutting or grinding. For ductile materials, Finnie predicted that the maximum erosion for an impact angle of 13° would vary as the square of the impact velocity. Subsequent experiments with $Si\,C$ particles impinging on pure aluminum showed good agreement with the model in that the maximum erosion occurred at approximately 15°. The data for normal (perpendicular) impact deviated from the model and the erosion dependence on velocity varied as $v^{2.4}$.

Since Finnie's original work, there have been many studies reported on particle erosion. An example of recent data (Magnée, 1995) obtained on erosion of aluminum and high chromium cast iron by alumina particles is shown in Figure 5.7 which gives the mass of material removed normalized with respect to the mass of the impacting particle. One notes that maximum erosion occurs near 15° for ductile materials (aluminum) and at 90° for brittle materials. The basic premise is that the cutting action of the particles is more important for ductile materials and deformation (or displacement) is more significant for brittle materials. The data also show that the erosion varies as the square of the impact velocity. Similar tests have been done on coatings for protecting jet engine parts and steam turbine components (Shanov et al., 1996)

Data of this type have been used in numerical models for erosion in ducts (Shimizu, 1993) and turbine blades (Tabakoff, 1982). An extensive summary on erosion in pneumatic conveying can be found in Marcus et al. (1990). Erosion in liquid-solid flows (slurries) is addressed in Wilson et al. (1992).

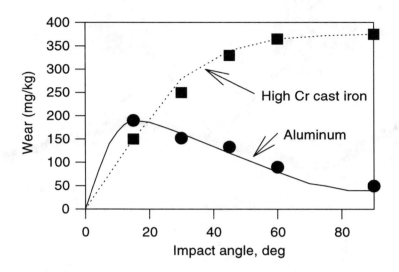

Figure 5.7: Erosion rate of aluminium and high chromium cast iorn by alumina particles impinging at 50 m/s.

5.1.4   van der Waals forces

Inter-particle forces due to particle wetness, electrostatic charges and van der Waals forces are known to cause adhesion of a particle to a wall or onto another particle. The van der Waals forces arise from molecular interaction between solid surfaces. This force becomes apparent when very smooth surfaces are brought into contact. To estimate the magnitude of van der Waals forces acting on solid bodies, the contribution of the many molecules constituting the surfaces must be considered. Hamaker (Hamaker, 1937) carried out calculations for various geometries. The force F between two infinite flat plates with separation z is expressed by

$$F = \frac{A}{6\pi z^3} \tag{5.40}$$

where F is force per unit area and A is referred to as the *Hamaker constant*. The minimum separation distance for two plates in air at standard conditions is 0.4 nm which is the mean free path of the air molecules.

The force between two spheres shown in Figure 5.8 is expressed by

$$F = \frac{Ad}{12z^2} \tag{5.41}$$

where

$$d = \frac{D_1 \cdot D_2}{D_1 + D_2} \tag{5.42}$$

and where z is the separation distance at the point of "contact". Czarnecki and Dabros (1980) suggest that the effect of roughness of the sphere surfaces modifies the van der Waals force by

$$F = \frac{Ad}{12(z+b)^2} \tag{5.43}$$

where $b = (b_1 + b_2)/2$ and b_1 and b_2 are the roughness heights of the two spheres. The force between a sphere and a flat plate can be obtained by putting the diameter of one sphere equal to infinity; i.e., $d = D$ in Equation 5.42.

Examples of the Hamaker constant for various materials are shown in Table 5.5.

Material	Hamaker constant $[J \times 10^{20}]$
Water	4.38
Polystyrene	6.15-6.6
Al_2O_3	15.5
Cu	28.4
Au	45.5

Table 5.5. Hamaker constant.

Example: Determine the maximum roughness of a 10-μm copper particle that would be suspended by the van der Waals force on the bottom side of a perfectly smooth flat copper plate in air.

Solution: The Hamaker constant for copper is 28.4×10^{-20} J. Equating the van der Waals force to the weight gives

$$\frac{AD}{12(z+b)^2} = \frac{\pi}{6}\rho_d D^3 g$$

Solving for $z + b$ one has

$$z + b = \frac{1}{D}\sqrt{\frac{A}{2\pi\rho_d g}}$$

The density of copper is 8900 kg/m^3. Using this value in the above equation, one finds $z + b = 7.2 \times 10^{-8}$ m. Taking z as 0.4×10^{-9} m, the value for the maximum roughness of the sphere is 1.43×10^{-7} m or 0.143 μm!

The Hamaker constant between different materials is given by

$$A_{12} = \sqrt{A_{11} \cdot A_{22}} \tag{5.44}$$

where A_{11} and A_{22} are the Hamaker constants for each material. The above relation is used for "dry" solids. If two solids is separated by a third material, the Hamaker constant is expressed by

$$A_{132} = \left(\sqrt{A_{11}} - \sqrt{A_{33}}\right)\left(\sqrt{A_{22}} - \sqrt{A_{33}}\right) \tag{5.45}$$

where A_{33} is the Hamaker constant of the third material. The Hamaker constant A_{33} for air is very small and thus Equation 5.45 is reduced to Equation 5.44. Comparing Equations 5.44 and 5.45, one finds that the attractive force between two solids becomes smaller in water than in air. In the above equation the sign of the van der Waals force between two solids can be negative, i.e., a repulsive force arises if A_{33} has a value between A_{11} and A_{22}.

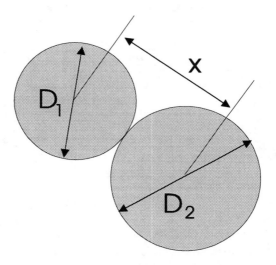

Figure 5.8: Two spherical particles in contact.

5.2   Particle-particle interaction

Particle-particle collision is negligible in dilute gas-particle flows. As the particle concentration becomes higher, particles collide with each other and the loss of particle kinetic energy due to inter-particle collision cannot be neglected. Fortunately, as long as the particulate phase is dispersed, it is sufficient to consider only simple binary collisions, not multiple collisions. Here, the case of two spheres shown in Figure 5.9 is considered. As briefly mentioned in Section 5.1.1 there are two models for the collision problem: the hard sphere model and the soft sphere model. The hard sphere model is described in Section 5.2.1 and the soft sphere model in Section 5.2.2.

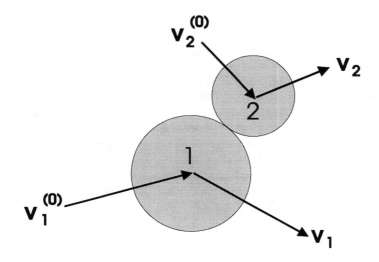

Figure 5.9: Particle-particle collision.

5.2.1 Hard sphere model

The hard sphere model is based on the impulsive force which is defined by the integral of the force acting on a particle versus time. If the particles are assumed to be rigid spheres, the impulse equations are given as follows,

$$m_1(\mathbf{v}_1 - \mathbf{v}_1^{(0)}) = \mathbf{J} \tag{5.46}$$

$$m_2(\mathbf{v}_2 - \mathbf{v}_2^{(0)}) = -\mathbf{J} \tag{5.47}$$

$$I_1(\boldsymbol{\omega}_1 - \boldsymbol{\omega}_1^{(0)}) = r_1 \mathbf{n} \times \mathbf{J} \tag{5.48}$$

$$I_2(\boldsymbol{\omega}_2 - \boldsymbol{\omega}_2^{(0)}) = r_2 \mathbf{n} \times \mathbf{J} \tag{5.49}$$

where $\mathbf{n}$ is the unit normal vector from particle 1 to particle 2 at the moment of contact and $\mathbf{J}$ is the impulsive force exerted on particle 1 (which also acts on particle 2 as the reaction force). The subscripts 1 and 2 refer to the two particles. The superscript (0) means values before collision. I is moment of inertia given by $I = (2/5)ma^2$. In the above equations, the particle mass m, size (radius $= r$), velocities before collision, $\mathbf{v}^{(0)}$ and positions before collision are given. The unknown variables are the impulsive force $\mathbf{J}$ and the post-collisional velocities $\mathbf{v}$. To solve the problem, the same assumptions for particle-wall collisions are used; namely,

1. particle deformation is neglected so, throughout the collision process, the distance between the particle centers of mass is constant and equal to the sum of particle radii,

2. the friction on sliding particles obeys Coulomb's friction law, and

3. once a particle stops sliding no further sliding occurs.

In the problem of particle-wall collision, the analysis is made by dividing the collision process into the compression and recovery periods. The reason for dividing periods in this way is that the coefficient of restitution is defined as the ratio of impulsive forces in the compression and recovery periods. Through such a definition of the restitution coefficient it is convenient to extend the analysis to nonspherical particles. In this section, the particle shape is limited to a sphere, and thus the ratio of the pre- and post-velocities, which is more familiar, is used as the definition of the restitution coefficient. Thus, the analytical procedure is somewhat simpler than that for a particle-wall collision.

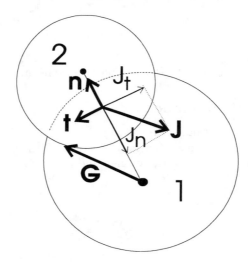

Figure 5.10: Relative motion of two spheres.

The relative velocities between particle centers before and after collision, $\mathbf{G}^{(0)}$ and $\mathbf{G}$, are

$$\mathbf{G}^{(0)} = \mathbf{v}_1^{(0)} - \mathbf{v}_2^{(0)} \tag{5.50}$$

$$\mathbf{G} = \mathbf{v}_1 - \mathbf{v}_2 \tag{5.51}$$

From this point, the analysis is made based on the relative particle motion shown in Figure 5.10. The relative velocity of the contact point before collision is

$$\mathbf{G}_c^{(0)} = \mathbf{G}^{(0)} + r_1\boldsymbol{\omega}_1^{(0)} \times \mathbf{n} + r_2\boldsymbol{\omega}_2^{(0)} \times \mathbf{n} \tag{5.52}$$

In the above equation, the second and third terms (vectors) of the right side lie in the tangential direction while the first term has normal and tangential components. The tangential component of the relative velocity $\mathbf{G}_c^{(0)}$ is given by

$$
\begin{aligned}
\mathbf{G}_{ct}^{(0)} &= \mathbf{G}_c^{(0)} - (\mathbf{G}^{(0)} \cdot \mathbf{n})\mathbf{n} \\
&= \mathbf{G}^{(0)} - (\mathbf{G}^{(0)} \cdot \mathbf{n})\mathbf{n} + r_1 \boldsymbol{\omega}_1^{(0)} \times \mathbf{n} + r_2 \boldsymbol{\omega}_2^{(0)} \times \mathbf{n}
\end{aligned}
\tag{5.53}
$$

The impulsive force $\mathbf{J}$ is divided into the normal and tangential components,

$$
\mathbf{J} = J_n \mathbf{n} + J_t \mathbf{t}
\tag{5.54}
$$

The unit vector $\mathbf{t}$ in the tangential direction is given by

$$
\mathbf{t} = \frac{\mathbf{G}_{ct}^{(0)}}{\left| \mathbf{G}_{ct}^{(0)} \right|}
\tag{5.55}
$$

The post-collisional relative velocity $\mathbf{G}$ is related to the pre-collisional velocity through the restitution coefficient

$$
\mathbf{n} \cdot \mathbf{G} = -e(\mathbf{n} \cdot \mathbf{G}^{(0)})
\tag{5.56}
$$

This equation is regarded as the definition of the restitution coefficient e. As in Section 5.1.1, the restitution coefficient is assumed to be a constant. From Equations 5.46, 5.47 and the definition of the relative velocities, $\mathbf{G}^{(0)}$ and $\mathbf{G}$, one has

$$
\mathbf{G} = \mathbf{G}^{(0)} + \frac{m_1 + m_2}{m_1 m_2} \mathbf{J}
\tag{5.57}
$$

Multiplying by the unit vector $\mathbf{n}$, one obtains

$$
\mathbf{n} \cdot \mathbf{G} = \mathbf{n} \cdot \mathbf{G}^{(0)} + \frac{m_1 + m_2}{m_1 m_2} \mathbf{n} \cdot \mathbf{J}
\tag{5.58}
$$

Substituting Equation 5.56 into the left side of the above equation results in

$$
-e(\mathbf{n} \cdot \mathbf{G}^{(0)}) = \mathbf{n} \cdot \mathbf{G}^{(0)} + \frac{m_1 + m_2}{m_1 m_2} J_n
\tag{5.59}
$$

From Equation 5.59, the normal component of the impulsive force, J_n, is given explicitly as

$$
J_n = -\frac{m_1 m_2}{m_1 + m_2}(1 + e)(\mathbf{n} \cdot \mathbf{G}^{(0)}) < 0
\tag{5.60}
$$

Assuming that the particles slide, the tangential component of the impulsive force, J_t, is

$$
J_t = f J_n < 0
\tag{5.61}
$$

from Coulomb's law for friction. The friction coefficient f and the coefficient of restitution are regarded as known parameters. By substituting the impulsive force $\mathbf{J}$ given by Equations 5.60 and 5.61 into Equations 5.46 to 5.49, all the post-collisional velocities are obtained. The results are

$$\mathbf{v}_1 = \mathbf{v}_1^{(0)} - (\mathbf{n} - f\mathbf{t})(\mathbf{n} \cdot \mathbf{G}^{(0)})(1 + e)\frac{m_2}{m_1 + m_2} \tag{5.62}$$

$$\mathbf{v}_2 = \mathbf{v}_2^{(0)} + (\mathbf{n} - f\mathbf{t})(\mathbf{n} \cdot \mathbf{G}^{(0)})(1 + e)\frac{m_1}{m_1 + m_2} \tag{5.63}$$

$$\boldsymbol{\omega}_1 = \boldsymbol{\omega}_1^{(0)} + \left(\frac{5}{2r_1}\right)(\mathbf{n} \cdot \mathbf{G}^{(0)})(\mathbf{n} \times \mathbf{t})f(1 + e)\frac{m_2}{m_1 + m_2} \tag{5.64}$$

$$\boldsymbol{\omega}_2 = \boldsymbol{\omega}_2^{(0)} + \left(\frac{5}{2r_2}\right)(\mathbf{n} \cdot \mathbf{G}^{(0)})(\mathbf{n} \times \mathbf{t})f(1 + e)\frac{m_1}{m_1 + m_2} \tag{5.65}$$

Equations 6.82 to 5.65 show the solutions corresponding to the case where the two spheres slide during the collision process. If the two spheres stop sliding, the results are different.

From the value for the tangential component of the relative velocity of the contact point it is determined whether or not the particles continue to slide. Consider the component based on the case in which the particles slide. The post-collisional relative velocity is given by

$$\mathbf{G}_c = \mathbf{G} + r_1\boldsymbol{\omega}_1 \times \mathbf{n} + r_2\boldsymbol{\omega}_2 \times \mathbf{n} \tag{5.66}$$

By using Equations 5.48, 5.49, 5.54 and 5.57, Equation 5.66 can be rewritten as

$$\mathbf{G}_c = \mathbf{G}_c^{(0)} + \left\{J_n\mathbf{n} + \left(\frac{7}{2}\right)J_t\mathbf{t}\right\}\frac{m_1 + m_2}{m_1 m_2} \tag{5.67}$$

Taking the scalar product of $\mathbf{t}$ (given by Equation 5.55) the above equation yields the tangential component of $\mathbf{G}_{ct}$ in the form

$$\mathbf{G}_{ct} = \frac{\mathbf{G}_{ct}^{(0)}}{\left|\mathbf{G}_{ct}^{(0)}\right|}\left\{\left|\mathbf{G}_{ct}^{(0)}\right| + \left(\frac{7}{2}\right)J_t\frac{m_1 + m_2}{m_1 m_2}\right\}. \tag{5.68}$$

The case for which the particles continue to slide during the collision process is given by the condition that the sign of the above parenthesis is positive; that is,

$$J_t > - \left(\frac{2}{7}\right)\frac{m_1 m_2}{m_1 + m_2}\left|\mathbf{G}_{ct}^{(0)}\right|. \tag{5.69}$$

The impulse J_t in the above inequality can be expressed in terms of the pre-collisional conditions through Equations 5.60 and 5.61. Rewriting the above relation one has

$$\frac{\mathbf{n} \cdot \mathbf{G}^{(0)}}{\left|\mathbf{G}_{ct}^{(0)}\right|} < \left(\frac{2}{7}\right)\frac{1}{f(1 + e)} \tag{5.70}$$

$\mathbf{G}^{(0)}$ is the pre-collisional relative velocity at the contact point which is given by Equation 5.50 and $\mathbf{G}_{ct}^{(0)}$ is given by Equation 5.53. If Equation 5.70 is not satisfied, the particle stops sliding. This case corresponds to the condition that the post-collisional relative velocity at the contact point is zero; i.e., the left side of Equation 5.68 is equal to zero. Therefore,

$$J_t = -\left(\frac{2}{7}\right)\frac{m_1 m_2}{m_1 + m_2}\left|\mathbf{G}_{ct}^{(0)}\right| \tag{5.71}$$

Substituting the impulsive force $\mathbf{J}$ given by Equations 5.60 and 5.71 into Equations 5.46 to 5.49, all the post-collisional velocities are obtained for the case where the particles stop sliding during the collision. The results are as follows,

$$\mathbf{v}_1 = \mathbf{v}_1^{(0)} - \left\{(1+e)(\mathbf{n}\cdot\mathbf{G}^{(0)})\mathbf{n} + \frac{2}{7}\left|\mathbf{G}_{ct}^{(0)}\right|\mathbf{t}\right\}\frac{m_2}{m_1 + m_2} \tag{5.72}$$

$$\mathbf{v}_2 = \mathbf{v}_2^{(0)} + \left\{(1+e)(\mathbf{n}\cdot\mathbf{G}^{(0)})\mathbf{n} + \frac{2}{7}\left|\mathbf{G}_{ct}^{(0)}\right|\mathbf{t}\right\}\frac{m_1}{m_1 + m_2} \tag{5.73}$$

$$\boldsymbol{\omega}_1 = \boldsymbol{\omega}_1^{(0)} - \frac{5}{7r_1}\left|\mathbf{G}_{ct}^{(0)}\right|(\mathbf{n}\times\mathbf{t})\frac{m_2}{m_1 + m_2} \tag{5.74}$$

$$\boldsymbol{\omega}_2 = \boldsymbol{\omega}_2^{(0)} - \frac{5}{7r_2}\left|\mathbf{G}_{ct}^{(0)}\right|(\mathbf{n}\times\mathbf{t})\frac{m_1}{m_1 + m_2} \tag{5.75}$$

5.2.2 Soft sphere model

The difference between the hard and soft sphere models lies in whether the basic Newtonian equations of motion are given in differential or integral form. The hard sphere model uses the integrated form. With this model the momentum difference between two points in time is equal to the impulsive force defined as the integral of the force versus time. The soft sphere model starts with the differential equations. Therefore the variations in momentum and displacement are obtained for arbitrary times as solutions of the differential equations.

When two particles collide, they actually deform. However, in the soft sphere model, the overlap displacement δ is assumed as shown in Figure 5.11 instead of considering deformation. The larger the displacement, the larger the repulsive force. In such a particle-particle interaction, the particles lose kinetic energy. When the two particles slide under the application of a normal force, a frictional force results. Considering these forces, the soft sphere model is composed of mechanical elements such as a spring, dash-pot and friction slider as shown in Figure 5.12. This model for contact forces was proposed by Cundall and Strack (1979) and is the same as the Voigt model used in the field of rheology. The spring simulates the effect of deformation and the dash-pot the damping effect. The coupler shown in Figure 5.12 connects the two particles when they are in contact but allows the particles to separate under the influence of repulsive forces. If the diameter of one particle is set equal to infinity, the model reduces to a particle-wall collision. In the soft sphere model, not only the relationship

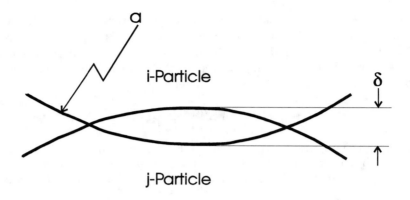

Figure 5.11: Displacement of two particles in contact.

between the pre- and post-collisional velocities can be obtained but also the forces acting on spheres during contact are obtained as well.

The effects of these mechanical components on particle motion appear through the following parameters;

stiffness k

damping coefficient η

friction coefficient f

The normal component of the contact force, $\mathbf{F}_{nij}$, acting on particle i is given by the sum of the forces due to the spring and the dash-pot; namely,

$$\mathbf{F}_{nij} = (-k_n\delta_n - \eta_{nj}\mathbf{G}\cdot\mathbf{n})\mathbf{n} \qquad (5.76)$$

where δ_n is the displacement of the particle caused by the normal force, $\mathbf{G}$ is the velocity vector of particle i relative to particle j ($\mathbf{G} = \mathbf{v}_i - \mathbf{v}_j$) and $\mathbf{n}$ is the unit vector in the direction of the line from the center of particle i to that of particle j .

In this section the suffixes i and j are used instead of 1 and 2, because the following analysis can be easily extended to cases where many particles are in contact with each other. The suffix i corresponds to 1 and the suffix j corresponds to 2 in Figure 5.10. In the case addressed here the spheres are assumed to have the same size (radius = a). The tangential component of the contact force, $\mathbf{F}_{tij}$, is given by

$$\mathbf{F}_{tij} = -k_t\delta_t - \eta_{tj}\mathbf{G}_{ct} \qquad (5.77)$$

where k_t and η_{tj} are the stiffness and displacement, respectively, in the tangential direction. In the above equations, the suffixes n and t designate the components in the normal and tangential directions. $\mathbf{G}_{ct}$ is the slip velocity at the contact point, which is given by

$$\mathbf{G}_{ct} = \mathbf{G} - (\mathbf{G}\cdot\mathbf{n})\mathbf{n} + a(\boldsymbol{\omega}_i + \boldsymbol{\omega}_j)\times\mathbf{n} \qquad (5.78)$$

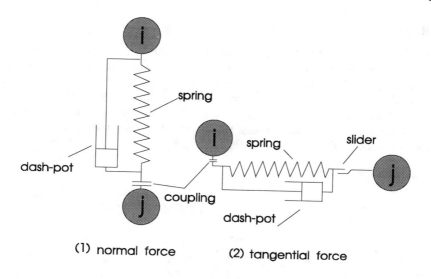

Figure 5.12: Spring-damper system to model contact forces.

This expression is the same as Equation 5.53 except that the two spheres have the same radius a. If the following relation is satisfied

$$|\mathbf{F}_{tij}| > f\,|\mathbf{F}_{nij}| \qquad (5.79)$$

particle i slides and the tangential force is given by

$$\mathbf{F}_{tij} = -f\,|\mathbf{F}_{nij}|\,\mathbf{t} \qquad (5.80)$$

instead of Equation 5.77. Equation 8.114 is the Coulomb-type friction law and $\mathbf{t}$ is the unit tangential vector defined by

$$\mathbf{t} = \frac{\mathbf{G}_{ct}}{|\mathbf{G}_{ct}|} \qquad (5.81)$$

In general, several particles are in contact with particle i at the same time as shown in Figure 5.13. Therefore the total force and torque acting on particle i are obtained by taking the sum of the above forces with respect to j.

$$\mathbf{F}_i = \sum_j (\mathbf{F}_{nij} + \mathbf{F}_{tij}) \qquad (5.82)$$

$$\mathbf{T}_i = \sum_j (a\mathbf{n} \times \mathbf{F}_{tij}) \qquad (5.83)$$

Stiffness

The next step after modeling of the contact forces is to determine the values for the stiffness k, the damping coefficient η and the friction coefficient f. Of these parameters, only the friction coefficient f is measurable and can be ascribed an empirical value. The *damping coefficient* η is deduced from the stiffness as discussed later. Therefore, the *stiffness* k must be determined first. Fortunately, the stiffness can be calculated using the *Hertzian contact theory* when physical properties such as the Young's modulus and Poisson ratio are known.

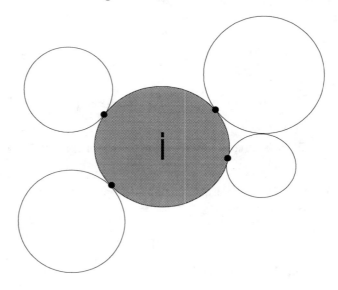

Figure 5.13: Particle with several contact forces.

According to Hertzian contact theory, the relationship between the normal force P_{na} and the normal displacement δ_n is

$$P_n = K_n \delta_n^{3/2} \tag{5.84}$$

For the case of two spheres of the same radius a, K_n is expressed by

$$K_n = \frac{\sqrt{2a}\,E_s}{3(1-\sigma_s^2)} \tag{5.85}$$

and K_n for contact between a sphere and wall is given by

$$K_n = \frac{\frac{4\sqrt{a}}{3}}{\frac{1-\sigma_s^2}{E_s}\cdot\frac{1-\sigma_w^2}{E_w}} \tag{5.86}$$

where E_s and σ_s are Young's modulus and Poisson's ratio for the particle material, respectively, and E_w and σ_w are the same parameters for the wall material.

Example: Calculate the normal force due collision of two 100-micron aluminum particles due to a displacement of 2 μm.

Solution: Poisson's ratio and modulus of elasticity for aluminum is 0.33 and 68 GPa, respectively. The spring constant for a Hertzian contact is

$$K_n = \frac{\sqrt{10^{-4}} \times 68 \times 10^9}{3(1 - 0.33^2)} = 0.234 \times 10^9 \text{ N/m}^{3/2}$$

The force due to a 2 μm displacement is

$$P_n = 0.254 \times 10^9 \times (2 \times 10^{-6})^{3/2} = 0.719 \text{ N}$$

Equation 8.118 implies that the relationship between the force and displacement is not linear but rather varies to the 3/2 power of the displacement. If the above results are applied to the model for contact forces given above, then Equation 5.76 is replaced by

$$\mathbf{F}_{nij} = (-K_n \delta_n^{3/2} - \eta_{nj} \mathbf{G} \cdot \mathbf{n})\mathbf{n} \tag{5.87}$$

The relationship between the tangential force P_t and the displacement δ_t has also been derived theoretically by Mindlin (1949) and Mindlin & Deresiewicz (1953). According to their theories, the force-displacement relationship depends on the normal displacement δ_n. Moreover, if the contact surface is allowed to slip, the expression for the relationship between the force and displacement becomes more complicated. For the case with no slip the result is

$$P_t = \frac{2\sqrt{2r_s}H_s}{2 - \sigma_s}\delta_n^{1/2} \cdot \delta_t \tag{5.88}$$

where H_s is the shear modulus, which is related to Young's modulus E_s and the Poisson ratio σ_s by

$$H_s = \frac{E_s}{2(1 + \sigma_s)}. \tag{5.89}$$

Equation 8.122 shows that the tangential force-displacement ratio is linear. Applying the above result to Equation 5.77, one finds the stiffness k_t to be

$$k_t = \frac{2\sqrt{2r_s}H_s}{2 - \sigma_s}\delta_n^{1/2} \tag{5.90}$$

The stiffness for contact between the sphere and wall becomes

$$k_t = \frac{8\sqrt{a}H_s}{2 - \sigma_s}\delta_n^{1/2} \tag{5.91}$$

This equation is based on the following assumption. When considering the tangential displacement in the contact between a sphere and the wall, the wall is regarded as a rigid body because elastic displacement of the wall in the tangential direction is much smaller than that of a sphere. Therefore, the wall properties are not included in Equation 5.91. However, the elastic displacement of the wall in the normal direction cannot be neglected.

As shown above, the stiffness can be determined from the material properties. This is the advantage of the model. Unfortunately, it is often difficult in practice to use the stiffness calculated by the Hertzian theory, because the time step Δt required for numerical integration becomes so small that an excessive amount of computational time is needed. The time step Δt should be less than one-tenth of the natural oscillation period $2\pi\sqrt{m/k}$ of a spring-mass system. Therefore, in many cases, a small value of stiffness is assumed for convenience of the calculation. In some cases the results based on a stiffness much smaller than an actual value are not very different from those based on a precise stiffness value. Fortunately, this is the case for fluid-particle multiphase flows because the fluid forces acting on the particles make the difference in stiffness a minor effect.

There are other cases where small values of stiffness are assumed. When the soft sphere model is applied to a sand or snow avalanche, it is not practical to take the smallest sand or snow particle as a discrete element but rather an assembly of particles of suitable size should be taken as one particle (virtual particle) because of the limits of memory size and computational time. In such a case, the stiffness of the actual particle is meaningless. The stiffness of the virtual particle is expected to be smaller than that of the actual particle. However, there is no theory available at present which yields the stiffness of the virtual particle.

Damping coefficient

Cundall & Strack (1979) proposed two expressions for the damping coefficient; namely,

$$\eta_n = 2\sqrt{mk_n}$$
$$\eta_t = 2\sqrt{mk_t}$$

(5.92)

which were derived for the critical damping condition of a single degree-of-freedom system consisting of a mass, spring and dash-pot. The reason for choosing the critical damping condition, as shown in Equation 5.92, comes from the requirement that bouncing motion after collision should be damped as soon as possible.

Another method for determining the damping coefficient is based on the idea that the damping coefficient should be related to the coefficient of restitution which is regarded as a physical property of the particle material. Fortunately the coefficient of restitution can be measured in a simple experiment. If the restitution coefficient is assumed constant and the relationship between spring force and displacement is linear, analytical expressions for the damping coefficient and restitution coefficient are obtained from the dynamics of an oscillating spring-mass-damper system. Tsuji et al. (1992) showed the damping coefficient η_n for a nonlinear spring is numerically related to the restitution coefficient by Equation 8.121. According to Tsuji et al., the damping coefficient η_n is expressed by

$$\eta_n = \alpha \sqrt{mK_n} \delta_n^{1/4} \tag{5.93}$$

where α is a constant related to the coefficient of restitution. The relationship between α and the coefficient of restitution e is shown in Figure 5.14. As for the damping coefficient η_t in the tangential direction, the same value as used for η_n is often used for η_t with no firm justification.

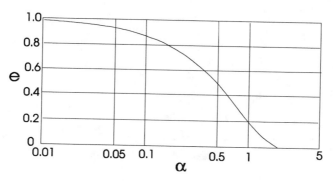

Figure 5.14: Relationship between α and coefficient of restitution. (Reprinted from *Powder Tech.*, 71, Tsuji et al., Lagrangian numerical simulation of plug flow of collisionless particles in a horizontal pipe, 246, 1992, with kind permission from Elsevier Science S.A., P.O. Box 564, 1001 Lausanne, Switzerland.)

Originally the soft sphere model was used to model the flow of granular materials without considering the affects of the interstitial fluid. Many papers have been published for flows in chutes and hoppers, as well as for vibrating beds, packing, attrition and pulverization. Williams & Mustoe (1993) and Mehta (1994) review recent developments in this area. Even though these models have only recently been applied to multiphase flows, it has been found that the soft sphere model is very useful for discrete particle simulation of dense phase flows such as fluidized beds (Tsuji et al., 1993) and dense phase pneumatic conveying (Tsuji et al., 1992). In this section, a description has been provided mainly for the model shown in Figure 5.12. However, other elements can be added, if necessary, as shown in Figure 5.15.

5.2.3 Hard sphere simulation of a soft sphere model

The primary problem with the implementation of the soft sphere model is the small time steps needed to calculate the motion during collision. Walton (1993) has proposed a hard sphere analogy of the soft sphere model which can be used if

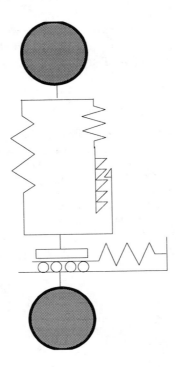

Figure 5.15: Partially latching spring model (Walton, O.R. and Braun, R.L., *J. Rheology,* 30, 949, 1986. With permission.)

only two particles are in contact at one time. This model is based on a normal coefficient of restitution, e, and a rotational restitution coefficient, β. First of all, a calculation is done assuming that sliding occurs and a predetermined maximum value of β is calculated,

$$\beta^* = -1 + f(1+e)(1+\frac{1}{K})\frac{\left|\mathbf{n}\cdot\mathbf{G}^{(0)}\right|}{\left|\mathbf{G}_{ct}^{(0)}\right|} \tag{5.94}$$

where $K = 4I/mD^2$. For a uniform density sphere, $K = 0.4$. If β^* exceeds the maximum value for β; namely β_o, then $\beta = \beta_o$. This corresponds to the rolling solution. Otherwise, $\beta = \beta^*$ which corresponds to the sliding solution. For computational convenience and to obviate the need to take square roots, Walton rewrote Equation 5.94 as

$$(\beta^* + 1)^2 = f^2(1+e)^2(1+\frac{1}{K})^2\frac{\left(\mathbf{n}\cdot\mathbf{G}^{(0)}\right)^2}{\mathbf{G}_{ct}^{(0)}\cdot\mathbf{G}_{ct}^{(0)}} \tag{5.95}$$

The condition for β then becomes

$$\text{If } (\beta^* + 1)^2 > (\beta_o + 1)^2 \quad \text{then } \beta = \beta_o$$

$$\text{Otherwise } \beta = \beta^*$$

Finally for two colliding particles of the same size, the post-collisional translational velocities are

$$\mathbf{v}_1 = \mathbf{v}_1^{(0)} - \tfrac{1}{2}(1+e)(\mathbf{n} \cdot \mathbf{G})\mathbf{n} - \frac{K}{2}\frac{1+\beta}{1+K}(\mathbf{G}_{ct}^{(0)})$$

$$\mathbf{v}_2 = \mathbf{v}_2^{(0)} + \tfrac{1}{2}(1+e)(\mathbf{n} \cdot \mathbf{G})\mathbf{n} + \frac{K}{2}\frac{1+\beta}{1+K}(\mathbf{G}_{ct}^{(0)})$$

(5.96)

and the rotational velocities are

$$\boldsymbol{\omega}_1 = \boldsymbol{\omega}_1^{(0)} - \frac{1+\beta}{a(1+K)}\mathbf{n} \times \mathbf{G}_{ct}^{(0)}$$

$$\boldsymbol{\omega}_2 = \boldsymbol{\omega}_2^{(0)} - \frac{1+\beta}{a(1+K)}\mathbf{n} \times \mathbf{G}_{ct}^{(0)}$$

(5.97)

The value for β_o has to be obtained from experiment. For plastic spheres, $\beta_o \simeq 0.35$.

This hard sphere analogy of the soft sphere collision dynamics greatly reduces the computational time for a soft sphere model. The model has been developed for collisions of particles of unequal size as well.

5.2.4 Fluid forces on approaching particles

Consider two spheres in a fluid approaching one another. As the distance between the spheres becomes smaller, the fluid pressure between the spheres becomes larger to move the fluid outward and the resulting force acts to prevent contact. This force can be deduced theoretically from the equations of fluid motion using the lubrication approximations for creeping flow. Here, the two spheres are assumed to be the same size so the flow is symmetric about the center plane between the spheres and axisymmetric along the line connecting the centers of the spheres. Although not addressed here, there is a model for the case where a liquid bridge is formed but the bridge does not affect the final result so long as the separation distance is small.

As shown in Figure 5.16, fluid motion is described by the cylindrical coordinates (r, θ, z). The basic equations to be used are the continuity equation and the Navier-Stokes equations with the lubrication approximation. In this case, the inertia terms and some of viscous stress terms can be neglected and the pressure is assumed to be a function only of r so the Navier-Stokes equations reduce to

$$\frac{dp}{dr} = \mu\frac{\partial^2 u_r}{\partial z^2}$$

(5.98)

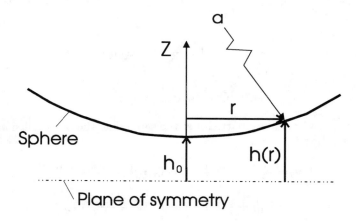

Figure 5.16: Coordinate system for two approaching particles.

where u_r is the velocity component in the r-direction. Integrating Equation 5.98 in the region $z = 0 - h$ with the boundary conditions of zero shear rate in the center plane ($\partial u_r/\partial z = 0$ at $z = 0$) and no slip at the wall ($u_r = 0$ at $z = h$) yields,

$$u_r = \frac{1}{2\mu}\frac{dp}{dr}(z^2 - h^2) \tag{5.99}$$

Conservation of mass over the region bounded by the radius r and the space between the symmetry plane and the sphere requires

$$2\pi r \int_o^h u_r dz + \pi r^2 \dot{h} = 0 \tag{5.100}$$

where $\dot{h}$ is the rate at which the sphere is approaching the symmetry plane. Substituting u_r from Equation 5.99 and integrating yields

$$\frac{dp}{dr} = \mu\left(\frac{3}{2}\right)\frac{r\dot{h}}{h^3} \tag{5.101}$$

The pressure difference $\Delta p = p - p_0$ is obtained by integrating Equation 5.101 from $r = r$ to $r = a$;

$$\Delta p = \mu\frac{3\dot{h}}{2}\int_a^r \frac{r}{h^3}dr \tag{5.102}$$

The force due to the pressure is obtained by integrating Equation 5.102 over the regions, $\theta = 0 - 2\pi$ and $r = 0 - a$ which results in

$$F = \int_0^{2\pi} \int_o^a \left[\mu \frac{3\dot{h}}{2} \int_a^r \frac{r}{h^3} dr \right] r d\theta dr = -\pi\mu \frac{3\dot{h}}{2} \int_0^a \frac{r^3}{h^3} dr \qquad (5.103)$$

The distance $h - h_0$ is approximated by

$$h - h_0 = \frac{r^2}{2a} \qquad (5.104)$$

By substituting Equation 5.104 into Equation 5.103 with the assumption of small h_0 one obtains

$$F = -\frac{3\pi\mu a^2 \dot{h}}{2h_0} \qquad (5.105)$$

This is the force acting on two approaching spheres. Note that the distance R corresponding to the liquid bridge disappears. The more general case in which the shear stress is expressed by a power law in the form

$$\tau = \mu \left(\frac{\partial u_r}{\partial z} \right)^n \qquad (5.106)$$

can be found in Adams & Edmondson (1987).

5.3 Summary

Particle-particle interaction is the key element in dense phase flows. The two models for particle-particle interaction are the hard sphere and the soft sphere models. With the hard sphere model the post-collisional velocities and rotations are determined as a function of the pre-collisional conditions, coefficient of restitution and coefficient of friction. The soft sphere model describes the particle history during the collision process and can be modeled with spring-damper arrangements. Particle-wall collisions are treated with the hard sphere model. Wall roughness can be simulated with a local wall slope at the surface.

Exercises

5.1. A 100-μm spherical particle on a smooth surface. The velocity components are shown on the figure. The initial angular velocity is zero. The coefficient of friction is 0.2. Find the coefficient of restitution, the post-collisional angular velocity and velocity component in the x-direction.

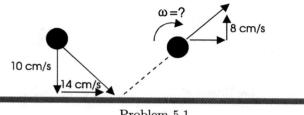

Problem 5.1

5.2. A spherical ball, 38 cm in diameter, impinges on a flat surface. The incident translational and angular velocities are

$$\mathbf{v} = (1.13\mathbf{i} - 1.62\mathbf{j} - 0.48\mathbf{k}) \text{ m/s}$$

$$\boldsymbol{\omega} = (0.1\mathbf{i} - 0.02\mathbf{j} + 0.05\mathbf{k}) \text{ rad/s}$$

The coefficient of restitution is 0.65 and the coefficient of friction is 0.3. Find the post-collisional velocity and rotation vector.

5.3. Use the soft sphere model to analyze the collision of two 600-μm spherical aluminum particles shown in the figure. The particles deform such that the overlap displacement is 25 μm. The coefficient of restitution is 0.8 and the coefficient of friction is 0.35. Determine if sliding occurs.

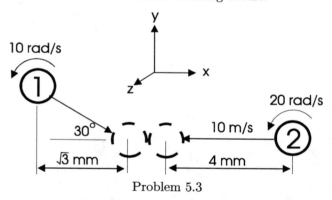

Problem 5.3

5.4. Two particles, with coefficient of restitution e, collide while traveling in the same direction as shown in the figure. Both particles are not rotating.

1. Using the momentum conservation, find the velocities of each particle after collision.

2. Simplify Equations 6.82 and 5.63 in the text and compare the results with those obtained in first part of this problem.

3. Particle 1 is 10 μm in diameter and is moving at 5 m/s. Particle 2 is 5 microns in diameter and is moving at 10 m/s. Both particles have a density of 2500 kg/m^3. The coefficient of restitution is 0.9. Find the velocities after collision.

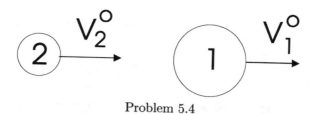

Problem 5.4

5.5. Show that

$$F = -\pi \mu_c \frac{3\dot{h}}{2} \int_0^R \frac{r^3}{h^3} dr$$

reduces to

$$F = -3\pi \mu_c \frac{a^2 \dot{h}}{2h_0}$$

for small h_0 and $h = r^2/2a + h_0$.

5.6. Determine the post-collisional translation velocities resulting from collision of a 2 mm particle with a smooth inclined wall. The inclination angle of the wall is $30°$. The precollision velocities are $v_x^0 = 5$ m/s, $v_y^0 = -3$ m/s and $v_z^0 = 1$ m/s. Assume $f = 0.4$ and $e = 0.8$. The initial angular velocity is zero.

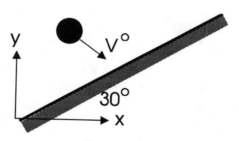

Problem 5.6

5.7. Assume two hard spheres collide with no initial rotation. Identify the condition (relations between **G**, **n** and **t**) for which the collision occurs without sliding. What are the translation velocities (as a function of $\mathbf{v}_1^0, \mathbf{v}_2^0, \mathbf{G}^0, e, m_1$ and m_2) and angular velocities after collision?

5.8. Low density polyethylene (LDPE) particles are pneumatically transported through a duct. All the particles are spheres with a diameter of 1 mm. Two particles, i and j, move through the duct in the manner illustrated in the figure. Find the magnitude of the force of collision for a normal displacement (δ_n) and tangential displacement (δ_t) of 1 μm. LDPE has the following properties: E = 30 GPa, $\sigma = 0.5$, $\rho = 920$ kg/m³, $f = 0.1$ and $e = 0.2$. Assume aerodynamic forces are insignificant and that the two particles move in the same plane.

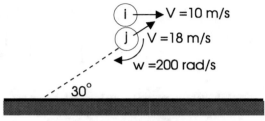

Problem 5.8

5.9. A particle drops vertically onto a flat plate. Just before collision with the wall, the ratio $a\omega_z/v_Y = -0.5$. The coefficient of restitution is 0.8 and the friction factor is 0.2. Find the angle with respect to the vertical with which the particle bounces from the surface.

5.10. Two particles of equal mass, m, collide as shown in the figure. The particles are glass beads, 50 μm in diameter. The velocity of bead 1 is 10 m/s and the velocity of bead 2 is 5 m/s at the point of collision when bead 2 is directly above bead 1. Bead 1 has a rotational velocity of 10 rad/s in the CW (clockwise) direction while bead 2 has a rotational velocity of 5 rad/s in the CCW (counter-clockwise) direction. Find the impulsive force, J, and the post-collisional velocities of both particles.

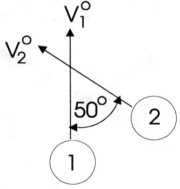

Problem 5.10

5.11. Find the size of an alumina particle which would be suspended (van der Waals force = weight) by 3-cm copper sphere. The roughness of the copper sphere is 1 μm and the roughness of the alumina sphere is 1% of its diameter.

5.12. As a particle approaches a plate, how far (in terms of h_o/a) from the plate is the force on the particle due to the plate being equal to Stokes drag?

Chapter 6

Continuous Phase Equations

Unlike the flow of a single phase liquid or gas, the carrier phase of a dispersed phase flow contains dispersed particles or droplets. For purposes of analysis, the ideal situation would be to solve the governing conservation (continuity, momentum and energy) equations for the carrier phase by accounting for the boundary conditions imposed by each and every particle or droplet in the field. This would provide a complete description of the carrier phase throughout the mixture. Computationally, this would require a grid dimension at least as small as the smallest particle in the field. Such a solution is beyond current computer capability. Solutions have been obtained in limited cases with a finite number of particles in a low Reynolds number (Stokes) flow (Brady 1993). Also numerical solutions have been obtained for flows in which the particles occupy no volume but produce a drag force on the flow (Elghobashi & Truesdell 1992, Squires & Eaton 1991). These solutions are also limited to low Reynolds numbers. In general, however, one must resort to the use of equations based on the average properties in a flow.

The purpose of this chapter is to introduce the averaging procedures and to present the equations in volume average form suitable for numerical model development. The various averaging approaches are first discussed. The effects of particles or droplets on control volume boundaries are presented. The conservation equations for a quasi-one-dimensional flow are developed first to show the essential features of the continuous phase equations. The latter portion of the chapter presents the more general multidimensional equations based on volume averaging. The formal volume averaging procedures are provided in Appendix B. The final portion of the chapter provides a brief discussion on turbulence modulation.

6.1 Averaging procedures

In essence there are three approaches to averaging the continuous phase equations: time, volume and ensemble averaging.

6.1.1 Time averaging

The time average is the result of averaging the flow properties over time at a point in the flow as shown in Figure 6.1. This type of measurement corresponds to hot-wire or laser-Doppler anemometry which has been used extensively to obtain average and fluctuational properties in single-phase flows. The time average of some property B of the fluid is defined as

$$\hat{B} = \frac{1}{T} \int_0^T B \, dt \qquad (6.1)$$

where T is the averaging time.

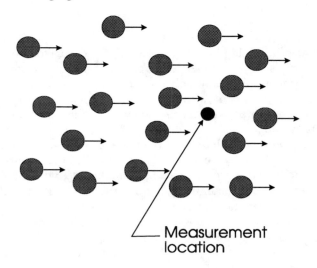

Figure 6.1: Measurement location in a multiphase flow field.

Assume that the velocity of each phase is measured as particles and fluid pass the measuring point. The signal may appear as shown in Figure 6.2. Obviously the averaging time must be large compared to the local fluctuation time, t', in order to define an average value. Yet, the averaging time must be smaller than the time associated with the system change, T'.

$$t' \ll T \ll T' \qquad (6.2)$$

In many transient flow systems, this condition may not be realizable. A true time average can only be obtained in a steady flow system and is given by

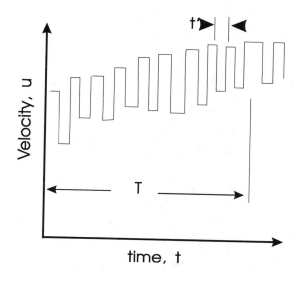

Figure 6.2: Velocity variation with time at measurement point.

$$\hat{u} = \lim_{T \to \infty} \frac{1}{T} \int_0^T u\, dt$$

A more detailed discussion of temporal averaging is provided by Ishii (1975).

6.1.2 Volume averaging

Volume averaging is carried out by averaging properties at an instant in time over a volume and ascribing the average value to a point in the flow. For example, the *volume averaged* property B would be defined as

$$\bar{B} = \frac{1}{V} \int_V B\, dV \tag{6.3}$$

where V is the averaging volume. Assume the distribution of the dispersed phase mixture appears as shown in Figure 6.3 where ℓ is the nominal distance between the particles and L is a distance which characterizes the spatial change in mixture properties. Obviously, in order to obtain a near stationary average (an average which does not change with a change in the size of the averaging volume), the averaging volume must be much larger than ℓ^3. However, in order that the average provide a local value for B in the field, the averaging volume must be much less than L^3. Thus the constraints on the averaging volume are

$$\ell^3 \ll V \ll L^3 \tag{6.4}$$

This constraint is essential to approximate spatial derivatives of $\bar{B}$ in the flow field. An accurate volume average is only possible for a homogeneous mixture.

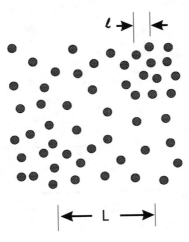

Figure 6.3: Spatial distribution of particles in averaging field.

Besides the volume average defined by Equation 6.3 there is also a *phase average* which is the average over the volume occupied by the phase.

$$\langle B \rangle = \frac{1}{V_c} \int_{V_c} B \, dV \qquad (6.5)$$

where V_c is the volume associated with the continuous phase. If B is the density of the fluid then $\langle \rho_c \rangle$ is the average material density of the fluid. If the density is constant then $\langle \rho_c \rangle = \rho_c$. The relationship between the volume averaged property and the phase average is $\bar{B} = \alpha_c \langle B \rangle$. The bulk density defined in chapter 2 is the same as the volume averaged density defined here.

A formalism has been developed for obtaining volume averaged equations for a mixture by averaging the differential equations for a single phase over a volume. This approach is used in Appendix B to obtain the continuity, momentum and energy equations for the carrier phase in a dispersed phase flow.

Another approach to volume and temporal averaging has been developed by Roco and Shook (1985) and has been applied to liquid-solid flows.

6.1.3   Ensemble averaging

Ensemble averaging avoids the shortcomings of time and volume averaging but is much more difficult to implement. Ensemble averaging is based on the probability of the flow field being in a particular configuration at a given time. For example, assume that the distribution of the fluid density over a region is measured many times. It is found that there are N different configurations and that the distribution in each configuration (realization) at a given time t is

$$\rho_c = f_\eta(x_i, t) \tag{6.6}$$

where η is one realization of the N configurations (ensemble). Assume that $n(\eta)$ is the number of times that configuration $f_\eta(x_i, t)$ occurred.. The ensemble average is then defined as

$$\langle \rho_c \rangle = \frac{\sum_N f_\eta(x_i, t) n(\eta)}{\sum_N n(\eta)} \tag{6.7}$$

In the limit of an infinite number of realizations, the above equation becomes

$$\langle \rho_c \rangle = \int_0^1 f(x_i, t, \mu) d\mu \tag{6.8}$$

where μ is the probability that the realization $f(x_i, t)$ will occur. Obviously ensemble averaging is not limited by volume or time constraints. Formulation of the equations for a particulate mixture has recently been published by Liljegren (1977).

6.2 Boundary particles

The approach to be used here to develop the equations for the carrier phase is the shell balance method. The conservation principles are applied to a finite size volume which is enclosed by an arbitrary boundary. The boundary may pass through several particles which are referred to as *boundary particles*. These particles are responsible for a blockage area of the carrier flow. Also these particles, or droplets, contribute to coupling effects such as mass transfer to and forces on the carrier phase fluid inside the control volume.

6.2.1 Blockage effects

Consider the portion of the boundary surface shown in Figure 6.4. The surface with area A cuts through several particles in the field. The area of the particles intersected by the surface is the *blockage* area, A_b. The objective here is to quantify the ratio of the blockage area to the total area (A_b/A) as a function of particle volume fraction.

Assume that all the particles are spherical and of the same size. The cross-sectional area of the cut where the surface passes through the particle is given by

$$A_c = \pi(a^2 - \epsilon^2) \tag{6.9}$$

where ϵ is the distance from the particle center to the surface. The centers of all the particles will lie in the volume defined by a distance $-a$ and $+a$ from the surface area.

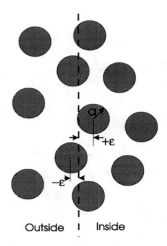

Figure 6.4: Boundary particles on a control surface.

There is a uniform probability that a particle will lie at a distance ϵ from the surface so the probability distribution function is

$$f(\epsilon) = \frac{1}{2a} \tag{6.10}$$

for $|\epsilon| \leq a$ and is equal to zero for $|\epsilon| > 0$. If there is a total of N particles intersected by the surface, the blockage area is given by

$$A_b = N \int_{-a}^{+a} \pi(a^2 - \epsilon^2) f(\epsilon) d\epsilon \tag{6.11}$$

Integrating this equation using the above distribution for $f(\epsilon)$ gives

$$A_b = N \frac{2\pi}{3} a^2 \tag{6.12}$$

The number of particles intersected by the surface is related to the number density, n, by

$$N = 2aAn \tag{6.13}$$

Substituting this value for N into Equation 6.12 results in

$$\frac{A_b}{A} = n \frac{4}{3} \pi a^3 \tag{6.14}$$

and the product of the number density and the volume of a single particle is the dispersed phase volume fraction so

$$\frac{A_b}{A} = \alpha_d \tag{6.15}$$

Thus the flow rate of the continuous phase through a cloud of particles can be expressed as

$$\dot{m} = \rho_c(1 - \alpha_d)uA = \rho_c\alpha_c uA \qquad (6.16)$$

where u is the phase[1] velocity of the continuous phase.

Another more direct approach to relate the blockage to the volume fraction is to realize that the bulk density of the carrier phase is the mass of the carrier phase per unit volume of mixture. Thus the product of the bulk density, the phase velocity and the cross-sectional area must yield the mass flow rate which is expressed by Equation 6.16 . The relationship between volume fraction and bulk density leads directly to Equation 6.16 with no restrictions on the dispersed phase size distribution and shape.

6.2.2 Property transfer

The boundary droplets also contribute to the mass, momentum and energy transfer to the continuous phase. The question arises as to how to account for these contributions for particles which are only partially inside the volume.

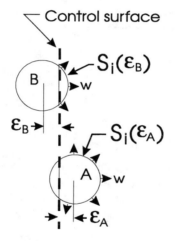

Figure 6.5: Mass transfer from boundary particles.

As an example, consider mass transfer from the evaporating boundary droplets shown in Figure 6.5. The center of droplet A is located at ε_A with respect to the boundary and the mass transfer to the carrier phase inside the computational volume is

$$\dot{m}_{A,i} = \int_{S_{A,i}} \rho_s w dS \qquad (6.17)$$

---

[1] As opposed to the "superficial velocity" which is the product of the phase velocity and continuous phase volume fraction, $\alpha_c u_c$.

where ρ_s is the density of the fluid at the droplet surface, w is the velocity of the fluid with respect to the surface and $S_{A,i}$ is the surface inside the computational volume. The surface S_i is a fraction of the total droplet surface S_d and can be expressed as $S_i = S_i(\varepsilon_A)$. Thus the integral becomes

$$\dot{m}_{A,i}(\varepsilon_A) = \int_{S_i(\varepsilon_A)} \rho_s w dS \tag{6.18}$$

and can be expressed as

$$\dot{m}_{A,i}(\varepsilon_A) = \dot{m} - \dot{m}_{A,o}(\varepsilon_A) \tag{6.19}$$

where $\dot{m}$ is total mass transfer from droplet and $\dot{m}_{A,o}(\varepsilon_A)$ is the mass transferred from the droplet to the outside of the computational volume. Droplet B has the same distribution of mass transfer over the surface as A (and, therefore, the same total mass transfer) and is located at ε_B. The mass transferred inside the computational volume from droplet B is

$$\dot{m}_{B,i}(\varepsilon_B) = \dot{m} - \dot{m}_{B,o}(\varepsilon_B) \tag{6.20}$$

Summing Equations 6.19 and 6.20 for the net mass transferred to inside the volume from both boundary droplets gives

$$\dot{m}_{A,i}(\varepsilon_A) + \dot{m}_{B,i}(\varepsilon_B) = 2\dot{m} - [\dot{m}_{B,o}(\varepsilon_B) + \dot{m}_{A,o}(\varepsilon_A)] \tag{6.21}$$

The position ε of complementary droplet B is defined such that $\dot{m}_{B,o}(\varepsilon_B) + \dot{m}_{A,o}(\varepsilon_A) = \dot{m}$ so the mass flux to the inside from the two complementary droplets is

$$\dot{m}_{A,i}(\varepsilon_A) + \dot{m}_{B,i}(\varepsilon_B) = \dot{m} \tag{6.22}$$

Thus a pair of complementary boundary droplets can be treated as one droplet completely inside the control volume.[2] This is valid as well for the integral of other scalar and vector properties over the surface of the boundary droplets.

6.3   Quasi-one-dimensional flow

A quasi-one-dimensional flow is a flow in which the flow variations in one direction are considered but the variations in the cross-stream direction are neglected. This model can be applied to a flow with gently sloping walls. The control volume used for the quasi-one-dimensional flow is shown in Figure 6.6. The control surface can be broken into two parts: the surface through the continuous phase at stations 1 and 2 and along the wall and the surface adjacent to the droplets inside the control volume. The continuous phase velocities at the stations 1 and 2 are u_1 and u_2, respectively. A shear stress of τ_w acts on the fluid at the wall. The average pressure acting on the continuous phase along the sloping wall is

---

[2] The complementary boundary droplets may not be symmetrically located from the boundary ($\varepsilon_A = -\varepsilon_B$) unless the distribution of the properties is uniform over the surface.

$\bar{p}$ which is the average pressure between stations 1 and 2; $\bar{p} = (p_1 + p_2)/2$. The mass rate of change of each droplet is $\dot{m}$ and the droplet velocity is v. There are boundary droplets on the control surface at stations 1 and 2. The number of droplets in the control volume plus the pairs of complementary droplets on the boundary is N. The volume of the control volume is V and is equal to the product of the average area, $\bar{A}$, and the length Δx.

Formal volume averaging procedures will not be used to develop the equations for quasi-one-dimensional flows. The deviations of the property values, such as density and velocity, from the phase averaged values will not be included. Formal volume averaging procedures are used for the multidimensional equations which are developed in Appendix B and presented in Section 6.4.

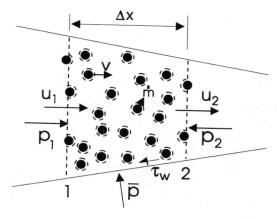

Figure 6.6: Quasi-one-dimensional control volume with dispersed phase elements.

6.3.1 Continuity equation

The continuity equation states that the rate of change of mass in the control volume plus the net efflux of droplet mass through the control surface is equal to zero.

$$\boxed{\begin{array}{c}\text{Rate of} \\ \text{mass} \\ \text{accumulation}\end{array}} + \boxed{\begin{array}{c}\text{Net} \\ \text{efflux} \\ \text{of mass}\end{array}} = 0$$

The continuous phase crosses the control surfaces at stations 1 and 2 and also passes through the surfaces surrounding every droplet, including the boundary droplets. The continuity for the continuous phase mass is

$$V\Delta_t(\rho_c\alpha_c) + \alpha_c\rho_c uA \mid_2 -\alpha_c\rho_c uA \mid_1 + \sum_{k}^{N} \dot{m}_k = 0 \qquad (6.23)$$

where the difference operator Δ_t is defined as

$$\Delta_t(\ ) = \frac{\Delta(\ )}{\Delta t}$$

and where $\dot{m}_k$ is the mass rate of change of droplet or particle k and the summation is carried out over all the surfaces inside the control volume. Two complementary boundary droplets are resolved as a single droplet inside the control volume. One notes here that if the droplet is evaporating ($\dot{m} < 0$), then mass flows into the continuous phase (influx). If all the droplets are evaporating at the same rate, the dispersed phase flux term becomes

$$\sum_{k}^{N} \dot{m}_k = N\dot{m} \qquad (6.24)$$

so continuity equation can be rewritten as

$$V\Delta_t(\rho_c \alpha_c) + \alpha_c \rho_c u A \mid_2 -\alpha_c \rho_c u A \mid_1 = -N\dot{m} = S_{mass} \qquad (6.25)$$

where S_{mass} is the mass source term. Obviously the mass source term is positive for evaporating droplets.

If the continuous phase is composed of several gaseous components such as nitrogen, oxygen or other chemical species, a continuity equation can be written for each species. The flux terms through the continuous phase and from the droplet surfaces would have a factor expressing the mass fraction of the species in question. Also a diffusional velocity would have to be included. These details will not be addressed here.

For convenience later, the difference in quantities between stations 1 and 2 will be expressed as

$$\Delta B = B_2 - B_1$$

so the shorthand form for the continuity equation is

$$V\Delta_t(\rho_c \alpha_c) + \Delta(\alpha_c \rho_c u A) = S_{mass} \qquad (6.26)$$

For models treating the particles as "points", the volume fraction of the continuous phase is taken as unity so the continuity equation becomes

$$V\Delta_t\rho_c + \Delta(\rho_c u A) = S_{mass} \qquad (6.27)$$

Dividing Equation 6.26 by the volume V yields

$$\frac{\partial}{\partial t}(\alpha_c \rho_c) + \frac{1}{\bar{A}}\frac{\Delta(\alpha_c \rho_c u A)}{\Delta x} = s_{mass} \qquad (6.28)$$

where s_m is the mass source per unit volume; i.e., $s_{mass} = S_{mass}/V$. In a continuum, the differential form of the continuity is obtained by taking the limit as $\Delta x \to 0$. However, in a two phase mixture the volume average will not be stationary as $\Delta x \to 0$ so a lower limit of $\Delta x'$ is defined. If $\Delta x'$ is much less than the distance associated with property changes in the flow then a differential formulation is valid as $\Delta x \to \Delta x'$. The differential form of the continuity equation is

$$\frac{\partial}{\partial t}(\alpha_c \rho_c) + \frac{1}{A}\frac{\partial}{\partial x}(\alpha_c \rho_c u A) = s_{mass} \tag{6.29}$$

If the continuous phase density is constant, the continuity equation becomes

$$\frac{\partial}{\partial t}(\alpha_c) + \frac{1}{A}\frac{\partial}{\partial x}(\alpha_c u A) = s_{mass}/\rho_c$$

Example: Find an expression for the mass source term per unit volume for droplets which evaporate according to the D^2-law as a function of $\bar{\rho}_d, \tau_m, t$.

Solution: The mass source term can be written as

$$S_{mass} = nVm\frac{\dot{m}}{m}$$

$$s_{mass} = \bar{\rho}_d\frac{\dot{m}}{m}$$

From the D^2-law,

$$m = \frac{\pi}{6}\rho_d D^2 \left(1 - \frac{t}{\tau_m}\right)^{3/2}$$

Taking the derivative with respect to time and dividing by the mass gives

$$\frac{\dot{m}}{m} = \frac{3}{2}\frac{1}{\tau_m}\left(1 - \frac{t}{\tau_m}\right)^{-1}$$

Thus, the source term becomes

$$s_{mass} = \frac{3}{2}\frac{\bar{\rho}_d}{\tau_m}\left(1 - \frac{t}{\tau_m}\right)^{-1}$$

6.3.2 Momentum equation

The momentum equation for the continuous phase can be expressed as the rate of change of momentum in the control volume plus the net efflux of momentum from the control volume is equal to the net forces acting on the continuous phase in the control volume.

Rate of momentum accumulation in control volume		Net efflux of momentum from control volume		Force on fluid in control volume
	+		=	

The momentum per unit mass is simply the velocity. Considering the momentum flux across stations 1 and 2 and through the surfaces surrounding the dispersed phase, the momentum equation is given by

$$V\Delta_t(\alpha_c\rho_c u) + \alpha_c\rho_c u^2 A \,|_2 \,-\alpha_c\rho_c u^2 A \,|_1 + \sum_{k}^{N} \dot{m}_k v_k = F \qquad (6.30)$$

where F is the force acting on the continuous phase. The assumption has been made that the mass flux is uniform over the droplet surface. If all the droplets have the same speed and evaporate at the same rate, the above equation can be written as

$$V\Delta_t(\alpha_c\rho_c u) + \alpha_c\rho_c u^2 A \,|_2 \,-\alpha_c\rho_c u^2 A \,|_1 + N\dot{m}v = F \qquad (6.31)$$

or

$$V\Delta_t(\alpha_c\rho_c u) + \Delta(\alpha_c\rho_c u^2 A) - S_{mass}v = F \qquad (6.32)$$

The forces acting on the fluid are the pressure forces on the boundary, the shear stress on the wall, the drag forces due to the dispersed phase, and the body forces on the fluid (such as the gravitational force).

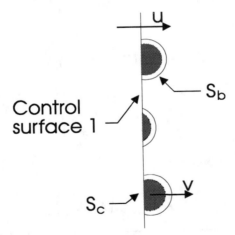

Figure 6.7: Boundary particles severed by control surface 1.

The boundary droplets require special consideration. Consider the boundary droplet on control surface 1 shown in Figure 6.7. The average pressure across the face is p_1. The force on the continuous phase due to pressure on the portion of the droplet inside the control volume is

$$\Delta \mathbf{F}_p = \int_{S_b} p_s \mathbf{n} dS \qquad (6.33)$$

where p_s is the pressure on the droplet surface and S_b is the droplet surface inside the control volume. The unit normal vector $\mathbf{n}$ is directed outward from the droplet surface. The pressure on the droplet surface is expressed as

$$p_s = p_1 + \delta p \tag{6.34}$$

so the force becomes

$$\Delta \mathbf{F}_p = p_1 \int_{S_b} \mathbf{n} dS + \int_{S_b} \delta p \mathbf{n} dS \tag{6.35}$$

The first integral reduces to

$$p_1 \int_{S_b} \mathbf{n} dS = p_1 S_s \mathbf{i} \tag{6.36}$$

where S_s is the area of the surface cut by the control surface and $\mathbf{i}$ is the unit vector in the flow direction. Summing the other integral over two complementary droplets on the control surface yields the negative of the force due to pressure (or "form" force) on a droplet completely inside the control volume in the same manner as discussed in Section 6.2.2. That is,

$$\overbrace{\int_{S_{b,A}} \delta p \mathbf{n} dS + \int_{S_{b,B}} \delta p \mathbf{n} dS}^{\text{complementary droplets}} = \int_{S_d} \delta p \mathbf{n} dS = -F_{D,form} \tag{6.37}$$

where $S_{b,A}$ and $S_{b,B}$ are the surface areas of the two complementary droplets inside the control volume and S_d is the surface area of a droplet completely inside the control volume. Summing the partial shear forces on the same complementary droplets will yield the shear force on a droplet completely inside the control volume. Finally the sum of the form and shear forces on the droplet results in the drag force. The droplet corresponding to the complementary pair of boundary droplets is included in the total number of droplets, N.

Adding all the pressure forces on the fluid due to the boundary droplets on control surface 1 yields $\alpha_d A_1 p_1$. The total force due to pressure on surface 1 is

$$\alpha_c A_1 p_1 + \alpha_d A_1 p_1 = p_1 A_1 \tag{6.38}$$

The corresponding force due to pressure at station 2 is $-p_2 A_2$.

The force in the flow direction due to the pressure on the sloping surface is $\bar{p}(A_2 - A_1)$ or $\bar{p}\Delta A$. Thus, the net force due to pressure is

$$F_p = p_1 A_1 - p_2 A_2 + \bar{p}\Delta A \tag{6.39}$$

or

$$F_p = -\Delta(pA) + \bar{p}\Delta A \tag{6.40}$$

Using the difference operator (like differentiation by parts),

$$\Delta(pA) = \bar{p}\Delta A + \bar{A}\Delta p$$

the pressure force can be expressed as

$$F_p = -\bar{A}\Delta p \tag{6.41}$$

The force due to wall friction is[3]

$$F_f = -\tau_w P \Delta x \tag{6.42}$$

where P is the perimeter (average) of the duct.

The force due to gravity is

$$F_g = \rho_c \alpha_c g \bar{A} \Delta x \tag{6.43}$$

where the gravity vector is in the flow direction.

Finally the force on the fluid due to dispersed phase drag, assuming all the dispersed phase elements move at the same velocity, is

$$F_D = -N \left[3\pi\mu D f(u - v) - V_d \frac{dp}{dx} \right] \tag{6.44}$$

where the unsteady forces have been neglected here but could be included for completeness if necessary. The force due to pressure gradient on the droplet can be expressed as

$$N V_d \frac{dp}{dx} = \alpha_d \bar{A} \Delta x \frac{\Delta p}{\Delta x} = \alpha_d \bar{A} \Delta p \tag{6.45}$$

which, when combined with Equation 6.41, modifies the net pressure force to be

$$F_p = -\alpha_c \bar{A} \Delta p \tag{6.46}$$

Summing all the forces and substituting them into Equation 6.30 results in the following form of the momentum equation.

$$V \Delta_t(\alpha_c \rho_c u) + \Delta(\alpha_c \rho_c u^2 A) = S_{mass} v$$

$$-\alpha_c \bar{A} \Delta p + 3\pi N \mu D f(v - u) - \tau_w P \Delta x + \alpha_c \rho_c g V \tag{6.47}$$

The drag force term due to the velocity difference can be more conveniently expressed as

$$N 3\pi\mu D f(v - u) = \bar{\rho}_d V \frac{f}{\tau_V}(v - u) = \beta_V V(v - u) \tag{6.48}$$

The final form of the momentum equation is

---

[3] The slope of the wall is small so the cosine of the wall angle is taken as unity.

$$V\Delta_t(\alpha_c\rho_c u) + \Delta(\alpha_c\rho_c u^2 A) = -\alpha_c\bar{A}\Delta p$$

$$S_{mass}v + \beta_V V(v-u) - \tau_w P\Delta x + \alpha_c\rho_c gV \tag{6.49}$$

The momentum source terms due to the dispersed phase can be identified as those due to mass transfer, $S_{mass}v$ and hydrodynamic drag, $\beta_V V(v-u)$.

For numerical models treating the dispersed phase as points, the momentum equation simplifies to

$$V\Delta_t(\rho_c u) + \Delta(\rho_c A u^2) = -\bar{A}\Delta p$$

$$+S_{mass}v + \beta_V V(v-u) - \tau_w P\Delta x + \rho_c gV \tag{6.50}$$

The differential form of the momentum equation is

$$\frac{\partial}{\partial t}(\alpha_c\rho_c u) + \frac{1}{A}\frac{\partial}{\partial x}(\alpha_c\rho_c u^2 A) = -\alpha_c\frac{\partial p}{\partial x}$$

$$+s_{mass}v + \beta_V(v-u) - \frac{1}{R_h}\tau_w + \alpha_c\rho_c g \tag{6.51}$$

where R_H is the *hydraulic radius* of the duct.

Example: Evaluate β_V for a mixture of 100 μm coal particles in air at standard conditions ($\mu_c = 1.8 \times 10^{-5}$ Ns/m^2, $\rho_c = 1.2$ kg/m^3). The particle volume fraction is 0.01, the material density of the coal is 1300 kg/m^3 and the velocity difference is 1 m/s. Assume the coal particles can be treated as spherical.

Answer: The bulk density of the coal particles is

$$\bar{\rho}_d = 0.01 \times 1300 = 13 \text{ kg/m}^3$$

The relative Reynolds number is

$$\text{Re}_r = 1 \times 10^{-4} \times 1.2/1.8 \times 10^{-5} = 6.7$$

The drag factor corrected to multiple particle effects (see Equation 4.85) is

$$f = (1 + 0.15\text{Re}_r^{2/3})\alpha_c^{-\beta}$$

where

$$\beta = 3.7 - 0.65\exp\left[-\frac{(1.5 - \log\text{Re}_r)^2}{2}\right] = 3.18$$

The drag factor is

$$f = 1.53 \times 0.99^{-3.18} = 1.58$$

The velocity response time is

$$\tau_V = \frac{\rho_d D^2}{18\mu_c} = 0.04 \text{ s}$$

Finally, evaluating β_V

$$\beta_V = \frac{\bar{\rho}_d}{\tau_V} f = 513 \text{ kg/m}^3\text{s}$$

6.3.3   Energy equation

The first law of thermodynamics states that the rate of change of energy in the control volume plus the net efflux of energy through the control surface is equal to the net heat transfer to the system minus the rate at which work is done by the system.

Rate of energy accumulation		Net efflux of energy		Heat transfer rate to system		Work rate done by system
	$+$		$=$		$-$	

The control volume used to develop the energy equation is shown in Figure 6.8. Energy is convected through the control surfaces at stations 1 and 2 as well as through the control surfaces surrounding the droplets. The energies consist of both the internal and external (kinetic) energy.[4] Work is done by flow work associated with motion across the control surfaces and the surfaces enclosing the droplets. Work is also done by the drag force on the droplets and by the body forces (gravity). There is no work associated with the wall shear stress since the velocity at the wall is zero. Heat is transferred through the control surfaces, through the wall and from the droplets to the continuous phase.

Relating the change in energy in the control volume plus the net efflux through the control surfaces to the rate of work and heat transfer gives

$$V\Delta_t \left[\alpha_c \rho_c (i_c + \tfrac{u^2}{2}) \right] + \Delta \left[\alpha_c \rho_c u(i_c + \tfrac{u^2}{2})A \right]$$

$$+N\dot{m}(i_s + \tfrac{v^2}{2}) = \dot{Q} - \dot{W} \tag{6.52}$$

where i_c is the specific internal energy of the continuous phase and i_s is the internal energy of the fluid at the control surface surrounding the droplet.

Work

First consider the flow work. The pressure at the control surface of the particles is decomposed into the average pressure in the continuous phase plus the

---

[4]The potential energy is not included here because the work against gravitational forces will be included in the work term.

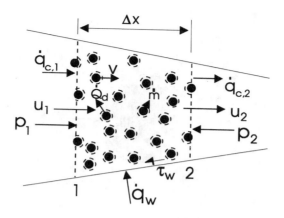

Figure 6.8: Quasi-one dimensional volume for energy equation.

deviation from that pressure as done previously in the development of the momentum equation, Equation 6.34. The flow work associated with a boundary droplet shown in Figure 6.7 is

$$\Delta \dot{W}_f = -\int_{S_b} (p_1 + \delta p) v \mathbf{n} \cdot \mathbf{i} dS \tag{6.53}$$

or

$$\Delta \dot{W}_f = -p_1 S_s v - v \int_{S_b} \delta p \mathbf{n} \cdot \mathbf{i} dS \tag{6.54}$$

Summing over all the boundary droplets on control surface 1 gives the flow work due to the droplet motion; namely, $-\alpha_d p_1 A_1$. Combining the second integral with that of a complementary droplet gives the product of the form force and velocity for a droplet completely inside the boundary. Adding the product of the shear stress and velocity on complementary droplets gives the product of the droplet velocity and drag force on a droplet inside the control volume which is included in the total number N.

The flow work associated with the flow across the 1 and 2 surfaces is

$$\dot{W}_f = \alpha_c p A u \mid_1^2 + \alpha_d p A v \mid_1^2 = \Delta \left(\alpha_c p A u \right) + \Delta \left(\alpha_d p A v \right) \tag{6.55}$$

The first term is due to the fluid flow while the second term is due to the motion of the particles which move at a different velocity than the fluid. The flow work associated with the mass flux from the droplet surface is

$$\dot{W}_f = N \dot{m} \frac{p_s}{\rho_s} \tag{6.56}$$

where p_s is the pressure at the droplet surface. If the droplet is evaporating, work is being done on the continuous phase so the sign on the work term is negative which is represented correctly here since $\dot{m}$ is negative.

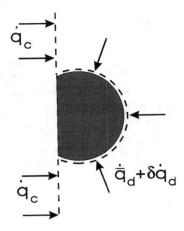

Figure 6.9: Heat transfer at a boundary particle.

The rate of work due to the change in droplet volume is

$$\dot{W}_f = -p_s N \dot{V}_d \tag{6.57}$$

The work rate associated with the drag force on the droplet is

$$\dot{W}_d = -Nv F_D \tag{6.58}$$

where F_D is the drag force *on* the droplet. This work term becomes

$$\dot{W}_d = -N\left[3\pi\mu_c D f(u-v)v - V_d v \frac{\Delta p}{\Delta x}\right] \tag{6.59}$$

Finally the work done against body (gravitational) forces is

$$\dot{W}_d = -\alpha_c \rho_c g V u \tag{6.60}$$

where g acts in the x-direction.

Heat transfer

Heat is transferred to the continuous phase through several means. There is heat transfer through the 1 and 2 surfaces of the control volume, heat transfer from the particles to the fluid and heat transfer through the walls.

The heat transfer across surfaces 1 and 2 can be broken down into two components; heat transfer through the fluid and heat transfer *through* the particles. The heat transfer across surface 1 with boundary droplets is depicted in Figure 6.9. The heat transfer by conduction through the carrier phase is $\dot{q}_f \alpha_c A_1$ where $\dot{q}_f$ is the heat transfer rate per unit area through the carrier fluid.

It is convenient to express the heat transfer on the surface of the boundary droplets as the sum of the average heat transfer rate over the surface of the droplet and the deviation therefrom.

$$\dot{q}_s = \widehat{\dot{q}}_s + \delta\dot{q} \tag{6.61}$$

The heat transfer rate to the fluid from the boundary droplet becomes

$$\dot{Q}_b = \int_{S_b} \widehat{\dot{q}}_s dS + \int_{S_b} \delta\dot{q} dS \tag{6.62}$$

The first integral when combined with a complementary droplet is the heat transfer to the fluid from a droplet completely inside the volume. The second integral represents the heat transfer rate *through* the droplet. This term, combined with the heat transfer through the carrier phase, gives the composite conductive heat transfer to the fluid in the volume through the surface 1. Thus

$$\dot{Q}_c = (\dot{q}_f \alpha_c + \dot{q}_d \alpha_d) A_1 = \dot{q}_{c,1} A_1 \tag{6.63}$$

where $\dot{q}_{c,1}$ is the *effective heat transfer* through both phases across surface 1. The effective heat transfer may be represented by a Fourier's law

$$\dot{q}_{c,1} = -k_{eff} \frac{\partial T_c}{\partial x} \mid_1 \tag{6.64}$$

where k_{eff} is the effective thermal conductivity which will be a function of the thermal conductivity of each phase and the local volume fraction. For low particle volume fractions, the effective thermal conductivity is

$$k_{eff} = \alpha_d k_d + \alpha_c k_c \tag{6.65}$$

where k_d and k_c are the thermal conductivity of the dispersed and continuous phases respectively. The net heat transfer to the continuous phase across stations 1 and 2 is

$$\dot{Q}_c = -A\dot{q}_c \mid_1^2 = -\Delta(A\dot{q}_c) \tag{6.66}$$

The heat transfer to the wall is expressed as

$$\dot{Q}_w = -\dot{q}_w P \Delta x \tag{6.67}$$

where $\dot{q}_w$ is the heat transfer to the wall per unit area. The minus sign reflects heat transfer from the fluid in the control volume. If the wall is insulated, $\dot{q}_w$ is zero. If the wall temperature is fixed, the heat transfer will depend on the temperature difference between the fluid and the wall and the Stanton number.

The heat transfer from the droplets to the fluid in the control volume is

$$\dot{Q}_d = NNu\pi k_c D(T_d - T_c) \tag{6.68}$$

It is assumed that there is no radiative energy absorbed by the continuous phase.

Substituting the expressions for flow work, Equations 6.55 and 6.56; the work rate associated with the expansion of the dispersed phase, Equation 6.57; the work done against droplet drag, Equation 6.59; the work done on body forces, Equation 6.60, the heat transfer to the fluid through the surfaces 1 and 2, Equation 6.66; the heat transferred to the walls, Equation 6.67; and the heat transferred from the droplets, Equation 6.68, into Equation 6.52 results in the following equation for total energy of the continuous phase.

$$V\Delta_t[\alpha_c\rho_c(i_c + \tfrac{u^2}{2})] + \Delta[\alpha_c\rho_c u(i_c + \tfrac{u^2}{2})] + N\dot{m}(i_s + \tfrac{v^2}{2}) =$$

$$- \overbrace{\Delta\,(\alpha_c p A u)}^{a} - \overbrace{\Delta\,(\alpha_d p A v)}^{b} - \overbrace{N\dot{m}\frac{p_s}{\rho_s}}^{c} + pN\dot{V}_d$$

$$-N\left[3\pi\mu_c D f(u-v)v - \overbrace{V_d v\frac{\Delta p}{\Delta x}}^{d}\right] + \alpha_c\rho_c gVu \tag{6.69}$$

$$-\Delta\,(A\dot{q}_c) - \dot{q}_w P\Delta x + N Nu \pi k_c D(T_d - T_c)$$

This equation can be simplified by combining terms. Terms a and c can be combined with the internal energy flux terms to yield the enthalpy flux,

$$\Delta[\alpha_c\rho_c u(i_c + \frac{u^2}{2})] + \Delta\,(\alpha_c p A u) = \Delta[\alpha_c\rho_c u(h_c + \frac{u^2}{2})] \tag{6.70}$$

$$N\dot{m}(i_s + \frac{v^2}{2}) + N\dot{m}\frac{p_s}{\rho_s} = N\dot{m}(h_s + \frac{v^2}{2}) \tag{6.71}$$

Term b can be written in difference form as

$$-\Delta\,(\alpha_d p A v) = -\alpha_d \bar{A} v\Delta p - p\Delta(\alpha_d A v) \tag{6.72}$$

and term d can be expressed as

$$N V_d v\frac{\Delta p}{\Delta x} = \alpha_d V v\frac{\Delta p}{\Delta x} = \alpha_d \bar{A}\Delta x v\frac{\Delta p}{\Delta x} = \alpha_d \bar{A} v\Delta p \tag{6.73}$$

which when combined with term b becomes

$$-\Delta\,(\alpha_d p A v) + \alpha_d \bar{A} v\Delta p = -p\Delta(\alpha_d A v)$$

The total energy equation assumes the form,

$$V\Delta_t[\alpha_c\rho_c(i_c + \tfrac{u^2}{2})] + \Delta\left[\alpha_c\rho_c u(h_c + \tfrac{u^2}{2})\right] - S_{mass}(h_s + \tfrac{v^2}{2}) =$$

$$-p\Delta(\alpha_d v A) + p_s N\dot{V}_d - \beta_V V(u-v)v + \alpha_c\rho_c gVu \tag{6.74}$$

$$-\Delta\,(A\dot{q}_c) - \dot{q}_w P\Delta x + \beta_T V(T_d - T_c)$$

where the coefficient β_V defined in the momentum equation, Equation 6.48, has been used here. The coefficient of the dispersed phase heat transfer term has been replaced by the more simple form

$$NNu\pi k_c D(T_d - T_c) = \frac{Nu}{2} \frac{\tilde{\rho}_d c_d}{\tau_T} V(T_d - T_c) = \beta_T V(T_d - T_c) \qquad (6.75)$$

where c_d is the specific heat of the dispersed phase. In most applications, the heat transfer through surfaces 1 and 2 is small compared to the enthalpy flux, $\Delta(A\dot{q}_c) << \Delta(\alpha_c\rho_c u A h_c)$. Also the heat transfer to the wall can be expressed as

$$\dot{q}_w = h(T_c - T_d) \qquad (6.76)$$

where h is the convective heat transfer coefficient which is a function of the Reynolds number and Prandtl number.

For numerical models based on "point" particles, the total energy equation reduces to

$$V\Delta_t[\rho_c(i_c + \frac{u^2}{2})] + \Delta\left[\rho_c u A(h_c + \frac{u^2}{2})\right] =$$

$$S_{mass}(h_s + \frac{v^2}{2}) - \beta_V V(u - v)v + \rho_c g V u \qquad (6.77)$$

$$-\Delta(A\dot{q}_c) - \dot{q}_w P\Delta x + \beta_T V(T_d - T_c)$$

The differential form of the total energy equation is

$$\frac{\partial}{\partial t}[\alpha_c\rho_c(i_c + \frac{u^2}{2})] + \frac{1}{A}\frac{\partial}{\partial x}[\alpha_c\rho_c A u(h_c + \frac{u^2}{2})] =$$

$$s_{mass}(h_s + \frac{v^2}{2}) - p\frac{1}{A}\frac{\partial}{\partial x}(\alpha_d v A) + p_s n\dot{V}_d - \beta_V(u - v)v \qquad (6.78)$$

$$+\alpha_c\rho_c g u - \frac{1}{A}\frac{\partial}{\partial x}(A\dot{q}_c) - \frac{\dot{q}_w}{R_H} + \beta_T(T_d - T_c)$$

Once again, some terms can be neglected to simplify the total energy equation in specific applications.

The source terms due to the dispersed phase can easily be identified: energy addition due to mass transfer, energy loss due to work done on the dispersed phase and heat transfer from the dispersed phase. The other terms, work done by body forces and heat transfer through the wall, exist in single-phase flows.

6.3.4 Thermal energy equation

The thermal energy equation is obtained by subtracting the equation for external energy from the total energy equation. The external energy equation is obtained by multiplying the momentum equation by the velocity. Multiplying Equation 6.57 by the velocity u gives

$$uV\Delta_t(\alpha_c\rho_c u) + u\Delta(\alpha_c\rho_c u^2 A) = -\alpha_c u\bar{A}\Delta p$$

$$S_{mass}uv + \beta_V Vu(v-u) - \tau_w uP\Delta x + \alpha_c\rho_c ugV$$

(6.79)

The left side of this equation can be rewritten as

$$uV\Delta_t(\alpha_c\rho_c u) + u\Delta(\alpha_c\rho_c u^2) =$$

$$V\Delta_t(\alpha_c\rho_c\tfrac{u^2}{2}) + \Delta(\alpha_c\rho_c u\tfrac{u^2}{2}) + \tfrac{u^2}{2}[V\Delta_t(\alpha_c\rho_c) + \Delta\,(\alpha_c\rho_c u)]$$

(6.80)

Using the continuity equation, Equation 6.26, for the last term in the above equation, allows one to express the external energy equation as

$$V\Delta_t(\alpha_c\rho_c\tfrac{u^2}{2}) + \Delta(\alpha_c\rho_c u\tfrac{u^2}{2}) + \tfrac{u^2}{2}S_{mass} = -\alpha_c u\bar{A}\Delta p$$

$$S_{mass}uv + \beta_V Vu(v-u) - \tau_w uP\Delta x + \alpha_c\rho_c ugV$$

(6.81)

Subtracting Equation 6.81 from the total energy equation, Equation 6.74, yields the thermal energy equation.

$$V\Delta_t(\alpha_c\rho_c i_c) + \Delta(\alpha_c\rho_c uh_c) = \alpha_c u\bar{A}\Delta p$$

$$+S_{mass}[h_s + \tfrac{(v-u)^2}{2}] - p\Delta(\alpha_d vA) + p_s N\dot{V}_d + \beta_V V(u-v)^2$$

$$+u\tau_w P\Delta x - \Delta\,(A\dot{q}_c) - \dot{q}_w P\Delta x + \beta_T V(T_d - T_c)$$

(6.82)

The term associated with the droplet drag, $\beta_V V(u-v)^2$, is always positive and represents the irreversible conversion of mechanical energy to heat. The term associated with wall friction, $u\tau_w P\Delta x$, can be written in terms of a shear stress coefficient

$$u\tau_w P\Delta x = \frac{1}{2}c_f\rho_c |u|\, u^2 P\Delta x$$

(6.83)

which is always positive and represents the irreversible conversion of mechanical energy to heat.

The differential form of the thermal energy equation is

$$\frac{\partial}{\partial t}(\alpha_c\rho_c i_c) + \frac{1}{A}\frac{\partial}{\partial x}(\alpha_c\rho_c uAh_c) = \alpha_c u\frac{\partial p}{\partial x}$$

$$+s_{mass}[h_s + \tfrac{(u-v)^2}{2}] - p\frac{1}{A}\frac{\partial}{\partial x}(\alpha_d vA) + p_s n\dot{V}_d$$

$$+\beta_V(u-v)^2 + \frac{1}{R_H}\,(u\tau_w - \dot{q}_w) - \frac{1}{A}\frac{\partial}{\partial x}(A\dot{q}_c) + \beta_T(T_d - T_c)$$

(6.84)

Example: Find the heating rate of air due to the conversion of mechanical energy to heat for 1-mm glass beads with a concentration of 10 in a flow with constant mass flux. The velocity difference is 10 m/s, the specific heat of the air is 1000 J/kgK and the viscosity of the air is 1.8×10^{-5} Ns/m^2

Solution: From Equation 6.82, one can write

$$\Delta(\alpha_c \rho_c u A h_c) = \beta V (u - v)^2$$

or

$$\bar{\rho}_c A u c_p \Delta T_c = \bar{\rho}_d A \Delta x f \frac{(u - v)^2}{\tau_V}$$

Recognizing that $\Delta x / u$ is a time interval Δt, one can write

$$\frac{\Delta T_c}{\Delta t} = \frac{\bar{\rho}_d}{\bar{\rho}_c} \frac{f}{\tau_V} \frac{(u - v)^2}{c_p}$$

The velocity response time is

$$\tau_V = \frac{\rho_d D^2}{18 \mu_c} = 7.72s$$

The relative Reynolds number is

$$\mathrm{Re}_r = \frac{|u - v| D}{\nu_c} = 667$$

and the drag factor is

$$f = 1 + 0.15 \mathrm{Re}_r^{0.67} = 12.7$$

Finally the rate of temperature increase is

$$\frac{\Delta T_c}{\Delta t} = \frac{10 \times 12.7 \times 10^2}{7.72 \times 1000} = 1.64 \text{ K/s}$$

6.3.5 Equation summary

The conservation equations for quasi-one-dimensional flow are summarized in Table 6.1. The coupling terms due to the dispersed phase are written on a separate line designated by the symbol, C. To make the table more compact, the following definitions for total conditions are used, $i_{c,t} = i_c + u^2/2$, $h_{c,t} = h_c + u^2/2$ and $h_{s,t} = h_s + v^2/2$. In the thermal energy equation, the term h_t is $h_s + (u - v)^2/2$. A summary of the differential form of the equations is provided in Table 6.2.

Continuity equation:

$$V\Delta_t(\bar{\rho}_c) + \Delta(\bar{\rho}_c u A) =$$
$$(C) \quad -N\dot{m}$$

Momentum equation:

$$V\Delta_t(\bar{\rho}_c u) + \Delta(\bar{\rho}_c A u^2) = -\alpha_c \bar{A}\Delta p + \bar{\rho}_c g V - \tau_w P \Delta x$$
$$(C) \quad \beta_V V (v - u) - N\dot{m}v$$

Total energy equation:

$$V\Delta_t(\bar{\rho}_c i_{c,t}) + \Delta(\bar{\rho}_c u A h_{c,t}) = -\Delta(\dot{q}_c A) + \bar{\rho}_c g u V - \dot{q}_w P \Delta x$$
$$(C) \quad -\beta_V V v(u - v) - N\dot{m}h_{s,t} + \beta_T V(T_d - T_c) - p\Delta(\alpha_d v A) + N\dot{V}_d p_s$$

Thermal energy equation:

$$V\Delta_t(\bar{\rho}_c i_c) + \Delta(\bar{\rho}_c u A h_c) = \alpha_c u \bar{A}\Delta p - \Delta(A\dot{q}_c) - \dot{q}_w P \Delta x + u\tau_w P \Delta x$$
$$(C) \quad +\beta_V V (u - v)^2 + \beta_T V (T_d - T_c) - p\Delta(\alpha_d v A) - N\dot{m}(h_t) + N\dot{V}_d p_s$$

Table 6.1. Summary of the difference form of the continuous phase equations for quasi-one-dimensional flow.

Continuity equation:

$$\frac{\partial}{\partial t}\bar{\rho}_c + \frac{1}{A}\frac{\partial}{\partial x}(\bar{\rho}_c u A) =$$
$$(C) \quad -n\dot{m}$$

Momentum equation:

$$\frac{\partial}{\partial t}(\bar{\rho}_c u) + \frac{1}{A}\frac{\partial}{\partial x}(\bar{\rho}_c u^2 A) = -\alpha_c \frac{\partial p}{\partial x} + \bar{\rho}_c g - \frac{1}{R_H}\tau_w$$
$$(C) \quad + \beta_v(v - u) - n\dot{m}v$$

Total energy equation:

$$\frac{\partial}{\partial t}(\bar{\rho}_c i_{c,t}) + \frac{1}{A}\frac{\partial}{\partial x}(\bar{\rho}_c h_{c,t} u A) = -\frac{1}{A}\frac{\partial}{\partial x}(A\dot{q}_c) + \bar{\rho}_c g u - \frac{1}{R_H}\dot{q}_w$$
$$(C) \quad - \beta_V v(u - v) - n\dot{m}h_{s,t} + \beta_T(T_d - T_c) - \frac{p}{A}\frac{\partial}{\partial x}(\alpha_d v A) + n\dot{V}_d p_s$$

Thermal energy equation:

$$\frac{\partial}{\partial t}(\bar{\rho}_c i_c) + \frac{1}{A}\frac{\partial}{\partial x}(\bar{\rho}_c h_c u A) = \alpha_c u \frac{\partial p}{\partial x} - \frac{1}{A}\frac{\partial}{\partial x}(A\dot{q}_c) + \frac{1}{R_H}(u\tau_w - q_w)$$
$$(C) \quad + \beta_V(u - v)^2 + \beta_T(T_d - T_c) - \frac{p}{A}\frac{\partial}{\partial x}(\alpha_d v A) - n\dot{m}h_t + n\dot{V}_d p_s$$

Table 6.2. Summary of differential form of continuous phase equations for quasi-one-dimensional flow.

6.4 Multidimensional flows

The essential features in the derivation of the basic equations for the continuous phase have been demonstrated for the quasi-one-dimensional flow. The purpose of this section is to present the conservation equations for multidimensional flows. The differential form of the equations are developed in Appendix B based on volume averaging. These equations will be presented and discussed here.

In Appendix B the properties are expressed as volume average or mass average. For example the mass average velocity, $\tilde{u}_i$, is defined as

$$\tilde{u}_i = \frac{\langle \rho_c u_i \rangle}{\langle \rho_c \rangle} \tag{6.85}$$

If the density of the continuous phase is constant, the volume average and mass average velocities are the same. If, however, the continuous phase density is nearly constant over the averaging volume, then $\tilde{u}_i \simeq \langle u_i \rangle$. For purposes here the mass average and volume average velocities will be considered equal and represented simply by u_i. All the other properties (pressure, internal energy etc.) will be the averaged properties but not designated by $\langle \rangle$ or a tilde.

6.4.1 Mass conservation

The general statement for mass conservation is that the net efflux of mass from a control volume plus the rate of accumulation of mass flow in the volume is zero. The continuity equation, Equation B64, is

$$\frac{\partial}{\partial t}(\alpha_c \rho_c) + \frac{\partial}{\partial x_i}(\alpha_c \rho_c u_i) = -\sum_{k}^{N} \dot{m}_k / V = s_{mass} \tag{6.86}$$

If all the droplets are evaporating at the same rate the mass source term simplifies to $s_{mass} = -n\dot{m}$. Of course, the use of a differential equation implies that continuous phase properties are averaged over a volume and the averaged properties are associated with point. Obviously if the size of the averaging volume is comparable to the dimensions of the system, the differential forms of the equations are inapplicable.

The continuity equation can be nondimensionalized to yield the scaling factors. The nondimensional distance is x_i' which is related to a characteristic distance λ by

$$x_i = \lambda x_i' \tag{6.87}$$

The velocity is normalized by a velocity U,

$$u_i = U u_i' \tag{6.88}$$

and the continuous phase bulk density by $\bar{\rho}_{co}$,

$$\alpha_c \rho_c = \bar{\rho}_c' \bar{\rho}_{co} \tag{6.89}$$

The continuity equation can be written in terms of the nondimensional variables as

$$\frac{\partial}{\partial t'}(\bar{\rho}'_c) + \frac{\partial}{\partial x'_i}(\bar{\rho}_c u'_i) = -n_o m \frac{\dot{m}}{m} \frac{\lambda}{\bar{\rho}_{co} U} \frac{n}{n_o} \tag{6.90}$$

where n_o is nominal number density. The ratio $\dot{m}/m$ is the reciprocal evaporation time $1/\tau_m$ so the equation can be rewritten as

$$\frac{\partial}{\partial t'}(\bar{\rho}'_c) + \frac{\partial}{\partial x'_i}(\bar{\rho}_c u'_i) = \frac{\bar{\rho}_{do}}{\bar{\rho}_{co}} \frac{\lambda}{\tau_m U} \frac{n}{n_o} \tag{6.91}$$

where $\bar{\rho}_{do}$ is the nominal dispersed phase bulk density. The coefficient of the mass source term is the mass coupling parameter defined in Chapter 2.

$$\Pi_{mass} = \frac{C}{St_{mass}}$$

If $\Pi_{mass} \ll 1$, mass coupling is unimportant.

If the continuous phase is a gas which consists of component chemical species, then the continuity equation for species A would be

$$\frac{\partial}{\partial t}(\alpha_c \rho_c \omega_A) + \frac{\partial}{\partial x_i}(\omega_A \rho_c \alpha_c u_i) = \omega_{A,s} s_{mass} + \frac{\partial}{\partial x_i}(\alpha_c \rho_c D_A \frac{\partial \omega_A}{\partial x_i}) + \dot{R}_A \tag{6.92}$$

where ω_A is the mass fraction of species A, $\omega_{A,s}$ is the mass fraction of species A at the droplet surface, D_A is the diffusion coefficient and $\dot{R}_A$ is the mass generation rate of species A per unit volume due to a chemical reaction. In this equation u_i is the mass averaged velocity for the mixture.

For numerical models which treat the dispersed phase as point elements, the volume fraction of the continuous phase, α_c, is set equal to unity.

6.4.2 Momentum conservation

The momentum equation states that the net rate of accumulation of momentum in the control volume plus the net efflux through the control surfaces are equal to the forces acting on the continuous phase. The momentum equation developed in Appendix B, Equation B70, is

$$\frac{\partial}{\partial t}(\alpha_c \rho_c u_i) + \frac{\partial}{\partial x_j}(\alpha_c \rho_c u_j u_i) = -\alpha_c \frac{\partial p}{\partial x_i} - \sum_k v_{k,i} m_k / V$$

$$+\alpha_c \frac{\partial}{\partial x_j}(\tau^e_{ij}) - \sum_k 3\pi \mu_c D_k f_k (u_i - v_{i,k})/V + \rho_c \alpha_c g_i \tag{6.93}$$

where $v_{k,i}$ is the velocity of droplet k. The unsteady drag forces have not been included here but could be added to the drag force term. If the droplets have the same local properties; that is, they evaporate at the same rate and have the same local velocity, the momentum equation becomes

$$\frac{\partial}{\partial t}(\alpha_c \rho_c u_i) + \frac{\partial}{\partial x_j}(\alpha_c \rho_c u_i u_j) = -\alpha_c \frac{\partial p}{\partial x_i} + v_i s_{mass}$$

$$+\alpha_c \frac{\partial}{\partial x_j}(\tau_{ij}^e) - \beta_V(u_i - v_i) + \rho_c \alpha_c g_i$$

(6.94)

where β_V is the factor defined by Equation 6.48.

In order to evaluate the shear stress terms, constitutive models are necessary to relate the shear stress to the properties of the conveying phase. As shown in Appendix B the effective volume averaged stress in the fluid is

$$\tau_{ij}^e = \langle \tau_{ij} \rangle - \alpha_c \langle \rho_c \delta u_i \delta u_j \rangle$$

(6.95)

where $\langle \rangle$ signifies a volume average and the operator δ means the difference between the local and the average velocity.. The last term derives from the momentum convection terms and appears because the momentum averaged velocity is not equivalent to a mass averaged velocity. It is directly analogous to *Reynolds stress* in a time-averaged formulation.

One approach is to use the constitutive equations for a single-phase flow; that is,

$$\tau_{ij}^e = \mu_c \left(\frac{\partial u_i}{\partial x_j} + \frac{\partial u_j}{\partial x_i}\right) - \frac{2}{3}\delta_{ij}\frac{\partial u_k}{\partial x_k}$$

(6.96)

It is unlikely that this approach is useful since the gradients in the average velocity do not capture the local gradients imposed by the presence of the particles. Also, as in single-phase turbulent flows, it is very likely that the Reynolds stress is much larger than the "laminar" stress.

There is little information available for the Reynolds stress in the continuous phase of a two phase mixture. There have been several approaches to modeling the Reynolds stress for single phase flows (Wilcox, 1993; Versteeg & Malalasekera, 1995). One approach has been to assume that the turbulence shear stress can be represented by

$$\tau_{12} = -\rho_c \widehat{u_1' u_2'} = \mu_e \left(\frac{\partial \hat{u}_1}{\partial x_2} + \frac{\partial \hat{u}_2}{\partial x_1}\right)$$

(6.97)

where μ_e becomes the *effective* viscosity and the velocities are the time averaged velocities. This is referred to as the *Boussinesq approximation*. The effective shear stress is related to the turbulence parameters of the flow. The same approach is used for the volume averaged equations with the volume averaged velocities replacing the time averaged values.

For several years, the $k - \varepsilon$ model (Launder & Spalding, 1978) has been used to characterize the turbulence in single-phase flows. By scaling arguments the turbulence viscosity is given by

$$\mu_T \propto \rho_c \ell u' \propto \rho_c \ell k^{\frac{1}{2}}$$

(6.98)

where ℓ is a mixing length and k is the kinetic energy associated with the turbulence fluctuations. In single phase flows it is difficult to develop an equation for the mixing length so the dissipation rate ε of the turbulence is used to relate ℓ to the turbulence kinetic energy; namely,

$$\ell \propto k^{\frac{3}{2}}/\varepsilon \tag{6.99}$$

The effective viscosity then becomes

$$\mu_e \propto \rho_c \frac{k^{3/2}}{\varepsilon} k^{\frac{1}{2}} \tag{6.100}$$

or

$$\mu_e = c_\mu \rho_c \frac{k^2}{\varepsilon} \tag{6.101}$$

where c_μ is an empirical constant. Of course the turbulence energy and dissipation rate will be affected by the presence of the dispersed phase. This is known as turbulence modulation and will be addressed in Section 6.4.6. However there is another issue that has to be addressed in a two phase flow. If the interparticle spacing, L, becomes comparable to or less than the mixing length (or dissipation length scale) of the continuous phase, then the interparticle spacing should establish the length scale in the flow. For example, the interparticle spacing for 100 μm coal particles flowing in air in a 2-cm diameter pipe becomes comparable to the mixing length when the loading is of order unity. For higher loadings the interparticle spacing should become the dominant length scale in the continuous phase. One might propose a new turbulence model in which the effective dissipation length scale would be

$$\ell_{eff} = \min(k^{3/2}/\varepsilon, L) \tag{6.102}$$

Considerable more work needs to be done to develop a reliable model for Reynolds stress in a dispersed phase flow.

Also the physical presence of the particles in the flow will create velocity disturbances which will contribute to the Reynolds stress. In fact, the Boussinesq assumption will not be valid because Reynolds stresses would be produced in the absence of a velocity gradient. A better approach may be to use Reynolds stress models which circumvent the need for the Boussinesq assumption.

Following the procedure used for the continuity equation the momentum equation can also be written using nondimensional variables.

$$\frac{\partial}{\partial t'}(\bar{\rho}_c' u_i') + \frac{\partial}{\partial x_j'}(\bar{\rho}_c' u_i' u_j') = -\frac{\partial p'}{\partial x_i'} + \frac{\bar{\rho}_{do}}{\bar{\rho}_{co}} \frac{\lambda}{\tau_m U} v_i' \frac{n}{n_o}$$

$$+\alpha_c \frac{\partial}{\partial x_j'}(\tau_{ij}^e/\bar{\rho}_{co} U^2) - \frac{\bar{\rho}_{do}}{\tau_V} f \frac{\lambda \bar{\rho}_d'}{\bar{\rho}_{co} U}(u_i' - v_i') + \bar{\rho}_c' g_i' \frac{g\lambda}{U^e} \tag{6.103}$$

Once again the mass coupling parameter appears with the mass coupling term. The coefficient of the coupling term associated with droplet drag becomes

$$\frac{\bar{\rho}_{do}}{\tau_V} \frac{\lambda}{\bar{\rho}_{co} U} = \frac{C}{St_{mom}} \tag{6.104}$$

One notes that the coefficient is the momentum coupling parameter presented in Chapter 2. However, as discussed in Chapter 2, the parameter has to be modified to recover the limiting case when the Stokes number approaches zero; that is[5],

$$\Pi_{mom} = \frac{C}{St_{mom} + 1}$$

The nondimensional parameter coefficient associated with the body (gravity) force term is the square of the reciprocal Froude number, $g\lambda/U^2 = (1/Fr)^2$. If the Froude number is large, gravitational forces are unimportant.

Developing a nondimensional form for the Reynolds stress term is more difficult because no constitutive model is currently available. If the velocity fluctuations are due to the relative velocity between the particles and the fluid, an estimate of the Reynolds stress would be

$$\alpha_c \langle \rho_c \delta u_i \delta u_j \rangle \sim \bar{\rho}_c (u - v)^2 \tag{6.105}$$

and the nondimensional stress term becomes

$$\frac{\partial}{\partial x_i'} \left(\alpha_c \langle \rho_c \delta u_i \delta u_j \rangle / \bar{\rho}_{co} U^2 \right) \sim \frac{(u - v)^2}{U^2} = O(1) \tag{6.106}$$

Thus the Reynolds stress term would be of the same order of magnitude as the convection terms. However, if the length scale for the velocity fluctuations is the interparticle spacing, the nondimensional stress term becomes

$$\frac{\partial}{\partial x_i'} \left(\alpha_c \langle \rho_c \delta u_i \delta u_j \rangle \right) \sim \frac{\lambda}{L} \frac{(u - v)^2}{U^2} = O(\frac{\lambda}{L}) \tag{6.107}$$

which suggests that the Reynolds stress may be very significant. Modeling Reynolds stress in dispersed phase flows is an important direction for future research.

6.4.3 Total energy equation

Energy conservation is a statement of the first law of thermodynamics for an open system; i.e., a control volume with mass and heat transfer across boundaries. The first law states that the rate of accumulation of energy in a control volume plus the net efflux of energy across control surfaces is equal to the rate of heat transfer to the system minus the work rate of the system on the surroundings. The differential form of the total energy equation developed in Appendix B, Equation B107, is

---

[5]It may be appropriate to include a representative value of the drag factor in the nondimensional grouping. The drag factor could be based on a particle Reynolds number using the characteristic velocity, U, and the nominal continuous phase volume fraction.

$$\frac{\partial}{\partial t}(\alpha_c \rho_c (i_c + u_i u_i/2) + \frac{\partial}{\partial x_j}[\alpha_c \rho_c u_j (h_c + u_i u_i/2)] = \alpha_c \rho_c g_i u_i$$

$$-\frac{1}{V}\sum_k \dot{m}_k(h_{s,k} + v_{i,k} v_{i,k}/2 + w'^2/2) + 3\pi \mu_c \sum_k f_k D_k(v_{i,k} - u_i)v_{i,k}/V$$

$$-p\frac{\partial}{\partial x_i}(\alpha_d v_i) + \tau_{ij}\frac{\partial}{\partial x_j}(\alpha_d v_i) + \frac{\partial}{\partial x_j}(\alpha_c u_i \tau_{ij}) + \sum_k \dot{V}_{d,k} p_s/V$$

$$+2\pi k_c \sum_k (\frac{Nu}{2})_k D_k(T_c - T_{d,k})/V + \frac{\partial}{\partial x_i}\left(k_{eff}\frac{\partial}{\partial x_i}T_c\right)$$

$$(6.108)$$

where the subscript s refers to the droplet surface. The effective thermal conductivity k_{eff} is the thermal conductivity for the mixture; that is, the composite heat transfer through the dispersed and continuous phase. A first-order model for k_{eff} would be

$$k_{eff} = \alpha_c k_c + \alpha_d k_d \qquad (6.109)$$

where k_c and k_d are the thermal conductivities of the continuous and dispersed phases, respectively. The value for k_c would be affected by the turbulence in the carrier phase.

If all the droplets evaporate at the same rate, move at the same local velocity and have the same local temperature, the total energy equation simplifies to

$$\frac{\partial}{\partial t}(\alpha_c \rho_c (i_c + u_i u_i/2) + \frac{\partial}{\partial x_j}[\alpha_c \rho_c u_j (h_c + u_i u_i/2)] =$$

$$\alpha_c \rho_c g_i u_i + s_{mass}(h_s + v_i v_i/2 + w'^2/2) + \beta_V (v_i - u_i)v_i$$

$$-p\frac{\partial}{\partial x_i}(\alpha_d v_i) + \tau_{ij}\frac{\partial}{\partial x_j}(\alpha_d v_i) + \frac{\partial}{\partial x_j}(\alpha_c u_i \tau_{ij}) + n\dot{V}_d p_s$$

$$(6.110)$$

$$+\beta_T (T_c - T_d) + \frac{\partial}{\partial x_i}\left(k_{eff}\frac{\partial}{\partial x_i}T_c\right)$$

Under certain conditions, several terms in the total energy equation can be neglected. The total enthalpy is given by $h_c + u_i u_i/2$. If the carrier phase is an ideal gas, the enthalpy is given by $c_p T_c$ where c_p is the specific heat at constant pressure. Comparing the kinetic energy and the enthalpy gives

$$\frac{\tilde{u}_i \tilde{u}_i}{2 c_p T_c} \sim \frac{U^2}{c_p T_c} = Ec \qquad (6.111)$$

which is referred to as the *Eckert number*, Ec. The Eckert number is proportional to the square of the Mach number so the kinetic energy can be neglected for low Mach number flows and $h_c + u_i u_i/2 \to h_c$. The same simplification applies to the total internal energy.

Using the same argument, the thermal coupling term due to mass exchange reduces to

$$s_{mass}(h_s + v_i v_i/2 + w'^2/2) \to s_{mass} h_s \qquad (6.112)$$

Also the work associated with droplet volume change can be compared with $s_{mass}h_s$.

$$\frac{n\dot{V}_d p_s}{s_{mass}h_s} = n\frac{\dot{m}}{\rho_d}p_s = -\frac{p_s}{h_s\rho_d} \sim \frac{\rho_s}{\rho_d} \tag{6.113}$$

where ρ_s is the density of gas at the droplet surface. Since $\rho_s/\rho_d << 1$ the volume expansion term is negligible.

The total energy equation can be nondimensionalized in the same manner as the continuity and momentum equations. An additional scaling parameter is needed for the thermal energy which is $c_p T_{co}$ where T_{co} is a representative temperature. The nondimensional form of the total energy equation becomes

$$\frac{\partial}{\partial t'}(\bar{\rho}'_c i'_c) + \frac{\partial}{\partial xi}(h'_c\bar{\rho}'_c u'_i) =$$

$$\frac{g\lambda}{U^2}\frac{U^2}{c_p T_{co}}\bar{\rho}'_c g'_i u'_i + \frac{\bar{\rho}_{do}}{\bar{\rho}_{co}}\frac{\lambda}{\tau_m U}\frac{h_{so}}{c_p T_{co}}\frac{n}{n_o}h'_s + \frac{\bar{\rho}_{do}}{\bar{\rho}_{co}}\frac{\lambda}{U\tau_V}\frac{U^2}{c_p T_{co}}\bar{\rho}'_d(v'_i - u'_i)v'_i$$

$$-p'\frac{\partial}{\partial x'_i}(\alpha_d v'_i) + \tau'_{ij}\frac{\partial}{\partial x'_j}(\alpha_d v'_i) + \frac{\partial}{\partial x'_j}(\alpha_c u'_i \tau'_{ij}) \tag{6.114}$$

$$+\frac{\bar{\rho}_{do}}{\bar{\rho}_{co}}\frac{\lambda}{U\tau_T}\frac{c_d}{c_p}(T'_c - T'_d) + \frac{k_c}{\lambda U\rho_c c_p}\frac{\partial}{\partial x'_i}\left(k'_{eff}\frac{\partial}{\partial x'_i}T'_c\right)$$

The work associated with the body force is scaled by Ec/Fr^2 so can be neglected for low Mach number flows. The scaling factor for the energy coupling due to mass transfer is

$$\frac{\bar{\rho}_{do}}{\bar{\rho}_{co}}\frac{\lambda}{\tau_m U}\frac{h_{so}}{c_p T_{co}} = \Pi_{mass}\frac{h_{so}}{c_p T_{co}} \tag{6.115}$$

The enthalpy at the droplet surface is the same order of magnitude as the latent heat of evaporation so

$$\Pi_{mass}\frac{h_{so}}{c_p T_{co}} \sim \Pi_{mass}\frac{h_L}{c_p T_{co}}$$

which is an energy coupling parameter also introduced in Chapter 2.

The work rate associated with the drag forces scales as

$$\frac{\bar{\rho}_{do}}{\bar{\rho}_{co}}\frac{\lambda}{U\tau_V}\frac{U^2}{c_p T_{co}} = \frac{C}{St_{mom}}Ec \tag{6.116}$$

so can be neglected in low Mach number flows.

The nondimensional scaling parameter associated with thermal coupling due to particle-fluid heat transfer is

$$\frac{\bar{\rho}_{do}}{\bar{\rho}_{co}}\frac{\lambda}{U\tau_T}\frac{c_d}{c_p} = \frac{C}{St_T}\frac{c_d}{c_p} \tag{6.117}$$

which is another energy coupling factor introduced in Chapter 2. The coupling parameter was modified to account for $St_T \to 0$ and expressed as

$$\Pi_{ener} = \frac{C}{St_T + 1}$$

The scaling parameter associated with conductive heat transfer through the mixture is

$$\frac{k_c}{\lambda U \rho_c c_p} = \frac{\mu}{\lambda U \rho_c} \frac{k_c}{\mu c_p} = \frac{1}{\text{Re Pr}} \qquad (6.118)$$

where Pr is the Prandtl number which is near unity for most gases. Thus the convective heat transfer can be neglected for high Reynolds number flows. This conclusion must be reconsidered for dense systems when there is significant heat transfer through the dispersed phase.

Thus, one finds that many terms can be neglected in the total energy equation for a dispersed phase flow if the Mach number is small. In this case the equation simplifies to

$$\frac{\partial}{\partial t}(\alpha_c \rho_c i_c) + \frac{\partial}{\partial x_i}[\alpha_c \rho_c u_i h_c] = s_{mass} h_s$$

$$-p\frac{\partial}{\partial x_i}(\alpha_d v_i) + \tau_{ij}\frac{\partial}{\partial x_j}(\alpha_d v_i) + \frac{\partial}{\partial x_j}(\alpha_c u_i \tau_{ij}) \qquad (6.119)$$

$$+\beta_T(T_c - T_d)$$

In a dilute phase flow with $\alpha_d/\alpha_c \ll 1$, the only source terms of importance are the mass coupling term and the particle-fluid heat transfer term.

6.4.4 Thermal energy equation

The thermal energy equation is obtained by subtracting the kinetic energy out of the total energy equation. The kinetic energy equation is obtained by multiplying the momentum equation by the velocity. Each component of the momentum equation is multiplied by the component carrier phase velocity in that direction and the three equations are combined to yield the kinetic energy equation. After some manipulation with the continuity equation, the equation for kinetic energy, as shown in Appendix B, becomes

$$\frac{\partial}{\partial t}\left(\alpha_c \rho_c \frac{u^2}{2}\right) + \frac{\partial}{\partial x_i}\left(\alpha_c \langle \rho_c \rangle u_i \frac{u^2}{2}\right) = \frac{u^2}{2}\sum_k \dot{m}_k/V - u_i \sum_k v_{i,k}\dot{m}_k/V$$

$$-\alpha_c u_i \frac{\partial}{\partial x_i}p + \alpha_c u_i \frac{\partial}{\partial x_j}\tau_{ij} \qquad (6.120)$$

$$-3\pi\mu_c u_i \sum_k D_k f_k(u_i - v_{i,k})_n/V + \alpha_c \rho_c g_i u_i$$

Subtracting this equation from the total energy equation results in the thermal energy equation. Assuming the droplets evaporate at the same rate, move at the same local velocity and have the same temperature, the thermal energy equation as derived in Appendix B assumes the form

$$\frac{\partial}{\partial t}(\alpha_c \rho_c i_c) + \frac{\partial}{\partial x_i}(\alpha_c \rho_c u_i i_c) - s_{mass}\left(h_s + \frac{|v - u|^2}{2} + \frac{w'^2}{2}\right) =$$

$$\beta_V |u_i - v_i|^2 - p\frac{\partial}{\partial x_i}(\alpha_d v_i + \alpha_c \langle u_i \rangle) + n\dot{V}_d p_s \tag{6.121}$$

$$+\tau_{ij}\frac{\partial}{\partial x_j}(\alpha_c u_i + \alpha_d v_i) + \beta_T(T_d - T_c) - \frac{\partial}{\partial x_i}k_{eff}\frac{\partial}{\partial x_i}T_c$$

Using the same scaling arguments presented above for the total energy equation, the thermal energy equation for low Mach number flows reduces to

$$\frac{\partial}{\partial t}(\alpha_c \rho_c i_c) + \frac{\partial}{\partial x_i}(\alpha_c \rho_c u_i i_c) - s_{mass} h_s =$$

$$-p\frac{\partial}{\partial x_i}(\alpha_d v_i + \alpha_c u_i) + \tau_{ij}\frac{\partial}{\partial x_j}(\alpha_c u_i + \alpha_d v_i) \tag{6.122}$$

$$+\beta_T(T_d - T_c)$$

If there is no mass transfer the thermal energy equation becomes

$$\frac{\partial}{\partial t}(\alpha_c \rho_c i_c) + \frac{\partial}{\partial x_i}(\alpha_c \rho_c u_i i_c) = \beta_V |u_i - v_i|^2$$

$$+\tau_{ij}\frac{\partial}{\partial x_j}(\alpha_c u_i + \alpha_d v_i) + \beta_T(T_d - T_c)$$

Further manipulations of the thermal energy equation are possible using the continuity equation to express the unsteady and convection terms in nonconservative form. This equation can also be formulated to give the time rate of change of the entropy of the continuous phase.

6.4.5   Equation summary

The conservation equations for multidimensional flow are summarized in Table 6.3.

Continuity equation

$$\frac{\partial}{\partial t}(\alpha_c \rho_c) + \frac{\partial}{\partial x_i}(\rho_c \alpha_c u_i) =$$
C $\quad -n\dot{m}$

Momentum equation

$$\frac{\partial}{\partial t}(\alpha_c \rho_c u_i) + \frac{\partial}{\partial x_j}(\alpha_c \rho_c u_j u_i) = -\alpha_c \frac{\partial p}{\partial x_i} + \alpha_c \frac{\partial}{\partial x_j}(\tau_{ij}) + \rho_c \alpha_c g_i$$
C $\quad -\beta_V(u_i - v_i) - n\dot{m}v_i$

Total energy equation

$$\frac{\partial}{\partial t}\left[\rho_c \alpha_c \left(i_c + \frac{u^2}{2}\right)\right] + \frac{\partial}{\partial x_j}[\alpha_c \rho_c u_j \left(h_c + \frac{u^2}{2}\right)] = \alpha_c \rho_c g_i u_i$$
$$+\frac{\partial}{\partial x_i}k_{eff}\frac{\partial}{\partial x_i}T_c + \tau_{ij}\frac{\partial}{\partial x_j}(\alpha_c u_i + \alpha_d v_i) + \alpha_c u_i \frac{\partial}{\partial x_j}\tau_{ij}$$
C $\quad -n\dot{m}\left(h_s + \frac{v^2}{2}\right) + \beta_V(v_i - u_i)v_i - p\frac{\partial}{\partial x_i}(\alpha_d v_i) +$
C $\quad +n\dot{V}_d p_s + \beta_T(T_c - T_d)$

Thermal energy equation:

$$\frac{\partial}{\partial t}(\rho_c \alpha_c i_c) + \frac{\partial}{\partial x_i}(\alpha_c \rho_c u_i i_c) = \frac{\partial}{\partial x_i}k_{eff}\frac{\partial}{\partial x_i}T_c$$
$$-p\frac{\partial}{\partial x_i}(\alpha_d v_i + \alpha_c u_i) + \tau_{ij}\frac{\partial}{\partial x_j}(\alpha_c u_i + \alpha_d v_i) +$$
C $\quad -n\dot{m}\left(h_s + \frac{|v_i - u_i|^2}{2}\right) + \beta_T(T_d - T_c)$
C $\quad +\beta_V|u_i - v_i|^2 + n\dot{V}_d p_s$

Table 6.3. Summary of differential form of conservation equations for the carrier phase for multidimensional flows.

6.4.6 Turbulence

Turbulence in a single-phase flow is generally characterized by the turbulence energy k and the dissipation rate ε. This is known as the *two-equation* model. The equations for k and ε are developed starting with Reynolds equations which are the equations for the fluctuating velocity components. The equation for k is obtained by taking the dot product of the Reynolds equations with the fluctuating velocity, producing an equation for kinetic energy. After numerous assumptions and clever physical reasoning, an equation is developed which relates the change in turbulence energy to terms representing the diffusion of turbulence energy, generation of energy and dissipation. The generation of turbulence energy results from velocity gradients in the flow and the dissipation from viscous effects.

The dissipation equation comes from taking the curl of the Reynolds equations which essentially gives the vorticity of the fluctuating velocity components. Once again the rate of change of dissipation is related to diffusion, generation and dissipation. There have been many modifications suggested to improve the $k - \varepsilon$ model but the essential features remain the same and it maintains a prominent role in the commercial fluid mechanics codes (Wilcox, 1993). More recently, more attention has been devoted to the turbulence models based on large eddy simulation as an improvement over the $k - \varepsilon$ model.

Turbulence modulation is the effect of particles or droplets on the turbulence of the carrier phase. Modulation is weak if the particle concentration is very low. Also, at very high concentrations, near the state of a packed bed, fluid turbulence is attenuated by the large viscous forces associated with the small Reynolds number based on the interparticle spacing. These limiting cases are not addressed here. While the study of the particle dispersion in turbulence has a long history, the effect of the particles on fluid turbulence has been the subject of studies conducted over the past ten years.

It was not until the 1970s that direct measurements of turbulence in the presence of particles were possible. Although these measurements were relatively new, practitioners in some engineering fields were already aware of the fact that the presence of particles can significantly change the rates of heat transfer and chemical reaction, which could not be explained except through the effect of the particles on the fluid turbulence. The drag reduction phenomenon observed in pipe flows with low solids concentrations is another example which suggested that the particle phase modified the carrier phase turbulence.

The primary obstacle in obtaining fluid turbulence data in particulate flows was the difficulty with making fluid property measurements in the presence of solid particles. Before the invention of laser Doppler velocimetry (LDV), hot-wire or hot-film anemometry were the only means to acquire a direct measurement of fluid turbulence. Hot-wire probes cannot be used in flows with solid particles but a conical probe coated with a hot-film is durable, to some degree, in solid-liquid flows. For this reason the early data on fluid turbulence in the presence of solid particles were obtained in liquid-solid flows in connection with sediment transport. Some researchers (Hetsroni & Sokolov, 1971) attempted to use a hot-wire probe in gas-droplet free-jets in the study of a two-fluid atomizer and obtained results showing that particles (liquid droplets) suppress the turbulence intensity of gas.

Invention of LDV (see Chapter 9) had a large impact on turbulence research in multiphase flows. Because of the availability of LDV, the number of studies on fluid turbulence in fluid-solid flows increased dramatically. Various configurations of particle-laden flows including pipe flows, channel flows, free jets, confined jets, impinging jets and boundary layers have been investigated using LDV. The distributions of turbulence intensities based on centerline velocities in a gas-solid flow in a vertical pipe (Tsuji et al., 1984) are shown in Figure 6.10. One notes that both suppression and enhancement of turbulence occur, depending on particle size, concentration and initial turbulence level. In general, small particles tend to attenuate the turbulence while larger particles augment the turbulence level. The effects of particles on fluid turbulence are similar to those of a grid or screen in a turbulent flow; that is, turbulence is generated or attenuated depending on the size of grid and turbulence intensity.

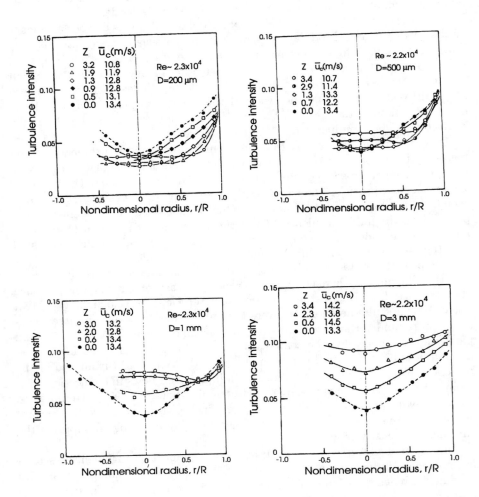

Figure 6.10: Variation of turbulence intensity for gas-particle flow in a vertical pipe with different particle sizes. (From Tsuji et al., *J. Fluid Mech.*, 139, 424-426, 1984. With permission.)

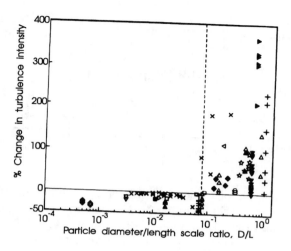

Figure 6.11: A summary of data for the change in turbulence intensity due to the presence of the dispersed phase. (Reprinted from *Intl. J. Multiphase Flow*, 15, Gore, R.A. and Crowe, C.T., The effect of particle size on modulating turbulence intensity, 280, 1989, with kind permission from Elsevier Science Ltd, The Boulevard, Langford Lane, Kidlington OX5 1GB, UK.)

Two papers (Gore & Crowe, 1989; Hetsroni, 1989) summarize the available data on turbulence modulation (until 1989). These papers suggest criteria for the suppression and enhancement of turbulence. One criterion is based on the length scale ratio D/L_e, where D is the particle diameter and L_e is the length scale characteristic of the most energetic turbulent eddies. A summary of the data is shown in Figure 6.11. It is observed that turbulence intensity is attenuated for D/L_e less than 0.1 while the turbulence level is increased for larger length scale ratios. Another criterion is based on the relative particle Reynolds number. This model suggests that particles with a low Reynolds number tend to suppress the fluid turbulence and particles with high particle Reynolds number tend to increase turbulence.

Although some qualitative trends have been observed for the effects of particles on the turbulence energy of the carrier phase, there is currently no general model that can be used reliably to predict carrier phase turbulence in particle-laden flows.

The presence of particles can affect the carrier phase turbulence in several ways (Crowe, 1993) such as:

- displacement of the flow field by flow around a dispersed phase element,

- generation of wakes behind particles,

- dissipation of turbulence transfer of turbulence energy to the motion of the dispersed phase,

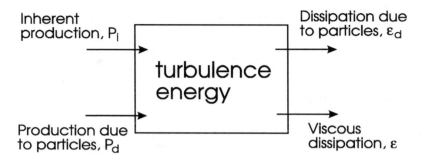

Figure 6.12: Schematic of model for turbulence modulation. (Reprinted from *Intl. J. Multiphase Flow*, 23, Kenning, V. M. and Crowe, C.T., Effect of particles on the carrier phase turbulence in gas-particle flows, 405, 1997, with kind permission from Elsevier Science Ltd, The Boulevard, Langford Lane, Kidlington OX5 1GB, UK)

- modification of velocity gradients in the carrier flow field and corresponding change in turbulence generation,

- introduction of additional length scales which may influence the turbulence dissipation,

- disturbance of flow due to particle-particle interaction.

Numerical models for turbulence modulation are still in development so no specific model has been adopted. The general approach is to modify the generation and dissipation terms in the $k - \varepsilon$ equations to account for the presence of the dispersed phase. Some approaches have been to use the volume averaged equations as single-phase flow equations and develop Reynolds equations for the averaged quantities. This approach leads to a fallacy (Crowe et al., 1996).

One approach which has been used to model the effect of particles on the carrier phase turbulence is direct numerical simulation (DNS). This approach is advantageous because it requires no Reynolds stress modeling. Squires and Eaton (1990) used DNS with spectral methods to study the effect of particles on isotropic, homogeneous turbulence at low Reynolds numbers. Energy had to be added to maintain the turbulence. They modeled the effect of the particles by including a point forces (drag) in the flow field. It was found that the presence of the particles increases the turbulence dissipation rate. Squires and Eaton also found that the particles tend to concentrate in regions of high strain. Elghobashi and Truesdell (1993) used a similar approach to model the effect of particles on turbulence. They also predicted that the rate of viscous dissipation is increased due to the presence of the particles.

Recently Liljegren (1996) has questioned the adequacy of modeling the effect of the particles as point forces in the flow in that the no slip condition at the particle surfaces is not correctly accounted for.

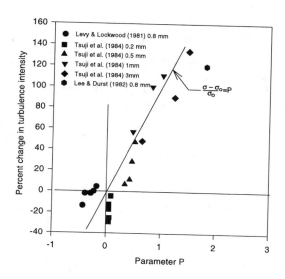

Figure 6.13: Comparison of experimental data for turbulence intensity modulation with model. (Reprinted from *Intl. J. Multiphase Flow*, 23, Kenning, V. M. and Crowe, C.T., Effect of particles on the carrier phase turbulence in gas-particle flows, 406, 1997, with kind permission from Elsevier Science Ltd, The Boulevard, Langford Lane, Kidlington OX5 1GB, UK.)

Sato and Hishida (1966) have reported experimental studies of turbulence modulation in liquid-solid flows. They found that small particles attenuate turbulence due to particle drag. They also developed a multiple-time scale model for particle-laden flows which predicts both turbulence attenuation and augmentation.

Also, recently, a simple model has been proposed by Kenning and Crowe (1996) in an attempt to explain the data presented in Figure 6.11. The basic idea of the model is shown in Figure 6.12. The intrinsic turbulence energy is the energy without particles. Additional turbulence energy can be generated by the particles or removed by transfer of the energy to particulate motion. Energy can also be removed by dissipation in the fluid.

For gas-particle flows it can be shown that the energy transferred to particulate motion is negligible. The energy added per unit mass to the flow because of the dissipation of the work associated with drag is

$$\dot{E}_d = \frac{f}{\tau_V} \frac{\bar{\rho}_d}{\bar{\rho}_c}(u - v)^2 \tag{6.123}$$

The dissipation of energy is given by

$$\varepsilon = \frac{k^{3/2}}{L_\varepsilon} \tag{6.124}$$

where L_ε is the dissipation length scale. However, in a two-phase mixture, several new length scales are introduced, the particle size and the interparticle spacing. If the interparticle spacing is less than the intrinsic length scale for the flow, then the interparticle spacing should influence the dissipation. Kenning & Crowe defined a "hybrid" length scale, L_h, which approached the correct limits and used this for the dissipation length scale. They showed that the change in turbulence intensity for particles transported by air in a vertical duct should vary as,

$$\frac{\sigma - \sigma_o}{\sigma_o} = P = \left[\frac{L_h}{L_i} + \frac{L_h}{k_i^{3/2}} \frac{f}{\tau_V} \frac{\tilde{\rho}_d}{\tilde{\rho}_c} (u - v)^2 \right]^{1/3} - 1 \tag{6.125}$$

where L_i and k_i are the dissipation length scale and turbulence energy in the flow without particles. This model is compared with the available data in Figure 6.13. A reasonable agreement is achieved.

Considerable more work needs to be done before viable models are found for carrier phase turbulence and for shear stress due to turbulence in fluid-solid flows.

6.5   Summary

The inability to model the local details of the continuous phase in a dispersed phase flow necessitates the use of averaging. The three general categories of averaging are time, volume and ensemble. The conservation equations based on volume averaging illustrate the influence of the dispersed phase on the carrier phase through the coupling terms and volume fractions of each phase. The deviations of the continuous phase velocity from the average velocity give rise to a Reynolds stress in the same manner as time averaging yields the Reynolds stress in a single-phase flow. Modeling the Reynolds stress in a dispersed phase flow is complicated by the length scales associated with the dispersed phase. Also the heat transfer through the mixture involves the heat transfer through both the continuous and dispersed phases. The presence of the particles or droplets in the field affect the turbulence of the carrier phase as a result of enhanced turbulence generation and dissipation.

Exercises

6.1. Consider a flow of evaporating droplets in a one-dimensional duct. The droplets evaporate according to the D^2-law and are always in kinetic equilibrium with the gas. The number flow rate $\dot{n}$ of the droplets is constant so

$$\dot{n} = nu_d A = nu_c A = const$$

where n is the droplet number density. The droplet volume fraction is sufficiently small that the continuous phase volume fraction can be taken as unity, $\alpha_c \simeq 1$.

a) Find an expression for the mass source term in the form

$$S_m = -n\dot{m} = f(Z_0, u_0/u, \rho_c, \tau_m, D/D_0)$$

where Z_0, u_0 and D_0 are the loading, velocity and droplet diameter at the beginning station in the tube. The evaporation time constant is τ_m and ρ_c is the gas density. Make use of the conservation of droplet number; namely, $n_0 u_0 = nu$.

b) Using the continuous phase continuity equation,

$$\frac{d}{dx}(\rho_c u) = S_m$$

find how the velocity varies with time in the form

$$\frac{u}{u_0} = f(Z_0, t/\tau_m)$$

and find the velocity ratio when evaporation is complete. When you complete the expression for S_m you will find a velocity u in the denominator which when combined with the continuity equation allows you to express

$$u\frac{d}{dx}(\rho_c u) = \rho_c \frac{du}{dt}$$

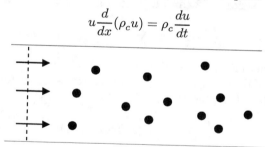

Problem 6.1

6.2. Two hundred-micron glass particles (sp. gr. = 2.5) are moving at 30 m/s through stagnant air at standard conditions and impinging on a plate as shown. The bulk density of the particles is 2 kg/m^3. What is the pressure gradient in the gas?

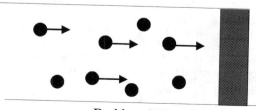

Problem 6.2

6.3. A fluid-particle flow is moving vertically in a tube as shown. The flow is steady and uniform. The particles are not accelerating. Find how the pressure gradient in the tube varies with the mixture density, $\rho_m = \alpha_c \rho_c + \alpha_d \rho_d$. How does this compare with the hydrostatic pressure gradient in a single-phase fluid. [Hint: It will be better to use the form of the momentum equation where the pressure gradient term is $-\partial p/\partial x$ instead of $-\alpha_c \partial p/\partial x$ so the force due to the particles contains all the surface forces on the particles (including the buoyancy force).]

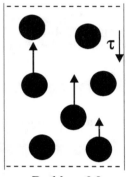

Problem 6.3

6.4. By performing the difference operation and using the continuity equation, show that Equation 6.32 can be written as

$$\rho_c \alpha_c V \frac{du}{dt} + \alpha_c \rho_c Au\Delta u - S_{mass}(v - u) = F$$

6.5. Show, in detail, how Equation 6.80 is obtained.

6.6. The rate of droplet evaporation can be expressed in terms of the Sherwood number as (see Chapter 4)

$$\dot{m} = Sh\pi\rho_c D_v D(\omega_\infty - \omega_s)$$

Assume that the Sherwood number, ω_s, is the same for all droplets in a volume and that the droplets have a Rosin-Rammler size distribution. Find the mass source term in the form

$$S_{mass} = V Sh\pi\rho_c D_v(\omega_\infty - \omega_s)f(D_{mm}, \sigma)$$

6.7. In Equation 6.74, the gravity term can be written as

$$\alpha_c \rho_c g V u = -\alpha_c \rho_c g A\Delta zu$$

where Δz is a distance increment in the direction opposite of gravity. Show that this term can be combined with the energy flux term to give

$$\Delta \left[\alpha_c \rho_c Au \left(h_c + \frac{u^2}{2} + gz \right) \right]$$

Note that in a steady flow with no mass transfer and heat transfer, Equation 6.74 would reduce to

$$\Delta \left[\alpha_c \rho_c A u \left(h_c + \frac{u^2}{2} + gz \right) \right] = -\beta V(u - v)v$$

Remark on the significance of this equation pertaining to Bernoulli's equation in a dispersed phase flow.

6.8. The equation for entropy in a flow system is

Accumulation of entropy in control volume	+	Net efflux of entropy across control surfaces	$\geq$	$\dot{Q}/T$

Show that the equation for entropy in a quasi-one-dimensional flow with no particles is

$$\rho \frac{\partial s}{\partial t} + \rho u \frac{\partial s}{\partial x} \geq -\frac{1}{A} \frac{\partial}{\partial x} \left(\frac{\dot{q} A}{T} \right) - \frac{\dot{q}_w}{T_w} \frac{1}{R_H}$$

where $\dot{q}_w$ is the heat transfer to the wall and s is the entropy per unit mass.

6.9. Using the thermal energy equation for a single-phase flow in a quasi-one dimensional duct,

$$\frac{\partial}{\partial t}(\rho_c i_c) + \frac{1}{A} \frac{\partial}{\partial x}(\rho_c u A h_c) = u \frac{\partial p}{\partial x} + \frac{u \tau_w - \dot{q}_w}{R_H} - \frac{1}{A} \frac{\partial}{\partial x}(\dot{q} A)$$

show that the equation for entropy of the fluid is

$$\rho_c \frac{Ds}{Dt} \geq -\frac{1}{A} \frac{\partial}{\partial x}\left(\frac{\dot{q} A}{T} \right) - \frac{\dot{q}_w}{T_w} + \frac{u \tau_w - \dot{q}_w}{R_H T} - \frac{1}{TA} \frac{\partial}{\partial x}(\dot{q} A)$$

6.10. Show that the thermal energy equation for a quasi-one dimensional flow in a duct with non-evaporation particles with constant material density can be expressed as

$$\frac{\partial}{\partial t}(\bar{\rho}_c i_c) + \frac{1}{A} \frac{\partial}{\partial x}(\bar{\rho}_c u A h_c) = \alpha_c \frac{Dp}{Dt}$$

$$+ \frac{u \tau_w - \dot{q}_w}{R_H} - \frac{1}{A} \frac{\partial}{\partial x}(\dot{q} A) + \beta_V (u - v)^2$$

6.11. Assume there was a Coulomb force acting on the particles in the flow direction. Would the momentum equation, Equation 6.55, have to be modified to include the Coulomb force on the particles? Discuss.

6.12. Starting with the differential form of the momentum equation for a quasi-one-dimensional flow, Equation 6.51, take the derivative by parts of the left side and subtract the continuity equation to show that

$$\alpha_c \rho_c \frac{\partial u}{\partial t} + \alpha_c \rho_c u \frac{\partial u}{\partial x} = -\alpha_c \frac{\partial p}{\partial x}$$

$$+ s_{mass}(v - u) + \beta_V(v - u) - \frac{\tau_w}{R_H} + \alpha_c \rho_c g$$

6.13. Would the total energy equation, Equation 6.74 be valid for flow in a quasi-one dimensional duct if particles were impacting and reflecting from the duct walls? Discuss.

Chapter 7

Droplet-Particle Cloud Equations

There are two generic approaches for the numerical simulation of a cloud of particles or droplets in a mixture: namely, the Lagrangian approach and the Eulerian approach. In the Lagrangian approach, the velocity, mass and temperature history of each particle (or a representative particle) in the cloud are calculated. The local cumulative motion and state of each particle in the cloud represents the spatial properties of the cloud. In the Eulerian approach, the cloud of particles is considered to be a second fluid which behaves like a continuum, and equations are developed for the average properties of the particles or droplets in the cloud. Each approach has its relative advantages and disadvantages depending on the nature of the flow.

The purpose of this chapter is to introduce the Lagrangian and Eulerian approaches and to show how the source terms introduced for mass, momentum and energy coupling are evaluated.

7.1 Lagrangian approach

The Lagrangian approach is applicable to both dilute and dense flows. In dilute flows, the time between particle-particle collisions is larger than the response time of the particles (or droplets) so the motion of the particles is controlled by the particle fluid interaction, body forces and particle-wall collisions. In a dense phase flow, the response time of the particles is longer than the time between collisions so particle-particle interaction controls the dynamics of the particles but is also influenced by the hydrodynamic and body forces as well as particle-wall interaction. If the flow is steady and dilute a form of the Lagrangian approach known as the *trajectory method* (Crowe et al., 1977) is easy to implement. If the flow is unsteady and/or dense, the more general *discrete element approach* is necessary.

7.1.1   Trajectory method

The trajectory approach can be explained best by reference to an example. Consider a nozzle spraying a liquid at a steady rate into the chamber shown in Figure 7.1. Assume, for now, that the flow is steady so the spatial distribution of the carrier flow properties is invariant with time. The flow field is subdivided into a series of computational cells as shown. The inlet stream is discretized into a series of starting trajectories. If the initial droplet velocity and mass are known, the droplet velocity can be calculated by solving the droplet motion equation (see Chapter 4) in the flow field; namely,

$$\frac{d\mathbf{v}}{dt} = \frac{\mathbf{F}_f}{m} + \mathbf{g} \tag{7.1}$$

where $\mathbf{F}_f$ is the fluid forces (form and friction forces) acting on the droplet of mass m and $\mathbf{g}$ is the gravity vector. The trajectory is obtained by integrating the velocity,

$$\mathbf{s} - \mathbf{s}_0 = \int_0^t \mathbf{v}\,d\tau \tag{7.2}$$

The integration scheme to be used depends on the desired accuracy and computational efficiency. This calculation would include both the drag and lift forces on the particle if the Magnus and Saffman lift forces are important in establishing the particle trajectories. The spin of the particle is determined from

$$\frac{d\boldsymbol{\omega}}{dt} = \frac{\mathbf{T}_f}{I} \tag{7.3}$$

where $\mathbf{T}_f$ is the moment applied due to the fluid forces and I is the moment of inertia about an axis through the particle centroid.

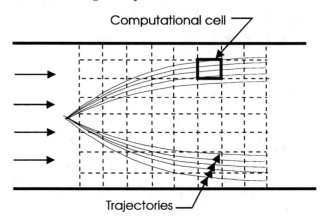

Figure 7.1: Droplet trajectories in a spray.

Concurrently, the droplet temperature history can be calculated using

$$\frac{dT_d}{dt} = \frac{1}{mc_d}(\dot{Q}_d + \dot{m}h_L) \tag{7.4}$$

where $\dot{Q}_d$ is the sum of both the convective and radiative heat transfer to the particle or droplet. If the Biot number of the droplet is large, this equation must be modified to account for the fact that the surface temperature is not the average temperature in the droplet.

The droplet mass transfer must also be calculated along the trajectory according to the relations given in Chapter 4. Of course, the droplet diameter must be adjusted to conform with the droplet mass (unless the application is the falling rate period of a drying porous particle).

Assume that the mass flow entering from the atomizer is discretized into j trajectories and the mass flow associated with each trajectory is $\dot{M}(j)$. Then the number flow rate along trajectory j would be

$$\dot{n}(j) = \frac{\dot{M}(j)}{\frac{\pi}{6}\rho_d D_0{}^3} \tag{7.5}$$

where ρ_d is the material density of the droplet and D_0 is the initial droplet diameter. If no droplet breakup or coalescence occurs, the number flow rate will be invariant along each trajectory.[1] Of course, more detail is possible by discretizing the starting conditions according to a size distribution as well. For example, if $\tilde{f}_m(D_s)$ is the fraction of particle mass associated with size D_s (see Chapter 3), then the number flow rate associated with size D_s on trajectory j would be

$$\dot{n}(j, D_s) = \frac{\tilde{f}_m(D_s)\dot{M}(j)}{\frac{\pi}{6}\rho_d D_s^3} \tag{7.6}$$

Obviously more detail requires more trajectories and increased computational time. In an axisymmetric flow, if the starting locations are discretized to a series of concentric rings, the mass flow rate on each ring must be weighted with the ring radius.

Once all the trajectories are calculated, the properties of the particle cloud in each computational cell can be determined. The particle number density is found using

$$n = \frac{\sum_{traj} \dot{n}\Delta t_j}{V} \tag{7.7}$$

where Δt_j is the time required for the particle to traverse the cell on trajectory j and V is the volume of the computational cell. The summation is carried out over all trajectories which traverse the cell. The particle volume fraction in each computational cell can be determined from

---

[1] The droplet size and velocity would not be known directly at the atomizer face. Thus the starting conditions must be taken at some point where data are available and beyond the dense spray region where droplet breakup and coalescence are predominant.

$$\alpha_d = \frac{\sum_{traj} \dot{n}_j \bar{V}_d \Delta t_j}{V} \tag{7.8}$$

where $\bar{V}_d$ is the average droplet volume along trajectory j in the cell. Other properties such as bulk density, particle velocity and temperature can be determined in the same way. Thus the properties of the cloud can be determined once all the trajectories have been calculated.

Example: The entering mass flow of 100-μm water droplets in an axisymmetric configuration is 1 kg/s. The droplets enter on concentric rings located at $r = 0.01, 0.02, 0.3, 0.04$ and 0.05 m. Find the number flow rate on each starting ring.

Solution: The total number flow rate of the droplets is

$$\dot{n}m = 1 \text{ kg/s}$$

The mass of each 100-μm droplet is 5.24×10^{-10} kg. The total number flow rate is

$$\dot{n} = \frac{1}{5.24 \times 10^{-10}} = 1.91 \times 10^9/s$$

The number flow rate is distributed according to the circumference of the starting ring. Thus the number on starting ring s is

$$\dot{n}_s = \frac{r_s}{\sum r_s} \dot{n}$$

The distribution is shown in the following table.

r_s	0.01	0.02	0.03	0.04	0.05
$\dot{n}_s \times 10^{-8}/s$	1.27	2.55	3.82	5.09	6.37

Particle or droplet wall collisions are included in the calculation by continuing the trajectory after wall collision, according to the models presented in Chapter 5. New velocities and spin rates are established depending on the nature of the collision. In the case of a droplet impact, the droplet may splatter on the surface and the trajectory is terminated or the trajectory is restarted with smaller droplets. In the case of annular mist flows, one would model reentrainment by initializing trajectories of droplets from the liquid layer on the wall. The specific conditions depend on the model selected for the problem. If droplet breakup or coalescence occurs, the number flow rate is modified according to

$$\dot{n}m_d \big|_{before} = \dot{n}m_d \big|_{after} \tag{7.9}$$

where the droplet mass before and after breakup or coalescence depends on the model used. The trajectory approach for droplet coalescence would not be advisable unless the agglomeration rate could be decoupled from the droplet number density.

The Lagrangian method has been the basis of many numerical simulations of gas-particle and gas-droplet flows.

7.1.2   Discrete element method

If the flow is unsteady and/or dense (particle-particle collisions important), the more general discrete element method is required. In this approach, the motion and position (as well as other properties) of individual particles, or representative particles, are tracked with time. Ideally, one would like to track each and every particle but this may not be computationally feasible. For a gas laden with 100-micron particles at a mass concentration of unity, there would be on the order of 10^9 particles per cubic meter. If the flow field of interest were one-tenth of a cubic meter, then 10^8 particles would have to be tracked through the field. This is impractical, so a smaller number of *computational* particles are chosen to represent the actual particles. For example, if 10^4 computational particles were chosen, then each computational particle would represent 10^4 physical particles. This computational particle is regarded as a *parcel* of particles. It is assumed then that the parcel of particles moves through the field with the same velocity and temperature, etc. as a single particle (physical particle). Of course, size distribution effects can be included by specifying parcels with a specific particle size. The parcel is identified as a *discrete element*. In some simulations, such as fluidized beds, it may not be possible to use parcels of particles without sacrificing the details necessary to simulate the system so the dynamics of each individual particle must be considered.

The equation for particle motion now assumes the form

$$\frac{d\mathbf{v}}{dt} = \frac{\mathbf{F}_f + \mathbf{F}_c}{m} + \mathbf{g} \qquad (7.10)$$

where $\mathbf{F}_c$ is the force due to particle-particle (and/or particle-wall) contact. Similarly, the rotation rate of the particle is determined from

$$\frac{d\boldsymbol{\omega}}{dt} = \frac{\mathbf{T}_f + \mathbf{T}_c}{I} \qquad (7.11)$$

where $\mathbf{T}_c$ is the torque due to contact forces.

Establishing the initial conditions for the discrete elements depends on the problem. For the example shown in Figure 7.1, a parcel could be the droplets emerging along a starting trajectory j in time interval Δt_p. Thus the number of droplets in the parcel would be $N_p = \dot{n}(j, D_s)\Delta t_p$, and the initial velocity would be determined from other information. For modeling a fluidized bed, the initial state may be all the particles at rest as a packed bed, and the interstitial gas flow initiates the motion.

The motion of each parcel over one time interval is obtained by integrating the particle motion equation. At the same time, the particle temperature, spin and other properties can be calculated. A field with a distribution of sample particle parcels is shown in Figure 7.2. During the time step, there may be particle-particle collisions which alter the trajectories and change the distribution of the parcels in each computational cell.

At every time step, the properties of the droplet cloud can be determined by summing over all the particles in a computational volume. For example, the

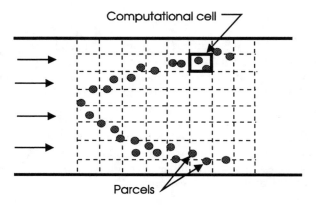

Figure 7.2: Distribution of droplet parcels in a spray field.

number density would be

$$n = \frac{\sum_p N_p}{V} \tag{7.12}$$

where the summation is carried out over all the parcels in the computational cell. Also the particle volume fraction would be

$$\alpha_d = \frac{\sum_p N_p V_{d,p}}{V_{cell}} \tag{7.13}$$

where $V_{d,p}$ is the volume associated with an individual particle and N_p is the number of particles in parcel p. The calculation of other properties such as bulk density and particle velocity is obvious. The distribution of the volume fraction of the solid phase is important in calculating the interstitial flow field. If each discrete element is an individual particle, then N_p in the above equations is unity.

7.1.3   Source term evaluation

The mass source term for a computational cell is simply the sum of the mass added by every droplet in the cell divided by the volume of the cell,

$$S_{mass} = -\sum_k \dot{m}_k \tag{7.14}$$

where $\dot{m}_k$ is the rate of change of the mass of droplet k. The minus sign indicates that a droplet losing mass is adding mass to the carrier phase. The mass source term using the trajectory approach is evaluated by

$$S_{mass} = -\sum_{traj.} \dot{n}_j \overline{\dot{m}} \Delta t_j \tag{7.15}$$

where $\overline{\dot{m}}$ is the average mass evaporation rate of the droplet during its traverse through the cell. The mass source term is evaluated using the discrete element approach by

$$S_{mass} = -\sum_p N_p \dot{m}_p \qquad (7.16)$$

where summation is carried out over all parcels which occupy the cell at the given time and $\dot{m}_p$ is the mass evaporation (or condensation) rate of the individual droplets in the parcel. The mass source term per unit volume is

$$s_{mass} = \frac{S_{mass}}{V} \qquad (7.17)$$

where V is the volume of the computational cell.

The momentum source term in the i-direction is given by

$$\mathbf{S}_{mom} = -\sum_k (\mathbf{F}_{f,k} + \mathbf{v}_k \dot{m}_k) \qquad (7.18)$$

where $\mathbf{F}_{f,k}$ is the fluid forces acting on the droplet and $\mathbf{v}_k$ is the velocity of droplet k. The force would include both a lift and drag force, but would not include the forces due to pressure gradient, shear stress gradient and body forces.[2] For the case in which the transient drag forces and lift forces are unimportant, the momentum source term for the trajectory approach becomes

$$\mathbf{S}_{mom} = \sum_{traj} \dot{n}_j \Delta t_j \left[m_j \frac{f_j}{\tau_{V,j}} (\mathbf{v}_j - \mathbf{u}) - \dot{m}_j \mathbf{v}_j \right] \qquad (7.19)$$

where $\mathbf{v}_j$ is the average velocity of droplets, f_j is the drag factor and $\tau_{V,j}$ is the velocity response time for the particles on trajectory j. The corresponding momentum source term for the discrete element approach is calculated by

$$\mathbf{S}_{mom} = \sum_p N_p \left[m_p \frac{f_p}{\tau_{V,p}} (\mathbf{v}_p - \mathbf{u}) - \dot{m}_p \mathbf{v}_p \right] \qquad (7.20)$$

where the subscript p refers to particles or droplets in the packet. Note that the body force due to gravity is not included in this expression.

The source term for the total energy equation using the trajectory approach is[3]

$$S_{ener} = -\sum_{traj} \dot{n}_j \Delta t_j \left[\dot{Q}_j + \dot{m}_j (h_{s,j} + \frac{|\mathbf{v}|_j^2}{2}) + \mathbf{F}_{f,j} \cdot \mathbf{v}_j \right] \qquad (7.21)$$

---

[2] The forces on the droplet due to pressure gradient and shear stress gradient can be included in the momentum source term if the pressure gradient term and shear stress term in the carrier phase equations do not have the α_c coefficient (see Chapter 6).

[3] The term associated with the dispersed phase volume change in the cell is not included here but can be if necessary.

where $h_{s,j}$ is the enthalpy of the carrier phase at the surface of the droplet on trajectory j and $\dot{Q}_k$ is the convective heat transfer to the droplet (radiative heat transfer is not included). The corresponding source term for the discrete element method is

$$S_{ener} = -\sum_p N_p \left[\dot{Q}_p + \dot{m}_p(h_{s,p} + \frac{|\mathbf{v}|_p^2}{2}) + \mathbf{F}_{f,p} \cdot \mathbf{v}_p \right] \qquad (7.22)$$

If the transient drag forces and lift forces are unimportant, the energy source term due to the dispersed phase is

$$S_{ener} = \sum_p N_p \left[\frac{Nu_p}{2} \frac{m_p c_d}{\tau_{T,p}}(T_{d,p} - T_c) - \dot{m}_p(h_{s,p} + \frac{|\mathbf{v}|_p^2}{2}) + m_p \frac{f_p}{\tau_{V,p}}(\mathbf{v} - \mathbf{u})_p \cdot \mathbf{v}_p \right]$$
$$(7.23)$$

The source terms for the thermal energy equation can be evaluated in the same fashion.

7.1.4   Particle dispersion

Calculation of particle motion is feasible, for the most part, in laminar flows using established formulas for forces acting on the particle. Very small particles (less than a micron) will be in velocity equilibrium with the carrier flow and will follow the fluid motion. However, there can still be diffusion due to Brownian motion which is diffusion associated with molecular impact on the particles. Diffusion due to Brownian motion is presented in Appendix C. The fundamental assumption in Brownian motion is that the time between molecular impacts is long compared to the particle response time, so molecular impacts can be regarded as random impulses, that is, there is no history of the previous impact. In this case the variance of the displacement varies linearly with time and gas temperature

$$\langle x^2 \rangle = 2kBTt \qquad (7.24)$$

where k is Boltzmann's constant and B is the mobility and depends on the particle size, fluid viscosity and the Cunningham correction factor (see Chapter 4).

The prediction of particle motion in turbulent flow fields is not straightforward because there are few detailed models for turbulence in single-phase flows, let alone multiphase flows. Moreover, the affects of turbulence on particle motion are significant except for massive particles which are unresponsive to turbulent velocity fluctuations.[4] Therefore, the problem of particle-fluid interaction in turbulent flows has been the subject of extensive research.

---

[4] If the particles are larger than the scale of turbulence, the fluid forces acting on the particle will be affected by the local fluid turbulence.

Particles disperse in a turbulent flow field due to fluctuating fluid forces. This phenomenon is similar to the Brownian motion where dispersion is caused by a random force but now the forces are continuous and vary in magnitude throughout the field. If particles are small enough to follow the instantaneous fluid motion, particle dispersion is equal to that of a fluid element. In general, however, particles do not follow the fluid motion due to their inertia. Moreover, the velocity nonequilibrium characteristic of large particles enables these particles to interact with several turbulent eddies which reduces the particle's residence time in each eddy and mitigates the influence of the eddy on the particle trajectory.[5] This effect is called the *crossing trajectory effect* (Csandy, 1963). Because of particle inertia, velocity nonequilibrium and the complexity of turbulent flows, the analysis of particle dispersion in turbulence is much more complicated than Brownian motion. The subject dates back to the pioneering work by G. I. Taylor (1921). Since that time, many studies of particle dispersion in homogeneous turbulence have been reported in the literature (Synder & Lumley, 1971; Reeks, 1977; Wells & Stock, 1983; Wang & Stock, 1992; Stock, 1994).

If the turbulent flow field is known, the calculation of particle dispersion evolves directly from the discrete element method. The instantaneous velocity field at each time level is included in the equation for motion of a computational particle (parcel). For example, the simplified form of the BBO equation for heavy particles is

$$\frac{d\mathbf{v}}{dt} = \frac{f}{\tau_V}(\mathbf{u} - \mathbf{v}) + \mathbf{g} \tag{7.25}$$

and $\mathbf{u}$ is the local instantaneous flow velocity of the carrier fluid and f is the drag factor introduced in Chapter 4. The temporal variation of the flow velocity through the field is responsible for particle dispersion.

In special cases numerical solutions for the carrier flow turbulence can be obtained. Direct numerical simulation (DNS) provides a direct solution to the Navier-Stokes equations with no empirical closure models. This scheme resolves the smallest scale turbulence (Komolgorov length scale) and provides (through a suitable interpolation scheme) the velocity at any point in the flow field. Because of the large number of grid points and times to achieve a statistically steady state, the scheme has been limited to small Reynolds numbers (Squires & Eaton, 1991; Elghobashi & Truesdell, 1992). However, the scheme provides a value for $\mathbf{u}$ which can be used in the equation of motion for the particles. Of course to apply the equation of motion, the particles should be smaller than the Komolgorov length scale.

Another approach for high Reynolds number flows in which the vorticity is concentrated in specific regions of the flow is the *discrete vortex method* (Chung & Troutt, 1988). This scheme is especially effective for free shear layers and wakes where large scale turbulent motion dominates the flow. In this scheme, the vorticity is discretized into discrete vortex elements which move through the

---

[5] A source of particle-fluid velocity difference is the terminal velocity due to gravity.

field and interact with each other. The instantaneous velocity at any point in the flow field is obtained by application of the Biot-Savart law.

More recently, *large eddy simulation* (LES) has been used to model the turbulence field (Squires & Wang, 1996). This technique involves both direct simulation and the Reynolds-averaging approach. The large scale structures are modeled directly while the smaller scales are assumed to homogeneous and isotropic, on average, and relatively independent of flow geometry. LES is attractive because it can be potentially used to model high Reynolds number, complex flows while maintaining acceptable accuracy.

In general, the instantaneous velocity at a point due to turbulence is unknown and approximate schemes must be used to simulate the effects of turbulence. These schemes fall into the general category of stochastic models. A review of these models is provided by Crowe et al. (1996).

A very common approach is to regard the turbulent flow as a collection of turbulent eddies with discrete velocities and life times (or scales). In the eddy life time model, originally proposed by Yuu et al. (1978), the fluid velocity encountered by a particle is constant during the time the particle spends in the eddy. The velocity is taken as the sum of the local time-averaged fluid velocity, $\overline{\mathbf{u}}$, and a fluctuation velocity, $\mathbf{u}'$ selected from a Gaussian distribution[6] with a variance proportional to the turbulence energy.

$$\mathbf{u} = \overline{\mathbf{u}} + \mathbf{u}' \qquad (7.26)$$

Using this velocity in the particle motion equation leads to the zigzag motion shown in Figure 7.3. Each arrow in the figure corresponds to an eddy encountered by the particle. The eddy life time is assumed equal to the Lagrangian integral time scale. Unfortunately the Lagrangian time scale T_L is not known *a priori* so it is usually deduced from the Eulerian properties through the following relationship

$$T_L = \frac{\ell}{\sqrt{k}} \qquad (7.27)$$

where ℓ is the Eulerian length scale and k is the local turbulence energy. The *Eulerian length scale* is defined by

$$\ell = \int_0^\infty R_E dr \qquad (7.28)$$

where R_E is the *Eulerian correlation function* of the fluctuating velocities. Measurement of the Eulerian correlation function is much easier to carry out than measurement of the Lagrangian correlation function so Eulerian length scales, as well as velocity profiles and RMS velocity fluctuation data, are available for a wide variety of flow fields.

A variety of other schemes to simulate particle dispersion due to turbulence have been proposed in the literature. Dukowicz (1980) suggested displacing

---

[6]The selection of a velocity through a random number generator is called the "Monte Carlo" method.

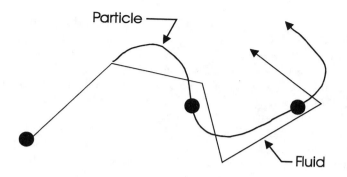

Figure 7.3: Particle motion in a turbulent velocity field.

the particle after every time step in the trajectory calculation by a distance selected randomly from a Gaussian distribution corresponding to a given particle dispersion coefficient. Lockwood et al. (1980) proposed augmenting at every time step the local fluid velocity by a value related to the local gas velocity gradient and turbulence energy. Jurewicz and Stock (1976) and later, Smith et al. (1981) added a *diffusional velocity* proportional to the gradient of the calculated particle bulk density to the calculated particle velocity. All these schemes fail to account for crossing trajectory effects.

Gosman and Ioannides (1981) introduced a scheme which was an improvement over Yuu et al.'s model because it accounted for the relative velocity between the turbulent eddy and the particle. This model was developed in concert with the two-equation $(k-\varepsilon)$ model for the turbulence of the carrier phase. They estimated the length scale of the turbulent eddy to be

$$\ell = C_\mu \frac{k^{\frac{3}{2}}}{\epsilon} \qquad (7.29)$$

and the eddy life time as

$$T_E = \frac{\ell}{\sqrt{\frac{2k}{3}}} \qquad (7.30)$$

where ϵ is the dissipation rate and C_μ is an empirical constant. The residence time of a particle in a eddy due to a relative velocity between the particle and the eddy is

$$T_R = \frac{\ell}{|\mathbf{u} - \mathbf{v}|} \qquad (7.31)$$

Thus, the interaction time between the fluid and the turbulent eddy is the minimum of the eddy life time and residence time,

$$T_I = \min(T_R, T_E) \qquad (7.32)$$

The fluctuation velocity associated with the turbulent eddy is selected randomly from a Gaussian distribution with a variance equal to $2k/3$. The particle is assumed to be in the eddy for the duration of the interaction time and then a new eddy with life time and scale corresponding to the local turbulence field is established and a new fluctuational velocity is selected with a random number generator. This scheme is used extensively in Lagrangian models because of its simplicity and robustness.

Several improvements of the Gosman and Ionnaides scheme have appeared in the literature. Berlemont et al. (1990) proposed a technique to avoid the step change in velocity when the particle passes from one eddy to another. The relative displacement between the eddy and the particle is tracked with time and the velocity in the eddy is assumed to change according to the velocity correlation for the Eulerian field. Zhou and Leschziner (1991) propose a similar approach but use a simpler technique for selection of the fluctuational velocity in the eddy. Lu et al. (1993) present a scheme whereby the relative displacement between the eddy and particle need only to be recorded over one time step. More recently Graham and James (1996) have evaluated the choice of eddy lifetime, eddy size and eddy velocity in homogeneous, isotropic and stationary turbulence and established criteria to ensure that long-time dispersion is predicted correctly.

Zhou and Yao (1992) proposed a *group modeling* scheme in which a volume occupied by the particles grows with time due to particle diffusion. The fraction of the volume in each computational cell represents the fraction of the particles in that cell. A similar model has been proposed by Litchford and Jeng (1991) and more recently by Chen and Pereira (1996). This approach breaks down in highly sheared flows and must be modified near walls. Other stochastic approaches are discussed in Crowe et al. (1996).

A shortcoming with the stochastic model for particle dispersion is that the velocity field established by adding the local turbulence fluctuational velocity and the mean carrier flow velocity does not satisfy the continuity equation (MacInnes and Bracco, 1992). This results in a nonphysical drift of the particles, especially near the edges of a mixing layer or jet.

Numerical simulations by Chien and Chung (1988), Martin and Meiburg (1994), Crowe et al. (1995) and others have shown that, under some situations, the stochastic approach is inadequate. The concept, first proposed by Crowe et al. (1985), for particle dispersion in a shear layer is illustrated in Figure 7.4. Particles which are very small will follow the fluid motion and the particles will disperse as fluid elements. On the other hand, large particles will not be affected by the fluid motion and will move in nearly straight lines. However, intermediate size particles will be centrifuged radially outward leading to a particle dispersion exceeding that of the fluid. The parameter establishing the degree of dispersion is the Stokes number,

$$St = \frac{\tau_V \Delta u}{\ell} \tag{7.33}$$

where Δu is the velocity difference across the shear layer and ℓ is the size of the turbulent structure. For $St \ll 1$, the particle will follow the fluid motion. For

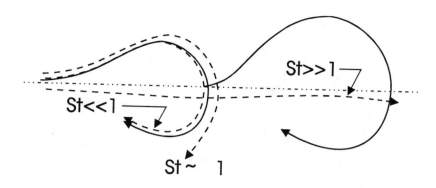

Figure 7.4: Effect of Stokes number on particle dispersion in a developing shear layer.

$St \gg 1$, the particles will move essentially independent of the fluid. However particles with $St \sim 1$ will tend to centrifuged toward the peripheries of the structures. In this case, the turbulence tends to concentrate the particles in bands at the peripheries of the vortex structures which is a *demixing* phenomenon (Crowe et al., 1988). Models treating turbulence as velocity fluctuations about a mean value will predict a concentration tending toward uniformity.

Numerical simulations of particle concentration in a wake (Tang et al., 1992) at four Stokes numbers are shown in Figure 7.5. One notes that the particles with $St \sim 1$ concentrate at the edges of the vortex structures.[7] Experimental results by Yang et al. (1995) corroborate this finding. The same phenomenon is also observed in droplet dispersion in forced jets (Ye & Richards, 1996).

7.2   Eulerian approach

In the Eulerian approach the particle or droplet cloud is treated as a continuous medium with properties analogous to those of a fluid. For example, the bulk density, or mass of particles per unit volume of mixture, is regarded as a continuous property. The particle velocity is the average velocity over an averaging volume. The purpose of this section is to develop and present the governing equations for the Eulerian approach. In that the continuous phase is a fluid, the use for the Eulerian approach for the dispersed phase is commonly referred to as the *two-fluid* or *Eulerian-Eulerian* approach. The terms will be used interchangeably here.

There are several levels of description for the Eulerian approach. If the Stokes number is sufficiently small, the particles and carrier fluid will have the same velocity (velocity equilibrium). If the characteristic time used in the definition of the Stokes number is a time representative of the carrier phase turbulence, a

---

[7]In this case the characteristic length used in the Stokes number is related to the width of the bluff body.

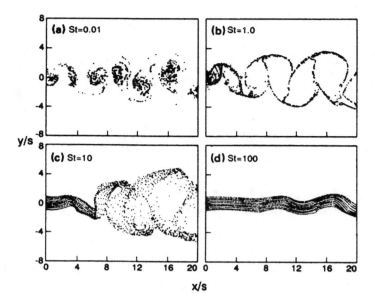

Figure 7.5: The effect of Stokes number on particle dispersion in the wake of a bluff body. (Reprinted with permission from Tang et al., Self organizing particle dispersion mechanism in free shear flows, *Phys. Fluids A*, 4, 2248, 1992. Copyright 1992 American Institute of Physics.)

small Stokes number implies that the particles will move with and disperse at the same rate as the carrier flow. In this case, the two-phase mixture can be regarded as a single phase with modified properties (density, thermal capacity, etc.) as presented in Chapter 2. If the Stokes number is based on some characteristic time of the flow field, a small Stokes number implies that the particles will move with the mean motion of the carrier flow but may not disperse at the same rate due to turbulence.

Of more practical interest is the situation where the velocities of the carrier fluid and particles are not the same. This could be the result of velocity gradients in the mean flow field, turbulent fluctuations and/or body forces acting on the particles. The local particle velocity is regarded as the average velocity of particles in an averaging volume

$$\langle \mathbf{v} \rangle = \frac{\sum_k \mathbf{v}_k}{N} \qquad (7.34)$$

where N is the number of particles in the volume. Another possibility would be the mass-averaged velocity defined by

$$\tilde{\mathbf{v}} = \frac{\sum_k m_k \mathbf{v}_k}{\sum_n m_k} \qquad (7.35)$$

where m_k is the mass of particle k in the averaging volume. This type of averaging is referred to as *Favre averaging*.

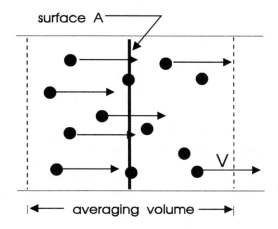

surface A

V

|◀── averaging volume ──▶|

Figure 7.6: Cloud of particles crossing a surface.

The flux of particles across a surface is an important aspect in the development of the conservation equations for the particle cloud. Consider a cloud of particles or droplets with number density n moving at velocity v across the surface A shown in Figure 7.6. The number flow rate across the surface is

$$\dot{N} = nvA \qquad (7.36)$$

and the corresponding mass flow rate would be

$$\dot{M}_d = nmAv = \bar{\rho}_d vA \qquad (7.37)$$

where m is the mass of an individual droplet and $\bar{\rho}_d$ is the local bulk density of the dispersed phase. If there is a distribution of droplet sizes and velocities, the mass flow rate is

$$\dot{M}_d = \sum_k \bar{\rho}_{d,k} v_k A \qquad (7.38)$$

where $\bar{\rho}_{d,k}$ is the bulk density and v_k is the velocity associated with size k and the summation is carried out over all discrete sizes.[8] It is convenient to define

---

[8] Due to turbulence and other factors, it is unlikely that all droplets of the same size will have the same velocity. In this case it is necessary to define the bulk density as a function of droplet velocity, position and time $\bar{\rho}_d(t, \mathbf{x}, \mathbf{v})$ so the the local bulk density at position $\mathbf{x}$ is given by

$$\bar{\rho}_d(\mathbf{x}, t) = \int \bar{\rho}_d(t, \mathbf{x}, \mathbf{v}) d\mathbf{v}$$

where the integration is carried out over all velocities (velocity space). The mass averaged velocity would be a mass average velocity as

$$\bar{\rho}_d \tilde{v} = \sum_n \bar{\rho}_{d,k} v_k \tag{7.39}$$

where $\bar{\rho}_d = \sum_k \bar{\rho}_{d,n}$ so the droplet mass flux is given by

$$\dot{M}_d = \bar{\rho}_d \tilde{v} A \tag{7.40}$$

The average is taken over the volume which surrounds the surface as shown in Figure 7.6.

7.2.1   Quasi-one-dimensional flow

The two-fluid form of the conservation equations for the dispersed phase in a quasi-one-dimensional duct are presented here. The control volume used for the derivation of the equations is shown in Figure 7.7.

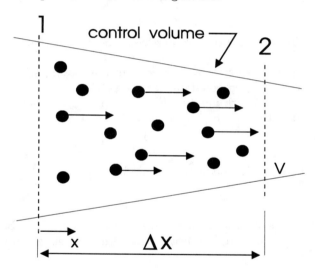

Figure 7.7: Particles/droplets in a quasi-one-dimensional duct.

Continuity equation

The continuity equation states that the accumulation rate of mass in the control volume plus the net efflux of mass through the control surface are equal to the

$$\bar{\rho}_d(t, x_i)\tilde{\mathbf{v}} = \int \bar{\rho}_d(t, \mathbf{x}, \mathbf{v})\mathbf{v}d\mathbf{v}$$

mass generated in the control volume. For a quasi-one-dimensional flow as shown in Figure 7.7, the continuity equation is expressed as[9]

$$V\Delta_t(\alpha_d\rho_d) + \Delta(\alpha_d\rho_d\tilde{v}A) = -S_{mass} \tag{7.41}$$

where the Δ operator is the same as used in Chapter 6. If all the droplets evaporated at the same rate, the mass source term would be

$$S_{mass} = -N_p\dot{m} \tag{7.42}$$

This equation has the same form as that for the continuous phase.

The corresponding differential form of the continuity equation is

$$\frac{\partial}{\partial t}(\alpha_d\rho_d) + \frac{1}{A}\frac{\partial}{\partial x}(\alpha_d\rho_d\tilde{v}A) = n\dot{m}$$

Momentum equation

The two-fluid momentum equation for a particle cloud can be obtained by summing the momentum equation for all the particles in a control volume. Consider the quasi-one-dimensional flow shown in Figure 7.5. The equation of motion for an isolated particle or droplet (with uniform mass flux from the surface) is written as[10]

$$m\frac{dv}{dt} = F_f + mg \tag{7.43}$$

where F_f is the fluid force on the droplet including the forces due to the pressure gradient. Adding $\dot{m}v$ to each side of this equation yields

$$\frac{d}{dt}(mv) = F_f + mg + \dot{m}v \tag{7.44}$$

The additional term represents the momentum associated with mass transfer from the droplet. Summing over all the droplets in the averaging volume and equating this value to the momentum change of the dispersed phase in the control volume yields the two-fluid momentum equation. If the dispersed phase elements have the same velocity, the momentum flux across the face of the control surface is

$$nmAv^2 = \bar{\rho}_dAv^2 \tag{7.45}$$

If there is a distribution of droplet sizes and velocities, the momentum flux must be calculated from $A\sum_k \bar{\rho}_{d,k}v_k^2$. Equating the momentum change of the droplets

---

[9]Boundary particles or droplets can also be included here. Combining complementary boundary droplet pairs is equivalent to one droplet inside the control volume.

[10]The cloud momentum equation ultimately obtained by starting from this equation will be valid if the particle-particle collisions are instantaneous. This equation will not be valid if the time over which the particle-particle contact force is operative is significant compared to the time between collisions.

in the control volume to the sum of the forces over all the droplets in the control volume gives

$$V\Delta_t(\bar{\rho}_d\tilde{v}) + \Delta(A\sum_k \bar{\rho}_{d,k}v_k^2) = \sum_k F_{f,k} + g\sum_k m_k + \sum_n \dot{m}_k v_k \qquad (7.46)$$

The second term on the right side can be written as $gV\bar{\rho}_d$. Breaking up the term representing surface forces on the droplets into the pressure gradient force and remaining terms,[11] the first term on the right side becomes

$$\sum_k F_{f,k} = -\frac{\Delta p}{\Delta x}\sum_k V_{d,k} + V\sum_k \beta_{V,k}(u - v_k) = -\bar{A}\alpha_d\Delta p + V\sum_k \beta_{V,k}(u - v_k)$$

$$(7.47)$$

The momentum flux term can be expanded in terms of the mass averaged velocity. If one defines δv_k as the deviation from the mass averaged velocity

$$v_k = \tilde{v} + \delta v_k \qquad (7.48)$$

then the momentum flux for a distribution of particle velocities becomes

$$\sum_k \bar{\rho}_{d,k}(\tilde{v} + \delta v_k)^2 = \tilde{v}^2\bar{\rho}_d + 2\tilde{v}\sum_k \bar{\rho}_{d,k}\delta v_k + \sum_n \bar{\rho}_{d,k}(\delta v_k)^2 \qquad (7.49)$$

By definition of the mass averaged velocity, the second term on the right is zero. Thus a mass averaged velocity is not equal to a momentum averaged velocity so an additional term appears. This is directly analogous to Reynolds stress in turbulent flows and shall be referred to here as a *dispersed phase Reynolds stress*

$$\tau_d^R = -\sum_k \bar{\rho}_{d,k}(\delta v_k)^2 \qquad (7.50)$$

These terms would obviously appear for flow with a distribution of particle sizes but would also appear in a flow with uniform size particles as a result of turbulent fluctuations. Also for particle-laden flow in a duct, this term would arise from particle wall collisions because the rebounding particles will produce a distribution of particle velocities in the duct. However in a fully developed one-dimensional duct flow the spatial change in τ_d^R would be zero. The Boussinesq approximation would not be applicable to model this stress.

Finally the momentum equation for the dispersed phase becomes

$$V\Delta_t(\bar{\rho}_d\tilde{v}) + \Delta\left(\bar{\rho}_d\tilde{v}^2 A\right) = -\alpha_d\bar{A}\Delta p + \Delta(\tau_d^R A)$$

$$+ V\sum_k \beta_{V,k}(u - v_k) + \bar{\rho}_d gV + \sum_k \dot{m}_k v_k$$

$$(7.51)$$

---

[11] The virtual mass, Basset and other forces are not considered here but must be included if significant.

The corresponding differential form of the momentum equation is

$$\frac{\partial}{\partial t}(\bar{\rho}_d \tilde{v}) + \frac{1}{A}\frac{\partial}{\partial x}\left(\bar{\rho}_d \tilde{v}^2 A\right) = -\alpha_d \frac{\partial p}{\partial x}$$

$$+ \sum_k \beta_{V,k}(u - v_k) + \bar{\rho}_d g - \sum_k \dot{m}_k v_k / V + \frac{1}{A}\frac{\partial}{\partial x}(A\tau_d^R) \tag{7.52}$$

If all the particles have the same size and evaporate at the same rate the dispersed phase momentum equation simplifies to

$$V \Delta_t(\bar{\rho}_d v) + \Delta\left(\bar{\rho}_d v^2 A\right) = -\alpha_d \bar{A} \Delta p$$

$$+ V \beta_V(u - v) + \bar{\rho}_d g V - S_{mass} v + \Delta(\tau_d^R A) \tag{7.53}$$

where now the velocity v is the average (not mass-averaged) velocity of the dispersed phase.[12] The corresponding differential form

$$\frac{\partial}{\partial t}(\bar{\rho}_d v) + \frac{1}{A}\frac{\partial}{\partial x}\left(\bar{\rho}_d v^2 A\right) = -\alpha_d \frac{\partial p}{\partial x}$$

$$+ \beta_V(u - v) + \bar{\rho}_d g - s_{mass} v + \frac{1}{A}\frac{\partial}{\partial x}(A\tau_d^R) \tag{7.54}$$

Example: Show how the quasi-one-dimensional momentum equation would be altered if there was a Coulomb force acting on the particles. The intensity of the electric field in the flow direction is E and the charge on a particle is q. There is no mass transfer.

Solution: The equation of motion for an individual particle is

$$\frac{d}{dt}(mv) = F_f + mg + qE$$

Adding all the particles in a computational cell using the same procedure outlined above yields the momentum equation in the form

$$V\Delta_t(\bar{\rho}_d \langle v \rangle_\mu) + \Delta\left(\bar{\rho}_d v^2 A\right) = \Delta(\tau_d^R A)$$

$$-\alpha_d \bar{A} \Delta p + V\beta_V(u - v) + \bar{\rho}_d g V + \bar{\rho}_d q E V$$

For the specific problem of quasi-one-dimensional flow in a duct, the walls will affect the particle velocity because the particles will collide with the wall and lose momentum on impact. Thus an effective friction term has to be included to account for this effect. There is voluminous literature on the friction factor for gas-particle flows in pipes (Klinzing, 1981). The momentum loss due to the wall will be quantified here by the *solids friction factor*, f_s, and is given by

$$\text{Momentum loss} = \frac{1}{2} f_s \bar{\rho}_d v |v| P \Delta x \tag{7.55}$$

---

[12] For convenience the average dispersed phase velocity, $\langle v \rangle$, is simply represented by v.

where P is the duct perimeter. Incorporating this momentum loss into the momentum equation yields the final form of the momentum equation for the dilute dispersed phase in a quasi-one-dimensional flow in a duct.

$$V\Delta_t(\bar{\rho}_d v) + \Delta\left(\bar{\rho}_d v^2 A\right) = -\alpha_d \bar{A}\Delta p + \Delta(\tau_d^R A)$$

$$+V\beta_V(u - v) + \bar{\rho}_d g V - S_{mass}v - \tfrac{1}{2}f_s\bar{\rho}_d v\,|v|\,P\Delta x \tag{7.56}$$

Expressions for the solids friction factor are given in the following table (Klinzing, 1981).

Investigator	Soilds friction factor, f_s
Stemerding	0.003
Reddy and Pei	$0.046/u_d$
Van Swaaij, Buurman and van Breugel	$0.080/u_d$
Capes and Nakamura	$0.048/u_d^{1.22}$
Konno and Saito	$0.0285\sqrt{gD}u_d^{-1}$
Yang, vertical	$0.00315\frac{\alpha_d}{\alpha_c}\left[\frac{\alpha_d u_t}{u_c - u_d}\right]^{-0.979}$
Yang, horizontal	$0.0293\frac{\alpha_d}{\alpha_c}\left[\frac{\alpha_d u_c}{\sqrt{gD}}\right]^{-1.15}$

Table 7.1. Empirical expression for the solids friction factor.[13]

In a dense flow, particle-particle collisions lead to flow properties of the particle cloud which are more reminiscent of a flowing fluid. For example, particle-particle collisions[14] give rise to propagation of information through the solid phase analogous to the speed of sound. The *powder modulus* is defined as (Gidaspow, 1994),

$$G = \left(\frac{\partial \sigma_s}{\partial \alpha_d}\right) \tag{7.57}$$

where σ_s is the normal stress applied to the powder and the sonic velocity is

$$C_s = \sqrt{\frac{G}{\rho_d}} \tag{7.58}$$

---

[13] The items in the table are listed in chronological order according to the date they were published.

[14] Particle collisions also contribute to the dispersed phase Reynolds stress because the collisions will affcet the deviations from the mass- averaged velocity.

Numerous empirical expressions have been proposed for G (Gidaspow, 1994). A useful formulation suggested by Bouillard et al. (1989) is

$$G = G_o \exp[-\kappa(\alpha_c - \alpha_{co})] \tag{7.59}$$

where G_o, κ and α_{co} are empirical constants. The value for G is often obtained from experiments on the velocity of wave propagation through a particle bed.

The momentum equation for nonreacting uniform size particles including the effect of particle-particle collisions as modeled with the powder modulus is

$$\frac{\partial}{\partial t}(\bar{\rho}_d v) + \frac{1}{A}\frac{\partial}{\partial x}\left(\bar{\rho}_d v^2 A\right) = -\alpha_d \frac{\partial p}{\partial x} + G\frac{\partial \alpha_c}{\partial x}$$
$$+\beta_V(u - v) + \bar{\rho}_d g - \frac{1}{2R_H} f_s \bar{\rho}_d v |v| \tag{7.60}$$

where the v is the averaged dispersed phase velocity. Note that the dispersed phase Reynolds stress is replaced by the term with the powder modulus.

Energy Equation

The thermal energy equation for a droplet was presented in Chapter 4; namely,

$$mc_d \frac{dT_d}{dt} = \dot{Q}_d + \dot{m}(h_s - h_d) \tag{7.61}$$

This equation can be written as

$$\frac{d}{dt}(mi_d) = \dot{m}i_d + \dot{m}(h_s - h_d) + \dot{Q}_d \tag{7.62}$$

where, for convenience, the internal energy, i_d, has been used in lieu of $c_d T_d$. The heat transfer term $\dot{Q}_d$ includes both the convective and radiative heat transfer. The enthalpy of the droplet can be written as

$$h_d = i_d + \frac{p_s}{\rho_d} + \frac{4\sigma}{D\rho_d} \tag{7.63}$$

where σ is the surface tension and D is the droplet diameter so the thermal energy equation can be expressed as

$$\frac{d}{dt}(mi_d) = -\dot{m}\frac{p_s}{\rho_d} - \frac{d}{dt}(\pi D^2 \sigma) + \dot{m}h_s + \dot{Q}_d \tag{7.64}$$

The term $\pi D^2 \sigma$ represents the energy associated with surface tension, E_σ, of a droplet. The first term on the right can also be written as $-p_s \dot{V}_d$. Finally the thermal energy equation for a single droplet is

$$\frac{d}{dt}(mi_d) = -p_s \dot{V}_d - \dot{E}_\sigma + \dot{m}h_s + Nu\pi k_c D(T_c - T_d) + \dot{Q}_r \tag{7.65}$$

where $\dot{Q}_r$ is the net radiative heat transfer to the particle discussed in Section 7.3. Summing over all the dispersed phase elements in the control volume results in

$$V\Delta_t(\bar{\rho}_d\tilde{i}_d) + \Delta(A\sum_k \bar{\rho}_{d,k}v_k i_{d,k}) = -\sum_k p_{s,k}\dot{V}_{d,k}$$

$$-\sum \dot{E}_{\sigma,k} + \sum_n \dot{m}_n h_{s,k} + \pi k_c \sum_k Nu_k D_k(T_c - T_d)_k + \sum_k \dot{Q}_{r,k}$$

(7.66)

where $\tilde{i}_d$ is the mass-averaged internal energy of the particles. If the dispersed phase elements do not have the same velocity and the mass averaged velocity is to be used in the convection term, a correction similar to that for the momentum equation must be applied; i.e.,

$$\sum_k \bar{\rho}_{d,k}v_k i_{d,k} = \bar{\rho}_d\tilde{v}\tilde{i}_d + \sum_k \bar{\rho}_{d,k}\delta v_k \delta i_{d,k} \qquad (7.67)$$

where $\delta i_{d,k}$ is the deviation from the mass averaged value. It is common practice in applications using the two-fluid model to utilize a number of equations for each particle size assuming that the same size moves with the same velocity. In this way the "diffusion" terms are eliminated, but more equations are needed to solve for the dispersed phase properties.

If all the droplets have the same mass, internal energy, evaporation rate and surface properties, the thermal energy equation simplifies to

$$V\Delta_t(\bar{\rho}_d i_d) + \Delta(A\bar{\rho}_d v i_d) = -p_s N\dot{V}_d$$

$$-N\dot{E}_\sigma + N\dot{m}h_s + \beta_T V(T_c - T_d) + N\dot{Q}_r$$

(7.68)

The corresponding differential form of the thermal energy equation is

$$\frac{\partial}{\partial t}(\bar{\rho}_d i_d) + \frac{1}{A}\frac{\partial}{\partial x}(A\bar{\rho}_d v i_d) = -p_s n\dot{V}_d$$

$$-n\dot{E}_\sigma + n\dot{m}h_s + \beta_T(T_c - T_d) + n\dot{Q}_r$$

Equation summary

The equations for the Eulerian-Eulerian formulation for quasi-one-dimensional flow are summarized in Table 7.2. It is assumed that all the droplets have the same size and evaporate at the same rate. The dispersed phase velocities are the average velocities. For dense flows, the Reynolds stress is set equal to zero and the powder modulus, G, is retained. For dilute flows, the powder modulus is set equal to zero and the Reynolds stress is retained. In general the Reynolds stress in dilute flow models is also neglected.

Continuity equation:

$$V\Delta_t(\bar{\rho}_d) + \Delta(\bar{\rho}_d vA) =$$
$$\text{(C)} \quad N\dot{m}$$

Momentum equation:

$$V\Delta_t(\bar{\rho}_d v) + \Delta\left(\bar{\rho}_d v^2 A\right) = -\alpha_d \bar{A}\Delta p + \bar{\rho}_d gV$$
$$\quad -\tfrac{1}{2}f_s\bar{\rho}_d v\,|v|\,P\Delta x + \Delta(A\tau_d^R) + G\bar{A}\Delta\alpha_c$$
$$\text{(C)} \quad +V\beta_V(u - v) + N\dot{m}v$$

Thermal energy equation:

$$V\Delta_t(\bar{\rho}_d i_d) + \Delta(A\bar{\rho}_d v i_d) = N\dot{Q}_r - N\dot{E}_\sigma$$
$$\text{(C)} \quad -p_s N\dot{V}_d + N\dot{m}h_s + \beta_T V(T_c - T_d)$$

Table 7.2. Summary of equations for the Eulerian formulation of the dispersed phase in a quasi-one-dimensional duct.

A summary of the differential form for the Eulerian formulation for the dispersed phase in a quasi-one dimensional duct is provided in Table 7.3. The equations are based on all the particles having the same size and evaporating at the same rate.

Continuity equation:

$$\frac{\partial}{\partial t}(\bar{\rho}_d) + \frac{1}{A}\frac{\partial}{\partial x}(\bar{\rho}_d vA) =$$
$$\text{C} \quad n\dot{m}$$

Momentum equation:

$$\frac{\partial}{\partial t}(\bar{\rho}_d v) + \frac{1}{A}\frac{\partial}{\partial x}\left(\bar{\rho}_d v^2 A\right) = -\alpha_d\frac{\partial p}{\partial x}$$
$$\quad -\frac{1}{2R_H}f_s\bar{\rho}_d v\,|v| + \bar{\rho}_d g + G\frac{\partial\alpha_c}{\partial x}$$
$$\text{C} \quad +\beta_V(u - v) + n\dot{m}v$$

Thermal energy equation:

$$\frac{\partial}{\partial t}(\alpha_d\rho_d i_d) + \frac{1}{A}\frac{\partial}{\partial x}(A\alpha_d\rho_d v i_d) = -n\dot{E}_\sigma + n\dot{Q}_r$$
$$\text{C} \quad -p_s n\dot{V}_d + n\dot{m}h_s + \beta_T(T_c - T_d)$$

Table 7.3. Summary of differential form of the conservation equations for the dispersed phase in a quasi-one-dimensional duct.

Mixture equations

The mixture equations can be obtained by adding the equations for the continuous phase (from Table 6.1) and the two-fluid formulation of the dispersed phase equations (Table 7.1). Adding the continuity equations for the continuous phase, Equation 6.26 to Equation 7.41 yields the continuity equation for the mixture,

$$V \Delta_t \left[(\bar{\rho}_d + \bar{\rho}_c) V \right] + \Delta \left[(\bar{\rho}_d v + \bar{\rho}_c u) A \right] = 0 \tag{7.69}$$

Defining a mixture velocity as

$$\rho_m U = \bar{\rho}_d v + \bar{\rho}_c u \tag{7.70}$$

the mixture continuity equation assumes the form,

$$V \Delta_t \rho_m + \Delta (\rho_m U A) = 0 \tag{7.71}$$

One notes that there is no source term and the equation is identical to the single-phase flow equation but based on mixture properties. If the velocity of both phases is the same (equilibrium flow), the continuity equation for the mixture is

$$V \Delta_t \rho_m + \Delta (\rho_m u A) = 0 \tag{7.72}$$

where u is the flow velocity.

The mixture momentum equation is obtained by adding the momentum equations for each phase, Equation 6.49 and the momentum equation in Table 7.2, to give

$$V \Delta_t (\rho_m U) + \Delta \left[A \left(\bar{\rho}_d v^2 + \bar{\rho}_c u^2 \right) \right] = -\bar{A} \Delta p + \Delta (A \tau_d^R)$$
$$+ \rho_m g V - \tfrac{1}{2} f_s \bar{\rho}_d v \, |v| \, P \Delta x - \tfrac{1}{2} \rho_c c_f u \, |u| \, P \Delta x + G \bar{A} \Delta \alpha_c \tag{7.73}$$

where c_f is the local skin friction coefficient for the continuous phase. For a dilute flow G can be set to zero. For a dense flow the Reynolds stress is set equal to zero

For a dilute equilibrium flow, the momentum equation of the mixture is

$$V \Delta_t (\rho_m u) + \Delta \left[A \left(\rho_m u^2 \right) \right] = -\bar{A} \Delta p$$
$$+ \rho_m g V - \tfrac{1}{2} \left(f_s \bar{\rho}_d + c_f \rho_c \right) u \, |u| \, P \Delta x \tag{7.74}$$

This equation has the same form as for a single-phase fluid except for the shear stress at the wall which contains the cumulative effect of the particles and the fluid.

The thermal energy equation for the mixture is obtained by adding the thermal energy equations for each phase, Equations 6.82 and 7.68 to yield

$$V \Delta_t (\bar{\rho}_d i_d + \bar{\rho}_c i_c) + \Delta [A (\bar{\rho}_d i_d v + \bar{\rho}_c h_c u)] = -N \dot{E}_\sigma$$
$$\alpha_c u \bar{A} \Delta p - p \Delta (\alpha_d v A) + (\beta_V V - N \dot{m} / 2) \left(u - v \right)^2 \tag{7.75}$$
$$+ \tfrac{1}{2} \rho_c c_f \, |u| \, u^2 P \Delta x - \Delta (A \dot{q}_c) - \dot{q}_w P \Delta x + N \dot{Q}_r$$

The enthalpy of the continuous phase in the convection term can be written as

$$\Delta(A\bar{\rho}_c h_c u) = \alpha_c u \bar{A} \Delta p + p\Delta(\alpha_c u A) + \Delta(A\bar{\rho}_d i_d v) \tag{7.76}$$

which, when substituted into Equation 7.75, gives

$$V\Delta_t(\bar{\rho}_d i_d + \bar{\rho}_c i_c) + \Delta[A(\bar{\rho}_d i_d v + \bar{\rho}_c i_c u)] = -N\dot{E}_\sigma$$

$$-p\Delta(\alpha_d v A + \alpha_c u A) + (\beta_V V - N\dot{m}/2)\frac{(u-v)^2}{2} \tag{7.77}$$

$$+\tfrac{1}{2}\rho_c c_f |u| u^2 P\Delta x - \Delta(A\dot{q}_c) - \dot{q}_w P\Delta x + N\dot{Q}_r$$

In this equation one can identify the sources of mechanical energy dissipation; namely, particle-fluid friction and wall friction. For an equilibrium flow, the thermal energy equation reduces to

$$V\Delta_t i_m + \Delta[i_m u A] = -N\dot{E}_\sigma - p\Delta(uA)$$

$$+\tfrac{1}{2}\rho_c c_f |u| u^2 P\Delta x - \Delta(A\dot{q}_c) - \dot{q}_w P\Delta x + N\dot{Q}_r \tag{7.78}$$

where $i_m = \bar{\rho}_d i_d + \bar{\rho}_c i_c$. Of course, the energy due to surface tension will not appear for a fluid-particle flow. The sources of thermal energy for an equilibrium flow are dissipation due to wall friction and heat transfer through the control surfaces. One also notes that the radiative heat transfer term appears in the equation since radiative energy is absorbed by the particles directly from the electromagnetic field and is not exchanged with the carrier phase.

7.2.2 Multidimensional flows

The Eulerian dispersed phase equation are obtained by summing the conservation equations for individual particles over all particles in the control volume to obtain the equations for a particle cloud. A cloud of particles or droplets in a Cartesian control volume is shown in Figure 7.8. The finite difference equations are then divided by the control volume and the limit is taken as the volume approaches the limiting value.[15] The details will not be provided here since the procedure is the same as for the quasi-one-dimensional flow. Index notation is throughout.

Continuity equation

Summing the rate of mass change of all the droplets in the computational cell and equating it to the net efflux of droplet mass through the control surfaces plus the rate of change of mass in the cell yields

$$\frac{\partial}{\partial t}(\alpha_d \rho_d) + \frac{\partial}{\partial x_i}(\alpha_d \rho_d \tilde{v}_i) = n\dot{m}_i \tag{7.79}$$

---

[15] As with the continuous phase equations, the limiting volume has to be sufficiently large to define average values and yet smaller than the flow system dimensions.

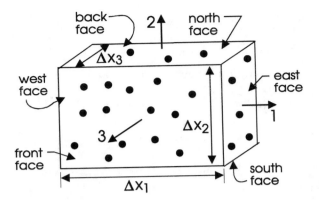

Figure 7.8: Three dimensional control volume for dispersed phase elements.

where $\tilde{v}_i$ is the mass-averaged velocity of the dispersed phase. If the simple average velocity is used, in lieu of the mass averaged velocity, the continuity equation has to be formulated differently. In this case the velocity in the averaging volume is expressed as

$$v_i = \langle v_i \rangle + \delta v_i \tag{7.80}$$

where δv is the deviation of the velocity from the averaged value. Similarly, the bulk density is written as

$$\bar{\rho}_d = \bar{\rho}_{d,o} + \delta\bar{\rho}_d \tag{7.81}$$

where $\bar{\rho}_{d,o}$ is the bulk density for the control volume shown in Figure 7.6 and $\delta\bar{\rho}_d$ is the deviation in bulk density in adjacent averaging volumes so the average mass flux is

$$\langle \bar{\rho}_d v_i \rangle = \langle \bar{\rho}_d \rangle \langle v_i \rangle + \langle \delta\bar{\rho}_d \delta v_i \rangle \tag{7.82}$$

The additional term can be regarded as a mass diffusion term. In a flow with homogeneous bulk density or uniform particle velocities, this term would be zero. Turbulence, however, will produce a distribution of particle velocities which will give rise to a net mass flux in nonhomogenous particle density fields.

The *gradient transport model* is used in the two-fluid formulation to simulate dispersion of particles in turbulent flows. With reference to Fick's law it is assumed that

$$\langle \delta\tilde{\rho}_d \delta v_i \rangle = -D_d \frac{\partial \bar{\rho}_d}{\partial x_i} \tag{7.83}$$

where D_d is the dispersion coefficient for the dispersed phase. The value for the dispersion coefficient has to be determined from experiment or through some auxiliary analysis. Picard et al. (1986) predicted a dispersion coefficient based on an early analysis of particle motion in turbulence reported by Tchen (1949). Adjustments were necessary to account for crossing trajectory effects. Rizk and Elghobashi (1989) employed a semi-empirical correlation suggested by Picard et al. Unfortunately, there are no simple analyses or models which will provide D_d as a function of particle properties and flow turbulence, so the choice of an appropriate value depends on finding or reducing a value from an experiment which is considered representative of the flow field to be simulated.

Analyses of a fundamental nature to evaluate the dispersion coefficient have been reported in the literature. Reeks (1977) considered the dispersion of fluid points in an isotropic, homogeneous and stationary turbulent field using an approximation scheme proposed by Phythian. He normalized the dispersion coefficient with turbulence intensity and length scale and calculated how the nondimensional coefficient varied with time as a function of particle response time. After a period, which depended on particle response time, the dispersion coefficient reached an asymptotic value which exceeded the value for a fluid point. Reeks (1993) has continued work in this area using phase space probability density functions and has considered flows with linear shear profiles. This work provides a rigorous basis for establishing dispersion coefficients but considerably more effort is needed to address industrially relevant turbulent flow fields.

A further problem exists with boundary conditions. If the particles bounce specularly from a wall, it is probably appropriate to set the gradient of the bulk density normal to the wall equal to zero. If the bouncing is not specular, another approximation must be made. If droplets impact the wall and there is no re-entrainment, setting bulk density equal to zero at the wall does not represent a realistic boundary condition for the bulk density field. The choice of suitable boundary conditions for bulk density in the two-fluid model is still an open question.

The final form for the two-fluid continuity equation using the volume averaged velocity is

$$\frac{\partial}{\partial t}(\alpha_d \rho_d) + \frac{\partial}{\partial x_i}(\alpha_d \rho_d \langle v_i \rangle) = \frac{\partial}{\partial x_i}\left(D_d \frac{\partial \bar{\rho}_d}{\partial x_i}\right) + n\dot{m} \qquad (7.84)$$

If Favre (mass averaging) is used, the diffusion term is eliminated.

Momentum equation

The momentum equation for a multidimensional flow can also be developed by summing over every droplet in the computational cell. The resulting differential equation is

$$\frac{\partial}{\partial t}(\alpha_d \rho_d \tilde{v}_i) + \frac{\partial}{\partial x_j}(\alpha_d \rho_d \tilde{v}_i \tilde{v}_j) = \sum_k \dot{m}_k v_{i,n}/V - \alpha_d \frac{\partial p}{\partial x_i}$$

$$\alpha_d \frac{\partial}{\partial x_i} \tau_{ij} + \sum_k \beta_{V,k}(u_i - v_i)_k + \alpha_d \rho_d g_i - \alpha_d \frac{\partial}{\partial x_j}\left(\sum_k \bar{\rho}_{d,k} \delta v_{i,k} \delta v_{k,j}\right) \qquad (7.85)$$

where $\tau_{c,ij}$ is the shear stress in the continuous phase and $\delta v_{i,k}$ is the deviation of the velocity of the k-th particle from the mass average velocity. The last term is analogous to a Reynolds stress,

$$\tau_{d,ij}^{R} = -\sum_{k} \bar{\rho}_{d,k} \delta v_{k,i} \delta v_{k,j} \qquad (7.86)$$

and has already been identified to as the dispersed phase Reynolds stress. This term arises because the mass averaged velocity is not a momentum averaged velocity. If all the droplets have the same mass and evaporate at the same rate the momentum equation reduces to

$$\frac{\partial}{\partial t}(\alpha_d \rho_d v_i) + \frac{\partial}{\partial x_j}(\alpha_d \rho_d v_i v_j) = -\alpha_d \frac{\partial p}{\partial x_i}$$

$$\alpha_d \frac{\partial}{\partial x_i}(\tau_{c,ij} + \tau_{d,ij}^{R}) + n\dot{m}v_i + \beta_V(u_i - v_i) + \alpha_d \rho_d g_i \qquad (7.87)$$

The dispersed phase Reynolds stress can arise from sources other than fluctuations due to turbulence. For example, in a gas-droplet flow with no turbulence, the Reynolds stress term would appear due to variations in particle velocity because of particle size. Such a situation would occur for a particle laden flow at the throat of a venturi where the smaller particles would tend to move at a velocity near the local fluid velocity while the larger particles would exhibit a larger velocity lag. The problem of Reynolds stress due to velocity variation because of particle size distribution could be circumvented by having a momentum equation for each particle size category. However, a Reynolds stress would still appear in a turbulent flow due to particle velocity fluctuations caused by the turbulent motion.

The dispersed phase Reynolds stress is usually modeled with the Boussinesq approximation (stress proportional to the rate of strain),

$$-\sum_{k} \bar{\rho}_{d,k} \delta v_{k,i} \delta v_{k,j} = \mu_s \left(\frac{\partial v_j}{\partial x_i} + \frac{\partial v_i}{\partial x_j} \right) - \frac{2}{3} \mu_s \frac{\partial v_k}{\partial x_k} \delta_{ij} \qquad (7.88)$$

where μ_s is an *solids viscosity*. It is very difficult to select a solids viscosity since the particle velocity fluctuations depend not only on local turbulence but also on the particle properties and particle history. Chung et al. (1986) related the solids viscosity to the eddy viscosity of the carrier fluid through a function which depended on the ratio of the Stokes number based on the integral time scale of the carrier phase turbulence. Rizk and Elghobashi (1989) simply used a constant ratio between the solids viscosity and the fluid eddy viscosity.

The problem with the Boussinesq approximation is that one can visualize a Reynolds stress without a gradient in the mean velocity field (no rate of strain) because the turbulent fluctuations of particles will give rise to a nonzero value for the dispersed phase Reynolds stress.

As with the dispersion coefficient for the Eulerian formulation, there is a continuing effort to develop expressions for the solids viscosity from fundamental principles. Reeks (1993) utilizes the equations for phase space probability density to evaluate the Reynolds stress and diffusion in a particle velocity field with a linear shear gradient. He finds that the Boussinesq assumption is inadequate and that the Reynolds stress is the result of two contributions, one from the carrier flow and the other from the linear shear gradient. Simonin et al. (1993) have also used the probability density function approach to model the dispersed phase Reynolds stress. Continued work through these and other fundamental analyses will provide a better foundation for the constitutive equations for the Eulerian approach.

The other problem is the choice of the correct boundary condition at a wall. Obviously the tangential component of the particle velocity adjacent to the wall need not be zero as for a fluid. One approach is to allow a slip velocity at the wall (Ding et al., 1993) similar to the procedures used for rarefied gas flows.

Because of the similarity of the two-fluid momentum equation to that for a single-phase flow, one is tempted to regard the particle phase as a continuous medium. However, in a dilute particle or droplet flow, information travels along particle trajectories and thus is parabolic in nature whereas a fluid is elliptic because information is transmitted in all direction by pressure waves. Thus the two-fluid model for dilute flows is always fraught with problems in trying to set up boundary conditions for the dispersed phase which are only appropriate for a continuous medium. One example is the tangential velocity at the wall discussed above. Moreover, for droplets impinging on a wall, the normal component of the droplet velocity is also not zero.

In the traditional two-fluid model for a dense phase flow, the force on the particles due to shear stress in the continuous phase is neglected and another term is added to account for the particle-particle interaction. The differential form of the momentum equation for nonevaporating dense phase flows is

$$\frac{\partial}{\partial t}(\alpha_d \rho_d v_i) + \frac{\partial}{\partial x_j}(\alpha_d \rho_d v_j v_i) = \frac{\partial \tau_{d,ij}}{\partial x_j}$$
$$-\alpha_d \frac{\partial p}{\partial x_i} - G\frac{\partial \alpha_d}{\partial x_i} + \beta_V(u_i - v_i) + \alpha_d \rho_d g_i \tag{7.89}$$

where G is the solids stress modulus obtained from empirical formulas and $\tau_{d,ij}$ is the dispersed phase shear stress from Equation 7.88. This model has been used to predict flow properties, including cluster formation, in circulating fluidized beds (Tsuo and Gidaspow, 1990).

An alternate form for the solids stress suggested by Harris and Crighton (1994) is

$$\tau = P_s \frac{\alpha_d}{\alpha_{d,o} - \alpha_d} \tag{7.90}$$

where P_s is a constant with units of pressure and $\alpha_{d,o}$ is the maximum solids volume fraction. For this formulation the solids modulus is

$$G = P_s \frac{\alpha_{d,o}}{(\alpha_{d,o} - \alpha_d)^2} \tag{7.91}$$

More recently the kinetic theory model has been used to derive relationships for the solids viscosity and other parameters for dense phase flows. An additional equation is included for the kinetic energy of the fluctuating motion of the particulate phase. Because of the similarities between particle-particle interactions and molecular interactions in a gas, the concepts from kinetic theory can be used to develop the governing equations for dense phase flows. This approach is nominally credited to Bagnold (1954) who derived an equation for repulsive pressure in uniform shear flow. Many others, particularly Savage (1983), have contributed to this approach since that time. Complete details of the derivations and applications to dense phase flows can be found in Gidaspow (1994). The basic concept is that particle-particle collisions are responsible for momentum and energy transfer in dense phase flow in the same way that molecular interactions are responsible for pressure wave propagation and viscosity in a single-phase fluid.

The kinetic energy associated with the particle velocity fluctuations is called the *granular temperature* and defined as

$$\Theta = \frac{1}{3} \langle C^2 \rangle \tag{7.92}$$

where C is the fluctuational velocity of the particle motion. Granular temperature can be produced by a shearing action in the granular flow and by hydrodynamic forces. Dissipation can occur through inelastic particle-particle and particle-wall collisions and dissipation in the fluid. Granular temperature can also be diffused in the same manner as heat. The stress term in the momentum equation based on kinetic theory becomes (Gidaspow, 1994)

$$\tau_{d,ij} = \left[-p_s + \xi_s \frac{\partial v_k}{\partial x_k} \right] \delta_{ij} + \mu_s \left(\frac{\partial v_j}{\partial x_i} + \frac{\partial v_i}{\partial x_j} \right) \tag{7.93}$$

where p_s is the solids pressure, ξ_s is the solids phase bulk viscosity and μ_s is the solids shear viscosity. These three parameters are functions of the granular temperature as well as the particle restitution coefficient, particle diameter, material density and volume fraction.

Several numerical models have been implemented for dense phase flows using the two-fluid models based on granular temperatures. These include flow in chutes, fluidized beds and sedimentation. Sinclair and Jackson (1989) have used the two-fluid model for modeling dense flows in vertical tubes. There are several advantages to using the two-fluid model for dense phase flows. The most significant advantage is that there is no need to consider the dynamics of individual particles, so large systems can be modeled. Also, the numerical formulations used for single-phase flows can be applied to the two-fluid equations for the solid phase. However, there is a level of empiricism which must be introduced in establishing the constitutive equations. Also, features such as particle-particle sliding, particle rotation and particle size distribution are not included

The granular temperature model should also extend to dilute flows. In this case, the hydrodynamic effects on particle oscillation in turbulence would have to be more accurately modeled. Extensions of the two-fluid model to dilute flows have been reported by Bolio and Sinclair (1996). In dense flows, the particle-particle contribution to particle fluctuation energy is more significant than that due to local turbulence in the continuous phase.

Thermal energy equation

The Eulerian thermal energy equation for the droplet cloud can be derived in the same way as the continuity and momentum equations by summing over the droplets in a control volume as shown in Figure 7.8. Following the development for the quasi-one-dimensional model, the multidimensional thermal energy equation is

$$\frac{\partial}{\partial t}(\alpha_d \rho_d \tilde{i}_d) + \frac{\partial}{\partial x_i}(\alpha_d \rho_d \tilde{v}_{in}\tilde{i}_d) = -\sum_k p_{s,k}\dot{V}_{d,k}/V - \frac{\partial}{\partial x_j}(\sum_k \tilde{\rho}_{d,k}\delta v_{j,k}\delta i_{d,k})$$

$$-\sum_k \dot{E}_{\sigma,k}/V + \sum_k \dot{m}_k h_{s,k}/V + \sum_k \beta_{T,k}(T_c - T_d)_k + \sum_k \dot{Q}_{r,k}/V$$

$$(7.94)$$

where $\tilde{i}_d$ is the mass-averaged droplet internal energy. The term $\sum_k \tilde{\rho}_{d,k}\delta v_{j,k}\delta i_{d,k}$ is analogous to the Reynolds stress term and has to be modeled in some fashion such as a gradient diffusion term. If the variations in droplet temperatures are small this term can be neglected.

If all the droplets have the same size, specific internal energy and evaporation rate, the thermal energy equation becomes

$$\frac{\partial}{\partial t}(\alpha_d \rho_d i_d) + \frac{\partial}{\partial x_i}(\alpha_d \rho_d v_i i_d) = -n p_s \dot{V}_d$$

$$-n\dot{E}_\sigma + n\dot{m}h_s + \beta_T(T_c - T_d) + n\dot{Q}_r$$

$$(7.95)$$

There may also be a heat transfer in dense flows due to particle-particle contact which is not included here.

A summary of the differential form of the two-fluid equations for the dispersed phase is given in Table 7.4. The equations are based on flows with uniform size particles which evaporate at the same rate. The expressions used for the shear stress may vary depending on the nature of the flow (dilute versus dense) and the sophistication of the model for the shear stress and particle interaction effects (traditional versus kinetic theory).

Continuity equation:

$$\frac{\partial}{\partial t}(\alpha_d \rho_d) + \frac{\partial}{\partial x_i}(\alpha_d \rho_d v_i) = \frac{\partial}{\partial x_i}\left(D_d \frac{\partial \bar{\rho}_d}{\partial x_i}\right)$$

C $\quad n\dot{m}$

Momentum equation

$$\frac{\partial}{\partial t}(\alpha_d \rho_d v_i) + \frac{\partial}{\partial x_j}(\alpha_d \rho_d v_j v_i) = \alpha_d \frac{\partial \tau_{ij}}{\partial x_j} - \alpha_d \frac{\partial p}{\partial x_i} + \alpha_d \rho_d g_i$$

C $\quad n\dot{m}v_i + \beta_V(u_i - v_i)$

Thermal energy equation:

$$\frac{\partial}{\partial t}(\alpha_d \rho_d i_d) + \frac{\partial}{\partial x_i}(\alpha_d \rho_d v_i i_d) = -n\dot{E}_\sigma + n\dot{Q}_r$$

C $\quad n\dot{m}h_s + \beta_T(T_c - T_d) - np_s\dot{V}_d$

Table 7.3 Summary of the differential form of the Eulerian equations for the dispersed phase.

7.3 Radiation modeling

Radiative heat transfer in the particle cloud is a very complex phenomenon and will not be discussed in detail here. Other references such as Smoot and Pratt (1979) provide a background for the subject.

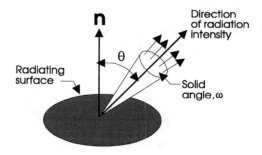

Figure 7.9: Intensity of radiation.

Radiation is the transport of energy by electromagnetic waves. Thermal radiation is the radiant energy emitted by a medium by virtue of its temperature and the corresponding wavelengths generally lie between 0.1 and 100 μm (Kreith, 1973). Radiation is quantified by intensity which is the energy emitted per unit area into a unit solid angle centered on the emitting surface and in a direction θ from the normal to the surface as shown in Figure 7.9. Integrating the radiation intensity over a hemisphere above the surface yields the total rate of radiation emission or *radiosity* which has units of heat transfer rate per unit area. If the surface is diffuse, the radiation is independent of the angle θ and the radiosity is simply

$$J = \pi I$$

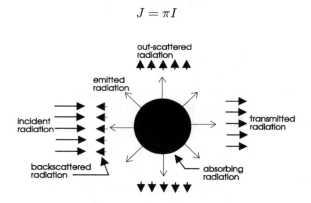

Figure 7.10: Absorption, scattering and emmision by a particle.

A very simplified description of the effect of a particle on the radiation intensity is shown in Figure 7.10. The incident radiative energy is scattered by the particle in all directions. Part of the incident beam is transmitted in the forward direction, part is backscattered in the opposite direction of the incident beam and part is out-scattered. The distribution of the scattered intensity depends on the parameter $\pi D/\lambda$ where λ is wave length. For $\pi D/\lambda < 1$, the scattering is defined as Rayleigh scattering while for $\pi D/\lambda > 5$ the scattering is known as *Mie scattering*. Analytic results are available for the scattering distribution for Rayleigh and Mie scattering (Van de Hulst, 1981) as well as numerical programs. Special treatment is needed to evaluate scattering in the intermediate range. For Mie scattering one finds that the ratio of the transmitted radiation (forward scattering) to the backscattering increases as $\pi D/\lambda$ increases. The particle affects the scattering distribution as well. Part of incident energy is also absorbed by the particle and energy is emitted by the particle as well.

The amount of energy emitted from the particle is

$$\dot{Q}_{emit} = \varepsilon_\lambda \pi D^2 I_b(\lambda, T_d) \tag{7.96}$$

where $I_b(\lambda, T_d)$ is the *black body radiation intensity* and ε_λ is the *emissivity* which is a function of the wavelength. For a gray body, the emissivity is assumed constant over all wavelengths. In this equation it is assumed that the radiation is diffuse (same in all directions). Integrating over all wave lengths, the total energy emitted by a gray particle is

$$I_{emit} = \varepsilon \pi D^2 \sigma T_d^4 \tag{7.97}$$

where σ is the *Stephan-Boltzmann* constant.

The amount of scattering is quantified by an efficiency factor, Q_s, also known as the Mie coefficient, and is a function of $\pi D/\lambda$, particle shape refractive index. The fraction of the incident beam energy which is scattered by a single spherical particle is $\pi \frac{D^2}{4} Q_s$. Thus the total energy scattered by N particles in a control volume is

$$I_{scattered} = I\frac{\pi}{4} \sum_{n=1}^{N} D_n^2 Q_{s,n} = K_s I \tag{7.98}$$

where I is the intensity of the incident beam and K_s is the scattering coefficient. If all the particles have the same size and efficiency factor, the scattered intensity is

$$I_{scattered} = I\frac{\pi}{4} D^2 Q_s n V = K_s I \tag{7.99}$$

This is called single particle scattering in that the cumulative effect is the sum of the scattering of an isolated, individual particle. As the number density of the particles increases, the scattering becomes considerably more complex and is known as multiple scattering.

In the same fashion the fraction of energy absorbed by a single particle is $\pi \frac{D^2}{4} Q_a$, where Q_a is the efficiency factor for absorption. The energy absorbed by all the particles in a control volume, provided all the particles have the same size and efficiency factor, is

$$I_{abs} = I\frac{\pi}{4} D^2 Q_a n V = K_a I \tag{7.100}$$

where K_a is the *absorption coefficient*.

Modeling radiation in a particle cloud involves setting up an energy balance that accounts for the scattering, absorption and emission including the contributions of the particles and the surrounding walls (or flames). Many simplifications have to be made to make the problem tractable to obtain a solution. The main area of difficulty is accounting correctly for the scattering. Several models have been developed and applied to industrial systems.

Hottel's zone method (Siegel & Howell, 1981) is the classic method. It provides a complete solution to the gray gas system. In this method, each zone (control volume) can exchange energy with all other zones. Thus for a system with N zones, N nonlinear equations have to be set up and solved simultaneously. Although this method is versatile and accurate, it is computationally intensive and expensive.

Another approach is the *Monte Carlo method* (Howell, 1968). In this method a bundle of intensities are propagated in all directions. Individual rays are followed for their duration as they undergo scattering, absorption and emission. This method provides a solution for the gray gases. It is also versatile and accurate but, as the zonal method, is expensive.

The flux method has gained popularity because of its relative simplicity and general applicability. In this method, the angular distribution of intensities is

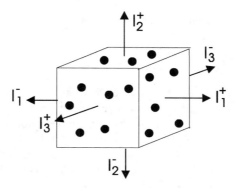

Figure 7.11: Control volume for six-flux model of radiative heat transfer.

replaced by discrete intensities in the coordinate directions. Chu and Churchill (1955) introduced the *six-flux model* in which there is a flux in the negative and positive sense of each coordinate direction. Consider the model shown in Figure 7.11. There is an incident beam in the positive 1-direction into the control volume. There is an out-scattering due to the particles in the volume as well as an in-scattering from the neighboring control volumes. Energy is also absorbed and emitted by the particles in the volume. The equation for radiant intensity change for the flux in the positive 1-direction is

$$\frac{dI_1^+}{dx_1} = -I_1^+ K_a - (1-f)K_s I_1^+ + bK_s I_1^- + sK_s(I_2^+ + I_2^- + I_3^+ + I_3^-) + \frac{K_a}{6}I_b(T_d)$$
(7.101)

The first term represents the energy absorbed. The second term is the out-scattering where f is the fraction associated with forward scattering. The third term is the backscattering from the flux in the negative 1-direction which yields a flux in the positive 1−direction and b is the fraction of the scattering associated with backscattering. The fourth term is the energy scattered in the 1-direction from the fluxes entering from the other directions and s is the fraction of the scattered flux in the 1-direction. Finally, the last term is the energy emitted by the particle. This equation can be solved using the same techniques as those used to solve the fluid equations.

The radiative energy absorbed per unit time by an individual particle is

$$\dot{Q}_r = I_t \frac{\pi}{4} D^2 Q_a$$
(7.102)

where I_t is the sum of the radiative intensity in all six directions.

A more recent model which provides improved accuracy over the flux model is the *discrete ordinate model* (Fiveland, 1988). In this case, finite difference equations are derived for discrete directions (not coordinate directions) of the radiative intensity. Improved accuracy is achieved by selecting more discrete directions but at the expense of increased computational time.

7.4 Summary

There are both advantages and disadvantages related to the discrete element and two-fluid approaches for the particle cloud equations. The discrete element approach provides the details of particle fluid interaction and does not require selection of an effective viscosity and thermal conductivity. If every particle could be included in the calculation, the discrete element approach provides a direct numerical simulation. However, this is not generally computationally feasible so computational particles must be used to represent a parcel of physical particles which compromises the accuracy of the model. The discrete element approach better represents the physics of the particulate flow field in that it correctly models the parabolic nature of dilute flows and allows for "natural" boundary conditions. The two-fluid model is not limited by the number of particles in the system and is amenable to the same calculational algorithm as used for the carrier phase. The fundamental problem with the approach is the selection of appropriate constitutive equations and boundary conditions. The kinetic theory for dense phase flows provides another relationship (granular temperature) which can be used to establish the constitutive relationships.

Exercises

7.1. Coal particles with a material density of 1300 kg/m^3 are injected into a flow. The particle distribution is log-normal with a mass median diameter of 120 μm and a standard deviation of 20 μm. The mass flow rate is 2 kg/s. Assume the particles are spherical. Assume the distribution is discretized into five sizes with equal mass fractions. Find the number flow rate associated with each size fraction.

7.2. The number flow rate of 100 μm particles along a trajectory is 10^5 per second. The average particle velocity in a computational volume is $\mathbf{v} = 6\mathbf{i} + 2\mathbf{j}$. The transit time of the particle in a computational cell is 0.001 seconds. The density of the carrier fluid is 1.2 kg/m^3 and the viscosity is 1.8×10^{-5} Ns/m^2. Find $\mathbf{S}_{mom}$ for this trajectory.

7.3. The convection term for the mixture momentum equation is

$$\Delta \left[A(\bar{\rho}_d v^2 + \bar{\rho}_c u^2) \right]$$

Expressing the velocity of each phase as a deviation about the mixture velocity,

$$\delta u = u - U$$
$$\delta v = v - U$$

rewrite the convection term in terms of the mixture velocity and an equivalent Reynolds stress.

7.4. Compare the surface energy to the internal energy of a 100-micron water droplet. For what droplet radius would both energies be the same?

7.5. Combine Equation 7.79 with the multidimensional continuity equation for the carrier phase in Chapter 6 to obtain the continuity equation for the mixture.

7.6. Explain why heat transfer due to radiation does not appear as a coupling term in the energy equation.

7.7. Combine Equation 7.87 with the multidimensional momentum equation for the continuous phase in Chapter 6 to yield the momentum equation for the mixture.

7.8. Using the procedures developed for the quasi-one-dimensional flows, derive the entropy equation first for a single, nonevaporating particle and then for a cloud of particles.

7.9. How would the multidimensional momentum equation, Equation 7.87, be extended to include a Coulomb force on the particles with charge q and electric field strength E_k.

7.10. For a quasi-one-dimensional flow with a Coulomb force on the particles, would the work associated with the Coulomb force appear in the thermal energy equation? Explain.

Chapter 8

Numerical Modeling

The current literature abounds with the improvement of current and the development of new numerical models for multiphase flows with droplets and particles. The state-of-the-art has advanced to the level that these computational models are not only fundamental components of multiphase-flow research but also find application to support engineering design. To adequately address the subject of numerical models for dispersed phase flows would require an extensive monograph. An earlier review of numerical models for dilute gas-particle flows (Crowe, 1982) provides some background. The purpose of this chapter is to introduce the principle ideas so the reader can understand the concepts while avoiding the specific details of each code. Numerical models for single-phase flows are addressed first using simple quasi-one-dimensional flow models which capture the essence of multidimensional modeling. The inclusion of the second phase is then illustrated. Eulerian-Lagrangian and Eulerian-Eulerian models are introduced for both dilute and dense flows.

8.1 Single-phase flows

There have been several computational models developed for the single-phase flows and can be reviewed in most any book on computational fluid mechanics. Many of the original developments in computational fluid mechanics for subsonic flows were the result of work at the Los Alamos Scientific Laboratories in New Mexico. These models (Harlow & Fromm, 1965) were based on finite difference formulations of the continuity and Navier-Stokes equations. The early work showed the need to use upwind differencing to obtain stable solutions of the equations at high Reynolds numbers. The work at Los Alamos also introduced the idea of using a formulation of the continuity equation based on pressure.

The interest in computational fluid mechanics was sparked by the publication of a book (Gosman et al., 1969) at Imperial College in London in which the *tank-and-tube* method was formulated. The flow field was conceptualized to consist of a field of "tanks" connected at adjacent interfaces by "tubes". This scheme used the stream function and vorticity as the dependent variables which reduced the number of dependent variables for a two-dimensional incompressible flow from three to two. Also, this scheme avoided the need to include the pressure as a dependent variable. The use of stream function and vorticity was not attractive because they are not generally used to describe a flow. The tank-and-tube concept introduced the idea of the finite volume method for formulating the conservation laws as introduced in Chapter 6. Further developments at Imperial College led to the adoption of the numerical scheme pioneered at Los Alamos using the *primitive* variables; namely, velocity and pressure. This required the introduction of a staggered grid to avoid nonrealistic solutions; that is, the velocity and pressures are defined at different locations. The general availability of the TEACH program in 1974 from Imperial College resulted in a widespread adoption of the finite volume approach. It underlies the basic structure of most commercial codes available today.

Further developments in turbulence modeling led to the two-equation models (Launder & Spalding, 1972) for turbulence in which the turbulence field is described by two variables. The most common variables are turbulence energy and dissipation rate discussed in Chapter 6. These two parameters provide the effective turbulence viscosity used in the conservation equations. Thus the primitive variable codes were extended to include two more dependent variables and adapted to modeling turbulent flow fields.

Work has continued to improve the finite volume codes. A key problem is the artificial viscosity introduced because of upwind differencing (Pantakar, 1980). That is, the use of upwind differencing results in numerical prediction corresponding to a flow with a viscosity higher than the actual viscosity. This problem is particularly severe in flows with stream lines which do not follow the directions of the grid lines. Other work has centered on improving the two-equation turbulence models or using other turbulence models (Wilcox, 1993). Also, a scheme to circumvent the need for a staggered grid using primitive variables was introduced by Rhie and Chow (1983). Still, the finite volume codes form the backbone of most codes for subsonic flows used in industry today.

Numerical schemes for compressible flows in which the density changes the result from pressure variations in the flow are generally based on conservative variables (not primitive variables). These dependent variables do not change as markedly as the primitive variables in high Mach number flows. By definition, they are invariant across a shock wave. Many compressible flow codes have been formulated using the conservative variables.

More accurate flow field modeling is obtained through higher spatial resolution with sufficient detail to include the smallest scale of turbulence. In this way, direct solutions of the Navier-Stokes equations (Rogallo & Moin, 1984; Hussaini & Zang, 1987) are obtained without the need to introduce an effective viscosity. These approaches are generally based on spectral methods because of the resolution achievable. At the present time, direct solutions (direct numerical simulations, DNS) are limited to small Reynolds numbers.

A more recent development in turbulence modeling has been the use of large eddy simulations (LES) in which the large-scale turbulence is solved for directly and the small-scale turbulence is modeled. This improves the prediction capability and removes the small Reynolds number restriction. LES is still in the early stages of development but shows considerable promise (Reynolds, 1990) as a scheme with sufficient accuracy for turbulence modeling with wide applicability. A recent review on the capabilities of LES is given by Rodi et al. (1997).

Another scheme which has been used to model large-scale turbulent structures in high Reynolds numbers flows is the discrete vortex method. In the scheme, the vorticity generated by a shear layer or bluff body is discretized into a series of vortex elements which interact as they are convected through the field (Leonard, 1985; Sarpkaya, 1989). Viscous effects are sometimes included by introducing diffusion of the vortices through a random walk.

It is well beyond the scope of this chapter to present a general multidimensional model for the carrier phase. In order to present the essential ideas and to have a model which can be used to illustrate the two-phase flow applications, a quasi-one-dimensional model will be introduced. A computational scheme for quasi-one-dimensional compressible flows using conservative variables will also be discussed to illustrate the essential features of numerical schemes based on conservative variables.

8.2 Dilute flows - Lagrangian models

In dilute flows particle and droplet motion is controlled by the hydrodynamic forces on the dispersed phase elements. Quasi-one-dimensional flows will be addressed first followed by more general multidimensional flows. Two formulations will be presented: primitive variable and conservative variable. The use of conservative variables is advantageous in compressible flows.

8.2.1 Quasi-one-dimensional flow

The conservation equations for a quasi-one-dimensional flow were derived and presented in Chapter 6. These equations, expressed in terms of the primitive variables which are the flow velocities, pressure and temperature, will be utilized here. The equations expressed in terms of conservative variables are presented later.

Primitive variable formulation - single-phase flow

First, consider the steady flow of an inviscid fluid in a tube as shown in Figure 8.1a. In this case the momentum equation is

$$\rho u \frac{du}{dx} = -\frac{dp}{dx} \tag{8.1}$$

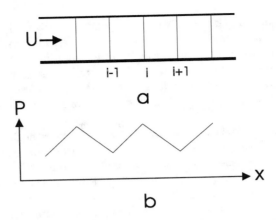

Figure 8.1: One-dimensional flow and saw-tooth pressure profile.

The central difference formulation for this equation using the grid system shown in the figure is

$$\rho u_i \left[\frac{u_{i+1} - u_{i-1}}{2\Delta x} \right] = \frac{p_{i-1} - p_{i+1}}{2\Delta x} \qquad (8.2)$$

A trivial solution to this finite difference equation is u and p are equal to a constant. However the solution for pressure shown in Figure 8.1b, the "saw-tooth" profile, also satisfies the difference equation for u equal to a constant because $p_{i+1} - p_{i-1} = 0$. This profile, however, would promote a flow because of the pressure gradient between adjacent nodal points. This leads to the introduction of the *staggered grid* arrangement[1] shown in Figure 8.2.

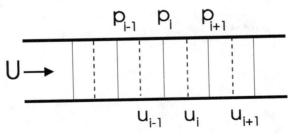

Figure 8.2: Staggered grid arrangement.

In this grid system the pressure and velocities are defined at different nodes with the velocity nodes lying between the pressure nodes. The finite difference equation for velocity now becomes

$$\rho u_i \left[\frac{u_i - u_{i-1}}{\Delta x} \right] = \frac{p_{i+1} - p_i}{\Delta x} \qquad (8.3)$$

---

[1] A lucid discussion of the need for a staggered grid can be found in Pantakar (1980).

The pressure nodes encompass the velocity nodes and the inconsistency noted with the conventional grid is avoided. The convection term in the above equation has utilized the upwind differencing scheme assuming the velocity is in the positive x-direction.

In order to illustrate the basic ideas underlying the primitive variable formulation consider the quasi-one-dimensional, steady flow in a duct illustrated in Figure 8.3.

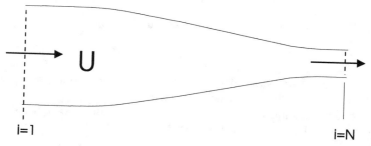

Figure 8.3: Quasi-one-dimensional duct.

Assume that the pressure is known at each end of the duct, and the distribution of the cross-sectional area is known through the duct. The fluid is incompressible.

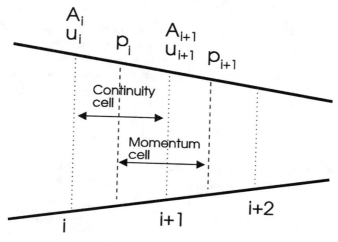

Figure 8.4: Continuity and momentum cells in quasi-one-dimensional duct.

The pressures are defined on the grid lines and the velocities are defined at points midway between the pressure nodes (grid lines) as shown in Figure 8.4. The velocity node i is displaced half a grid spacing downstream from the pressure node i. This represents the staggered grid arrangement. Consider the area enclosed by the velocity nodes i and $i + 1$ indicated in the figure. This control volume will be referred to as the *continuity cell*. The area enclosed by the pressure nodes is the *momentum cell*. The continuity equation for a steady flow in the absence of a dispersed phase is given by Equation 6.26 with $\alpha_c = 1$ and $S_{mass} = 0$; namely,

$$\Delta(\rho_c A u) = 0 \tag{8.4}$$

Applying the continuity equation to the continuity cell yields

$$\rho_c u_i A_i - \rho_c u_{i+1} A_{i+1} = 0 \tag{8.5}$$

where A_i is the cross-sectional area of the duct at the velocity node i. For convenience the mass flux $\rho_c u A$ will be represented by $\dot{M}_c$.

The momentum equation for a steady flow in the absence of a dispersed phase is given by Equation 6.49 with the unsteady term set equal to zero, $\beta_V = 0$, $\alpha_c = 1$ and $S_{mass} = 0$.

$$\Delta(\dot{M}_c u) = \bar{A}\Delta p - \tau_w P \Delta x + \rho_c g A \Delta x \tag{8.6}$$

Applying the momentum equation to the momentum cell between nodes i and $i + 1$ gives

$$\dot{M}_c(u_{i+1} - u_i) = (p_i - p_{i+1})A_{i+1} - \frac{1}{8}(f_c \rho u A \Delta x)_{i+1} + \rho_c g \Delta x A_{i+1} \tag{8.7}$$

where f_c is the Darcy-Weisbach friction factor for the fluid (later the continuous phase) and P is the duct perimeter. Notice that the specific momentum, namely the velocity, is not evaluated at the face of the momentum cell but at the velocity node upstream from the face. This corresponds to upwind differencing.

The wall friction term can also be expressed in terms of the mass flux by

$$\frac{1}{8}(f_c \rho u A \Delta x)_{i+1} = \dot{m}\left(\frac{f_c u \Delta x}{2D_H}\right)_{i+1} \tag{8.8}$$

where D_H is the hydraulic diameter. Substituting this equation for friction force into the momentum equation and solving for the velocity u_{i+1} gives

$$u_{i+1} = \frac{1}{F_{i+1}}\left(u_i + \frac{A_{i+1}}{\dot{M}_c}(p_i - p_{i+1}) + \frac{A_{i+1}}{\dot{M}_c}\rho_c g \Delta x\right) \tag{8.9}$$

where $F_i = 1 + (f_c \Delta x / 2D_H)_i$. This equation represents a recursion relationship for u_i once the mass flow rate and the pressure are known.

The continuity equation is formulated in terms of a pressure correction. One begins by assuming that the velocity is a function of the pressure gradient only or, in other words, a function of the two pressures on each side of the velocity node.

Applying a Taylor series expansion to estimate the affect of pressure change one has

$$u_{i+1} = u_{i+1}^* + \frac{\partial u_{i+1}}{\partial p_i}(p_i - p_i^*) + \frac{\partial u_{i+1}}{\partial p_{i+1}}(p_{i+1} - p_{i+1}^*) \qquad (8.10)$$

where the superscript $*$ represents the values from the previous iteration. Taking the indicated derivatives of the momentum equation, Equation 8.9, one finds

$$\frac{\partial u_{i+1}}{\partial p_i} = \frac{A_{i+1}}{\dot{M}_c F_{i+1}} \qquad (8.11)$$

and

$$\frac{\partial u_{i+1}}{\partial p_{i+1}} = -\frac{A_{i+1}}{\dot{M}_c F_{i+1}} \qquad (8.12)$$

Thus the new velocity can be expressed as a function of the previously calculated velocity and the pressure change by

$$\qquad (8.13)$$

$$u_{i+1} = u_{i+1}^* + \frac{A_{i+1}}{\dot{M}_c F_{i+1}}(\Delta p_i - \Delta p_{i+1}) \qquad (8.14)$$

where $\Delta p_i = p_i - p_i^*$. The corresponding equation for u_i is

$$u_i = u_i^* + \frac{A_i}{\dot{M}_c F_i}(\Delta p_{i-1} - \Delta p_i) \qquad (8.15)$$

Substituting the expressions for velocity back into the continuity equation yields the following equation for pressure change

$$\frac{A_i^2}{\dot{M}_c F_i}\Delta p_{i-1} - \left(\frac{A_i^2}{\dot{M}_c F_i} + \frac{A_{i+1}^2}{\dot{M}_c F_{i+1}}\right)\Delta p_i + \frac{A_{i+1}^2}{\dot{M}_c F_{i+1}}\Delta p_{i+1} = \rho u_{i+1}^* A_{i+1} - \rho u_i^* A_i$$

$$\qquad (8.16)$$

In this problem the pressure at the beginning and end of the nozzle are fixed so Δp_1 and Δp_N are zero. This equation can be solved using the TDMA.[2] Notice that the right-hand side of the equation is the continuity equation; when continuity is satisfied this term is equal to zero. Thus the pressure changes calculated from Equation 8.16 correct the pressure field in a direction to satisfy the continuity equation. Once the continuity equation is satisfied, the pressure changes are zero and no further correction is needed. The reformulation of the continuity equation in terms of pressure is called the *SIMPLE* method (Patankar, 1980) which stands for "Semi-Implicit Method for Pressure-Linked Equations".

The solution procedure is carried out by the steps illustrated in the flow diagram shown in Figure 8.5. First values for $\dot{M}_c$ and p_i are assumed. Using the value for $\dot{M}_c$ and the cross-sectional area at node 1 the velocity at node u_1 is

---

[2]TDMA is the tridiagonal algorithm which is a procedure for solving a system of equations with a tridiagonal matrix (main diagonal and its two adjacent diagonals). A description of this algorithm can be found in Anderson et al. (1984).

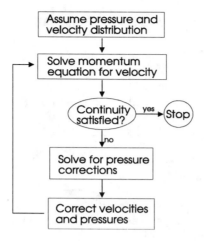

Figure 8.5: Flow diagram for numerical solution.

calculated. The momentum equation is then applied to calculate the remaining velocities. The equation for the velocity is relaxed and expressed as

$$u_{i+1} = (1 - \omega)u_{i+1}^* + \frac{\omega}{F_{i+1}}[u_i^* + \frac{A_{i+1}}{\dot{M}_c}(p_i^* - p_{i+1}^*) + \frac{A_{i+1}}{\dot{M}_c}\rho_c g \Delta x] \qquad (8.17)$$

where ω is the relaxation factor (usually $\omega = 0.5$). The new velocity field calculated using this equation becomes the u_i^* field which is substituted into the equation for pressure change. The TDMA is used to obtain the pressure changes and the velocities are corrected by applying Equations 8.13 and 8.15 and the pressure is updated by adding Δp to the previous pressure. The momentum equation is then solved again for the velocity distribution and the process is continued until convergence is achieved. Convergence is usually established by

$$\sum_{i=1}^{N-1} |\rho_c A_{i+1} u_{i+1}^* - \rho_c A_i u_i^*| \le \epsilon \qquad (8.18)$$

where ϵ is a preselected convergence criterion. The main features of the above equations are the use of the staggered grid and the pressure-change equation to satisfy continuity. This algorithm carries over to two- and three-dimensional flows.

The above model using the pressure-velocity variables can be extended to compressible flows as well. In this case the energy equation is needed to provide the gas temperature and the gas density is obtained from the equation of state. Introducing the energy equation into the formulation makes convergence more difficult because of the sensitivity of the mass flow to density changes. For this reason the conservative variable approach introduced in the next section is recommended for flows with significant compressibility effects.

The same scheme can be used to obtain a numerical solution for unsteady flows. Assume, for example, the fluid in the duct shown in Figure 8.3 is initially at rest and a pressure gradient is applied across the duct. The flow will accelerate with time until the steady-state condition is achieved. For an incompressible fluid, the continuity equation remains unchanged, but the momentum equation requires an additional term to account for the momentum change with time in a computational cell; namely, the unsteady term in Equation 6.49. The momentum equation now becomes

$$\rho_c A_{i+1} \frac{\Delta x}{\Delta t}(u_{i+1} - u_{i+1}^o) + \dot{M}_c(u_{i+1} - u_i) =$$

$$(p_i - p_{i+1})A_{i+1} - \frac{1}{8}\left(f_c\rho_c u^2 P\right)_{i+1}\Delta x + \rho_c g A_{i+1}\Delta x \tag{8.19}$$

where u_{i+1}^o is the velocity at the previous time level and Δt is the time step. The solution for u_{i+1} assumes the form

$$u_{i+1} = \frac{\dot{M}_c u_i + (p_i - p_{i+1})A_{i+1} + \rho_c A_{i+1}\frac{\Delta x}{\Delta t}u_{i+1}^o}{\dot{M}_c F_{i+1} + \rho_c A_{i+1}\frac{\Delta x}{\Delta t}} \tag{8.20}$$

Using this equation the pressure formulation of the continuity equation has the same form as Equation 8.16 but with different coefficients. The first term becomes

$$\frac{\rho_c A_i^2}{\dot{M}_c F_{i+1} + \rho_c A_{i+1}\frac{\Delta x}{\Delta t}}\Delta p_{i-1}$$

and equivalent changes appear in the other terms.

The solution procedure is the same as for the steady flow case. An initial pressure distribution and inlet velocity are assumed. In this case the initial velocity, u_i^o, would be zero everywhere. The velocity and pressure would be solved for using the same procedure as before. Once continuity is established, the solution at a new time level is sought with the initial velocity being the result from the previous time step. Solutions at new time levels are obtained in the same way until the prescribed time limit is reached. In the example considered here, the solution will approach the steady-state solution. This is an alternate approach to obtaining a solution for a steady flow problem.

If the flow is compressible, further modifications are needed. The continuity equation has to be extended to include the transient mass change within a control volume (accumulation term). Also the energy equation would have to include a transient term for temporal change of energy inside the control volume. These extensions will not be addressed here. The treatment of flows with significant compressibility effects may be better modeled using conservative variables discussed below.

Rhie and Chow (1983) introduced a method to circumvent the need for a staggered grid. Consider thenonstaggered grid system shown in Figure 8.6. The steady flow continuity equation for cell i is

$$\rho_c u A \mid_e - \rho_c u A \mid_w = 0 \tag{8.21}$$

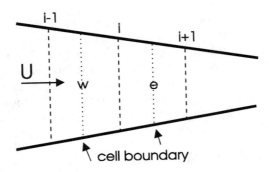

Figure 8.6: Nonstaggered grid system in a quasi-one-dimensional duct.

The momentum equation now uses a central difference formulation for the pressure gradient,

$$\dot{M}_c(u_i - u_{i-1}) = \frac{1}{2}(p_{i-1} - p_{i+1})A_i - \frac{1}{8}\left(f_c\rho_c u^2 A\Delta x\right)_i + \rho_c g\Delta x A_i \qquad (8.22)$$

which is a 2Δ center difference. For convenience, the momentum equation is written as

$$u_i = -\frac{A_i}{\dot{M}_c}\frac{dp}{dx}\Big|_i \Delta x + B_i \qquad (8.23)$$

where B_i is the remaining term. The velocity on the east face is the average of the velocity at the i and $i+1$ nodes,

$$u_e = \frac{u_i + u_{i+1}}{2} = -\frac{A_e}{\dot{M}_c}\frac{1}{2}\left(\frac{dp}{dx}\Big|_i + \frac{dp}{dx}\Big|_{i+1}\right)\Delta x + \bar{B} \qquad (8.24)$$

where $\bar{B}$ is the average between the two nodes. When this equation is used in the continuity equation, an oscillation develops in the equation for pressure change. Rhie and Chow suggested that

$$u_e + \frac{A_e}{\dot{M}_c}(p_{i+1} - p_i) = \frac{u_i + u_{i+1}}{2} + \frac{A_e}{\dot{M}_c}\frac{1}{2}\left(\frac{dp}{dx}\Big|_i + \frac{dp}{dx}\Big|_{i+1}\right)\Delta x \qquad (8.25)$$

which yields the mass flux through the east face in the form

$$\rho A u\,|_e = \rho A_e \frac{u_i + u_{i+1}}{2} - \rho\frac{A_e^2}{4\dot{M}_c}(-p_{i-1} + 3p_i - 3p_{i+1} + p_{i+2}) \qquad (8.26)$$

A similar equation can be developed for the west face. The same equation, Equation 8.16, is used for the pressure change with the right side replaced by the mass flux given by the above equation. This approach essentially removes the pressure oscillations attributed to the 2Δ difference in the pressure gradient.

Primitive variable formation - multiphase flow

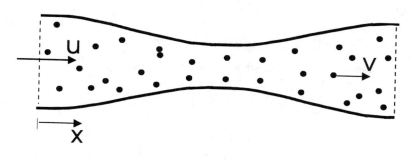

Figure 8.7: Particles flowing in a qausi-one-dimensional duct.

Consider the fluid-particle flow in the nozzle shown in Figure 8.7. In order to minimize the algebra and to show the fundamental ideas, it is assumed that the fluid is incompressible.

The continuity equation including the multiphase flow effects, Equation 6.26, is expressed as

$$(\alpha_c \rho_c A u)_{i+1} - (\alpha_c \rho_c A u)_i = -N\dot{m} \tag{8.27}$$

where N is the number of droplets in the control volume

The momentum equation for the quasi-one-dimensional flow of the carrier phase with the suspended droplets is

$$\dot{M}_{c,i+1}u_{i+1} - \dot{M}_{c,i}u_i = \alpha_{c,i+1}A_{i+1}(p_i - p_{i+1}) - \tau_w P \Delta x + S_{mom,i+1} + \alpha_c \rho_c g V_{i+1} \tag{8.28}$$

where $S_{mom,i+1}$ is the momentum source term at the $i + 1$ station and is due to the drag force of the particles on the fluid. Note that the mass flux of the continuous phase is not constant with mass addition to the flow. The source term can be expressed as

$$S_{mom,i+1} = A\Delta x \beta_V (v - u)\mid_{i+1} -N\dot{m}v_{i+1} \tag{8.29}$$

where v is the particle velocity and β_V is the proportionality constant for the steady-state drag force. The unsteady drag terms have not been included. To make the analysis less cumbersome it is assumed that there is no mass coupling. Incorporating this term into the momentum equation, Equation 8.28, one obtains[3]

---

[3]This equation has been written for a uniform size particle field. For a distribution of particle sizes, the source term becomes

$$S_{mom,i+1} = A_{i+1}\Delta x \left(\sum_k \beta_{V,k} v_k \mid_{i+1} -u_{i+1} \sum_k \beta_{V,k} \right)$$

where subscript k refers to a size class of particles.

$$\dot{M_c}(u_{i+1} - u_i) = \alpha_{c,i+1} A_{i+1}(p_i - p_{i+1}) + A_{i+1}\Delta x \beta_V (v - u)\mid_{i+1}$$

$$-\dot{M_c}\frac{\Delta x}{2}(f_c u \frac{1}{D_H})_{i+1} + g\rho_c \Delta x \left(\alpha_c \bar{A}\right)_{i+1}$$

(8.30)

The recursion equation for the velocity takes the form

$$u_{i+1}(1 + f_c\frac{\Delta x}{2D_H})_{i+1} =$$

$$u_i + \frac{1}{\dot{M_c}}\alpha_{c,i+1} A_{i+1}(p_i - p_{i+1}) + \frac{1}{\dot{M_c}}[A\Delta x\beta_V(v - u)]_{i+1}$$

(8.31)

One notes that the carrier fluid velocities appear in the source terms. It is convenient for stability, especially at high loadings, to split the source term and rewrite the equation in the form

$$u_{i+1}\left[1 + \left(f_c\frac{\Delta x}{2D_H}\right)_{i+1} + \frac{1}{\dot{M_c}}(A\Delta x\beta_V)_{i+1}\right]$$

$$= u_i + \frac{1}{\dot{M_c}}\alpha_{c,i+1} A_{i+1}(p_i - p_{i+1}) + \frac{1}{\dot{M_c}}(nA\Delta x\beta_V v)_{i+1}$$

(8.32)

At high particle concentrations, one notes that the gas velocity will approach the particle velocity which is the expected trend. The factor $\frac{1}{\dot{M_c}}A\Delta x\beta_V$ can be written as

$$\frac{1}{\dot{M_c}}(A\Delta x\beta_V)_{i+1} = \frac{1}{\dot{M_c}}\dot{N}m_d\left(f\frac{\Delta t}{\tau_V}\right)_{i+1} = Z\left(f\frac{\Delta t}{\tau_V}\right)_{i+1}$$

(8.33)

where Z is the loading (ratio of particle to gas mass flow rates), $\dot{N}$ is the number flow rate of the particles, Δt is the residence time of particles in the computational cell and f is the ratio of the drag coefficient to Stokes drag. The ratio of times $\frac{\Delta t}{\tau_R}$ can be regarded as the Stokes number associated with the particle residence time in the cell. The recursion formula for u_{i+1} becomes

$$u_{i+1} = \frac{1}{E_{i+1}}\left[u_i + \frac{\alpha_{c,i+1} A_{i+1}(p_i - p_{i+1})}{\dot{M_c}} + Z\left(f\frac{\Delta t}{\tau_R}v\right)_{i+1}\right]$$

(8.34)

where the factor E_i is

$$E_i = 1 + \left(f_c\frac{\Delta x}{2D_H}\right)_i + Z\left(f\frac{\Delta t}{\tau_R}\right)_i$$

(8.35)

This equation is the same form as that for a single-phase flow so the pressure form of the continuity can be carried out in the same way. The final form is

$$(\alpha_c\rho_c uA)^*_{i+1} - (\alpha_c\rho_c uA)^*_i = -\frac{\alpha^2_{c,i}\rho_c A^2_i}{\dot{M_c}E_i}\Delta p_{i-1}$$

$$+ \left[\frac{\alpha^2_{c,i+1}\rho_c A^2_{i+1}}{\dot{M_c}E_{i+1}} + \frac{\alpha^2_{c,i}\rho_c A^2_i}{\dot{M_c}E_i}\right]\Delta p_i - \frac{\alpha^2_{c,i+1}\rho_c A^2_{i+1}}{\dot{M_c}E_{i+1}}\Delta p_{i+1}$$

(8.36)

In order to complete the equation set, values for the carrier phase volume fraction, particle velocity and cell residence time are needed. These are determined from the particle equations by application of the trajectory approach or the discrete element approach. The trajectory approach is applicable to steady flow problems so will be illustrated here.

The particle velocity is obtained by integrating the particle cloud equations developed in Chapter 7 (Equation 7.54) and can formulated as

$$v \frac{dv}{dx} = -\frac{1}{\rho_d} \frac{dp}{dx} + \frac{f}{\tau_V}(u - v) - \frac{f_s}{2} \frac{|v| v}{D_H} + g \qquad (8.37)$$

Many schemes are available to integrate this equation such as the fourth-order Runge-Kutta methods, Adams-Bashforth methods and so on. These will not be discussed here but can be found in any textbook on numerical methods.

A useful formulation for this problem where one wishes to find the velocity at specific distances, not times, is formulated by expressing the particle motion equation in implicit finite difference form as

$$v_{i+1} \frac{v_{i+1} - v_i}{\Delta x} = \frac{f}{\tau_V}(u_{i+1} - v_{i+1}) - \frac{f_s}{2} \frac{|v_i| v_{i+1}}{D_H} + \gamma \qquad (8.38)$$

where γ represents the sum of the gravity and buoyant forces. Expressing this equation as a quadratic equation for v_i one has

$$v_{i+1}^2 - v_{i+1}\left(v_i - \frac{f}{\tau_V}\Delta x - \frac{f_d |v_i|}{2D_H}\Delta x\right) - \left(\frac{f}{\tau_V}u_{i+1}\Delta x + \gamma\Delta x\right) = 0 \quad (8.39)$$

and solving for v_{i+1} yields

$$v_{i+1} = B + \sqrt{B^2 + C} \qquad (8.40)$$

where

$$B = \tfrac{1}{2}\left(v_i - \frac{f}{\tau_R}\Delta x - \frac{f_d|v_i|}{2D_H}\Delta x\right)$$
$$C = \frac{f}{\tau_R}u_{i+1}\Delta x + \gamma\Delta x$$

Integrating the particle momentum equation through the nozzle yields information on the particle residence time and the average velocity in each computational cell. The particle volume fraction is obtained from

$$\alpha_d = \frac{\dot{M}_d}{\rho_d A v} \qquad (8.41)$$

where $\dot{M}_d$ is the mass flow rate of the dispersed phase. The carrier phase volume fraction is calculated from the dispersed phase volume fraction.

The solution procedure is outlined in the flow diagram shown in Figure 8.8.

First, a velocity and pressure field are assumed throughout the nozzle. Although it is not necessary, it is advantageous to have the initial flow assumptions

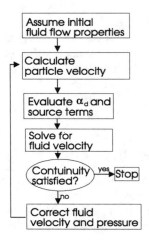

Figure 8.8: Flow diagram for solution scheme including two-phase coupling effects.

satisfy the continuity equation. The particle velocity equations are then integrated to provide the fluid volume fraction and parameters in the momentum source term. The continuous phase velocity distribution is calculated using the recursion formulas for the carrier phase. The pressure formulation of the continuity equation is used to correct the pressure field and the pressures and carrier flow velocities are updated. The computation returns to the beginning of the loop where the particle motion equations are integrated again. The loop continues until the continuity equation is satisfied to within a specified convergence range. In this way, two-way coupling is included. For high levels of loadings, convergence is accelerated if the coupling terms are not updated on every iteration.

Another approach may be to solve first only the carrier flow equations and obtain the velocity and temperature distribution with no particles present. The particle equations would then be called and the solution scheme would proceed as above.

As an example this scheme is applied to predict the flow rate of a mixture of 1-mm particles in water passing through a flow nozzle. The inlet diameter is 4 cm and the outlet is 2 cm. A 10-kPa pressure differential is applied across the nozzle. The particles are spherical and have a density of 2500 kg/m^3. The predicted fluid and particle velocity distributions in the nozzle are shown in Figure 8.9. As expected, the particle velocity is less than the water velocity with the 5-mm particles showing the greater velocity lag.

The effect of loading on the flow rate of the water with 1-mm particles and the total flow rate is shown in Figure 8.10. As the loading is increased, the water flow rate is decreased but the total flow rate $[\dot{m}_t = \dot{m}_c(1 + Z)]$ increases with the loading ratio.

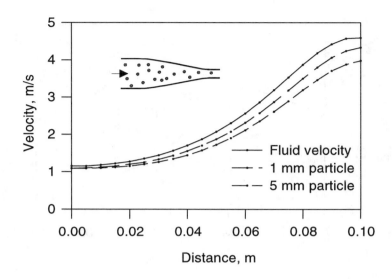

Figure 8.9: Particle and fluid velocity distribution in flow nozzle.

Essentially the same procedure is used for the discrete element approach. In this case, a parcel of particles is identified and tracked through the field. As an example, consider the problem of a duct flow shown in Figure 8.7, which is initially operating at steady flow conditions and particles are introduced into the duct at time $t = 0$ and at a constant rate thereafter. The particles are introduced as *parcels* which represent many particles but move at the same speed as an individual particle. At every time step, a new parcel is introduced into the duct as shown in Figure 8.11. At time level 3 there are three parcels in cell 1: namely, b, c and d. Parcel a has already passed through cell 1.

The same procedure for convergence is used at each time level and diagrammed in Figure 8.12.

A *time-splitting* procedure is required to account for the coupling terms at the most appropriate time level. The parcels in the duct are moved to new positions corresponding to half a time step by integrating the particle motion equations and assuming the velocity field is the initial velocity field. The particle temperature is also calculated at the half-time step. The momentum and energy source terms, as well as the particle volume fraction, are then evaluated for each cell corresponding to the conditions at the half-time step. For example, the momentum source term for cell i would be

$$S_{i,mom} = A_i \Delta x \sum_k \beta_{V,k}(v - u)_k \qquad (8.42)$$

where the summation is carried out over all parcels which are located in the

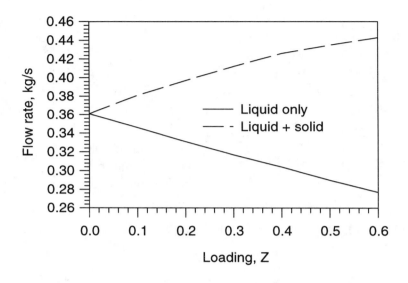

Figure 8.10: Effect of loading on the flow rate of water alone and on the total flow rate.

computational cell. These source terms can be split as done in Equation 8.32 to enhance stability.

The equations for the carrier flow are then solved over a complete time step using the source terms evaluated at the half-time step. The procedure used for single-phase flow is employed here except the coefficients are different to reflect the coupling terms. This completes the solution for the current time step, and the same procedure is used for the new time step. However, it may be desirable to obtain a better solution by re-evaluating the two-phase coupling terms using the average carrier phase velocities and temperatures to relocate the particles at the half step and recalculate the source terms. These new values are then used to find carrier flow properties over a complete time step.

Many computational schemes for unsteady flows use a two-step process referred to as the *predictor* step followed by the *corrector* step. With these schemes, it is convenient to use the particle source terms evaluated on the half-time step in the predictor step. The initial field is used to move the particles from the initial location by half a time step. The source terms are then evaluated and used in the corrector step. There are a variety of methods available to incorporate the source terms.

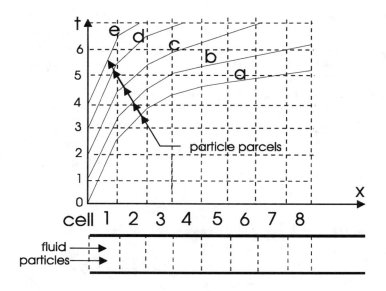

Figure 8.11: Particle parcels introduced into a duct.

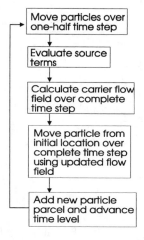

Figure 8.12: Flow diagram for unsteady flow.

Single-phase flow - conservative variable formulation

The pressure-velocity formulation performs well for incompressible fluids but does not converge easily for compressible flows in which the density changes can be large. A robust method for compressible flows utilizes *conservative* variables which do not vary as rapidly as the primitive variables in a compressible flow (Sharma & Crowe, 1978).

Consider the steady flow through the duct shown in Figure 8.3. The computational cell is shown in Figure 8.13. In this formulation a staggered cell arrangement is not necessary. The continuity equation is

$$\rho_c u A = \dot{M}_c = const \tag{8.43}$$

A new (conservative) variable is defined as

$$X = \rho_c u A \tag{8.44}$$

where X in this case is constant throughout the duct.

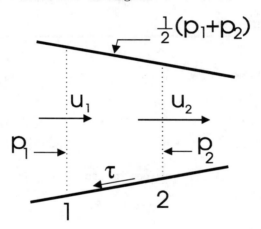

Figure 8.13: Computational cell for conservative variables.

The momentum equation for the computational cell is

$$\dot{M}_c(u_2 - u_1) = (p_1 A_1 - p_2 A_2) - \frac{p_1 + p_2}{2}(A_1 - A_2) - F_f + B \tag{8.45}$$

where the pressure force term has been decomposed into the pressure forces acting on faces 1 and 2 and on the wall. The force F_f is the friction at the wall and can be represented as in Equation 8.7 with the Darcy-Weisbach friction factor as

$$F_f = \frac{f_c}{8}P\Delta x \rho u^2 = \frac{f_c}{2}\frac{\Delta x}{D_H}Xu \tag{8.46}$$

where P is the duct perimeter and D_H is the hydraulic diameter. The body force due to gravity is

$$B = \rho_c g A \Delta x \tag{8.47}$$

where g is the acceleration due to gravity in the flow direction.

Defining the conservative variable for momentum as

$$Y = \dot{M}_c u + pA = \left(\rho_c u^2 + p \right) A \tag{8.48}$$

the momentum equation, Equation 8.45, can now be expressed as

$$Y_2 - Y_1 = -\frac{p_1 + p_2}{2}(A_1 - A_2) - F_f + B \tag{8.49}$$

The terms on the right side are responsible for changing the momentum and can be regarded as momentum source terms.

The energy equation for the computational cell becomes

$$\dot{M}_c \left(h_c + \frac{u^2}{2} \right)_2 - \dot{M}_c \left(h_c + \frac{u^2}{2} \right)_1 = \dot{Q}_w \tag{8.50}$$

where h_c is the enthalpy of the flowing fluid and $\dot{Q}_w$ is the heat transfer from the wall which can be expressed as

$$\dot{Q}_w = St_Q X \frac{4\Delta x}{D_H}(T_w - T) \tag{8.51}$$

where St_Q is the *Stanton number*. Introducing the conservative variable,

$$Z = \dot{M}_c(h_c + \frac{u^2}{2}) \tag{8.52}$$

the energy equation can be written as

$$Z_2 = Z_1 + \dot{Q}_w \tag{8.53}$$

The conservative variables are useful because they do not change rapidly in a compressible flow. In fact they are constant across a shock wave which is another way of saying that these combinations of properties are *conserved* across a shock wave.

However, once one has solved for the conservative variables in a duct flow, it is necessary to recover the primitive variables. This procedure is referred to as *decoding*. This can be accomplished easily for a calorically perfect ideal gas in which the enthalpy is expressed as

$$h_c = c_p T_c = \frac{k}{k - 1} R T_c \tag{8.54}$$

The conservative variable Z can then be written as

$$Z = X \left(\frac{k}{k - 1} \frac{p}{\rho_c} + \frac{u^2}{2} \right) \tag{8.55}$$

Using the definition of Y, the pressure is given by

$$p = \frac{Xu - Y}{A} \tag{8.56}$$

Substituting this equation into Equation 8.55 results in the following quadratic formula for the velocity.

$$u^2 - 2\frac{k}{k+1}\frac{Y}{X}u + 2\frac{k-1}{k+1}\frac{Z}{X} = 0 \tag{8.57}$$

Solving for u yields

$$u = \frac{k}{k+1}\frac{Y}{X}\left[1 \pm \left(1 - 2\frac{XZ}{Y^2}\frac{k^2-1}{k^2}\right)^{\frac{1}{2}}\right] \tag{8.58}$$

The two solutions of the quadratic equation correspond to subsonic (negative sign) and supersonic (positive sign) flow. At sonic conditions

$$2\frac{XZ}{Y^2}\frac{k^2-1}{k^2} = 1 \tag{8.59}$$

and the sonic velocity is

$$u = \frac{k}{k+1}\frac{Y}{X} \tag{8.60}$$

The solution procedure is straightforward, as shown in the flow diagram in Figure 8.14 for a quasi-one-dimensional flow. The conservative variables at the inlet are evaluated. The process of converting the primitive variables to conservative variables is called *encoding*. The values for the conservative variables at the end of the first step, Δx, are sought. There is insufficient information to explicitly solve for Y_2 and Z_2 since the pressure at station 2 and the fluid velocity and temperature in the computational cell are unknown. An iterative approach is used. As a first estimate one takes $p_2 = p_1$, $u = u_1$ and $T = T_1$. The source terms are then evaluated and values for the conservative variables at station 2 are calculated (in this example, X remains constant). The velocity, u_2, is determined from Equation 8.58 which, in turn, can be used to determine the density ρ from Equation 8.44 and then the pressure, p_2, from Equation 8.56. Finally, the temperature is obtained by using Z_2. This new pressure is used to evaluate the pressure source term and the average velocity and temperature in the cell are used for the friction force and heat transfer term. New values for the conservative variables are calculated and the procedure is continued until the pressure no longer changes with continued iterations. Few iterations are needed for convergence. Once the solution is obtained for one cell, the procedure is repeated for the adjacent downstream cell.

This method differs from the one based on primitive variables in that the inlet conditions are the initial (including mass flow) conditions and the program provides the outlet pressure. In the code based on primitive variables, the pressure boundary conditions are specified and the solution provides the mass

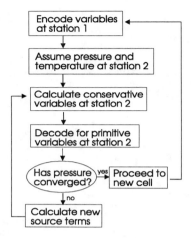

Figure 8.14: Flow diagram for conservative variables.

flow. Thus, if the pressure boundary conditions are specified for the conservative variable approach, the inlet mass flow must be determined by iteration until the correct exit pressure is established.

Another interesting feature of the conservative variable code is the behavior for sonic and supersonic flow conditions in the quasi-one-dimensional nozzle. Consider the flow through the de Laval nozzle shown in Figure 8.15. If the initial velocity selected is sufficiently small, sonic conditions will not be achieved at the throat and subsonic flow will persist through the nozzle, as shown in the figure. However, if the initial velocity is too high, the condition represented by Equation 8.59 will be encountered before the flow reaches the minimum area and a solution is not possible.[4] The initial velocity has to be found by iteration such that the sonic condition is achieved at the minimum area (or very close thereto). After this condition is established, the calculation is repeated with the sign on the discriminant in Equation 8.58 changed from minus to plus at the minimum area. The flow thereafter is supersonic.

If the exit pressure is higher than or equal to the back pressure (pressure of the environment), no further changes are needed. This corresponds to an underexpanded ($p_e > p_b$) or ideally expanded ($p_e = p_b$) nozzle. If the exit pressure is less than the back pressure, a normal shock must exist in the expansion section. Once again the location of the shock must be determined by iteration. The sign on the discriminant in Equation 8.58 is changed (corresponding to the presence of a normal shock wave) at a location in the nozzle which yields an exit pressure equal to the back pressure. The location is determined by some iterative technique such as binary search. The flow now exits from the nozzle at a subsonic Mach number.

The conservative variable approach can also be extended to unsteady quasi-

---

[4] The discriminant in Eqn. 8.58 becomes negative.

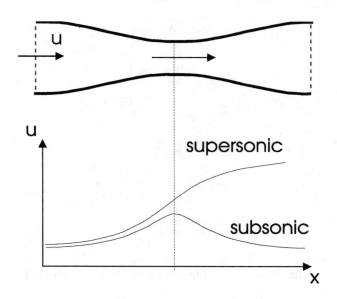

Figure 8.15: Flow velocity in the de Laval nozzle.

one-dimensional flows. In this case, the equations in differential form become

$$\frac{\partial}{\partial t}(\rho_c A) + \frac{\partial}{\partial x}(\rho_c u A) = 0 \tag{8.61}$$

$$\frac{\partial}{\partial t}(\rho_c u A) + \frac{\partial}{\partial x}(\rho_c u^2 A + pA) = -p\frac{dA}{dx} - \frac{f}{8}\rho_c u^2 P + \rho_c g A \tag{8.62}$$

$$\frac{\partial}{\partial t}\left[\rho_c A(i_c + \frac{u^2}{2})\right] + \frac{\partial}{\partial x}\left[\rho_c u A\left(h_c + \frac{u^2}{2}\right)\right] = St_Q \rho_c u P c_p(T_w - T) \tag{8.63}$$

These equations are often written in more compact form as

$$\frac{\partial \mathbf{F}}{\partial t} + \frac{\partial \mathbf{G}}{\partial x} = \mathbf{H} \tag{8.64}$$

where $\mathbf{F}, \mathbf{G}$ and $\mathbf{H}$ are vectors defined by

$$\mathbf{F} = \left\{ \begin{array}{c} \rho_c A \\ \rho_c u A \\ \rho_c A(i_c + \frac{u^2}{2}) \end{array} \right\} \qquad \mathbf{G} = \left\{ \begin{array}{c} \rho_c u A \\ \rho_c u^2 A + pA \\ \rho_c u A\left(h_c + \frac{u^2}{2}\right) \end{array} \right\} \tag{8.65}$$

and

$$\mathbf{H} = \left\{ \begin{array}{c} 0 \\ -p\frac{dA}{dx} - \frac{f}{4}\rho_c u^2 P + \rho_c gA \\ St_Q \rho_c u P c_p (T_w - T) \end{array} \right\} \tag{8.66}$$

There are several procedures available in the literature to integrate these equations (Anderson et al., 1978). The procedures generally involve finding the values for the vector $\mathbf{F}$ at a new time level and decoding the new $\mathbf{F}$ for the primitive variables to evaluate $\mathbf{G}$ and $\mathbf{H}$ for the next time step. These methods will not be covered in this chapter.

Obviously, the same approach can be extended to multidimensional flows. In this case, the shear stress terms have to be modeled with some appropriate stress-rate of strain relationship such as the Boussinesq equation. In some applications with significant compressibility effects, the effects of shear stress are insignificant.

Multiphase flow - conservative variable formulation

The use of conservative variables to model multiphase flows follows directly from that for the single phase. Consider, as before, the quasi-one-dimensional flow of gas-particle flow in a duct shown in Figure 8.7. Evaporating droplets would represent a mass source to the continuous phase. The presence of droplets or particles would give rise to momentum and energy source terms. Once the source terms are evaluated the procedure follows that for the single phase.

The carrier phase continuity equation for a steady flow is (see Table 6.1)

$$\Delta \left(\alpha_c u A \rho_c \right) = S_{mass} \tag{8.67}$$

and can be written as

$$X_2 = X_1 + S_{mass} \tag{8.68}$$

where the mass flux, X, now incorporates the continuous phase volume fraction,

$$X = \alpha_c u A \rho_c \tag{8.69}$$

The source term for uniformly evaporating droplets of a single size is given by

$$S_{mass} = -nA\Delta x \dot{m} \tag{8.70}$$

which also can be expressed as

$$S_{mass} = -\dot{N}\Delta t \dot{m} \tag{8.71}$$

where $\dot{N}$ is the number flow rate in the duct. This form may be more convenient since the number flow rate of the droplets would not change in a steady flow.

The momentum equation for the continuous phase for steady flow is

$$\Delta \left(\alpha_c \rho_c A u^2 \right) = -\alpha_c \bar{A} \Delta p + \alpha_c \rho_c gV - \tau_w P \Delta x + \beta_V V \left(v - u \right) + S_{mass} v \tag{8.72}$$

This equation can be rewritten as

$$\Delta\left(Xu\right) = -\Delta(\alpha_c Ap) - \frac{p_1+p_2}{2}\Delta A + \alpha_c\rho_c gV$$

$$-\tau_w P\Delta x + \beta_V V\left(v-u\right) + S_{mass}v \tag{8.73}$$

or as

$$Y_2 = Y_1 - \frac{p_1+p_2}{2}\Delta A + \alpha_c\rho_c gV$$

$$\beta_V V\left(v-u\right) - -\tau_w P\Delta x + S_{mass}v \tag{8.74}$$

where Y is the conservative variable,

$$Y = Xu + \alpha_c pA \tag{8.75}$$

For convenience, the unsteady forces are not included here but it would not be difficult to include them. The drag force term can be expressed more directly as

$$\bar{A}\Delta x\beta_V(v-u) = \dot{M}_d\Delta t\frac{f}{\tau_V}(v-u) \tag{8.76}$$

For a distribution of particle sizes, the drag force term is summed over all the size classifications. The friction term is the same as for a single-phase flow.

The total energy equation for the carrier phase for steady flow with no mass transfer is given in Table 6.1 as[5]

$$\Delta[X(h+\frac{u^2}{2})] = \bar{\rho}_c guV - \dot{q}_w P\Delta x - V\beta_V v(u-v) + V\beta_T(T_d - T_c) \tag{8.77}$$

The conservative variable Z is defined as $X(h+\frac{u^2}{2})$ so the energy equation can be expressed as

$$Z_2 = Z_1 + \bar{\rho}_c guV - \dot{q}_w P\Delta x - V\beta_V v(u-v) + V\beta_T(T_d - T_c) \tag{8.78}$$

The particle-fluid heat transfer term becomes

$$V\beta_T(T_d - T) = \dot{M}_d\Delta t\frac{Nu}{2\tau_T}c_d(T_d - T_c) \tag{8.79}$$

where c_d is the specific heat of the dispersed phase. This completes the equation set for the carrier fluid. One notes that the equations are the same as for the single-phase flow except for additional source terms.

The solution procedure, shown in the flow diagram in Figure 8.16, follows that for a single-phase flow. First, inlet flow conditions are selected and the solution proceeds from cell to cell. The pressure and velocities are not known

---

[5] The heat conduction through the fluid in the flow direction is neglected since this is small compared to the convective energy flux when the Péclet number is large.

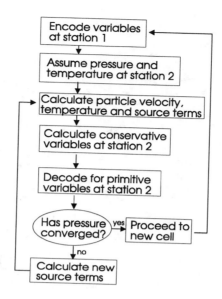

Figure 8.16: Flow diagram for conservative variable formulation with coupling.

a priori in the computational cell, so the upstream pressure and carrier velocity are chosen. The particle velocity and temperature are predicted at the downstream edge of the cell by integrating the particle momentum and energy equation. The volume fraction of the dispersed phase is obtained from the particle mass flow equation,

$$\dot{M}_d = \alpha_d \rho_d v A \tag{8.80}$$

For a distribution of particle sizes, the volume fraction for each size would have to be obtained and added to yield the overall dispersed phase volume fraction. Once the conservative variables are estimated at the downstream edge of the computational cell the equations are decoded to give the primitive variables. Then the pressure (or velocity) is compared with the previous iteration. When the change is sufficiently small, convergence has been achieved and the computation moves to the adjacent downstream cell.

A program utilizing this algorithm is provided in Appendix D. The geometry used in the program is a converging-diverging nozzle (venturi geometry) but the program is applicable to any quasi-one-dimensional configuration.

An example application of the conservative variable formulation is the flow of hot particles through the converging-diverging nozzle. The inlet diameter of the nozzle is 10 cm and the throat/inlet diameter ratio is 0.5. The particles have a diameter of 75 μm and a material density of 2500 kg/m^3. The initial gas velocity is 20 m/s and the initial gas temperature is 293 K. The initial temperature of the particles is 1000 K. The loading is unity.

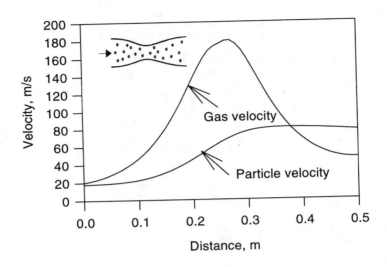

Figure 8.17: Gas and particle velocity distributions through nozzle.

The predicted gas and particle velocity distributions are shown in Figure 8.17. As the gas velocity accelerates, the particle velocity increases due to the increased drag force. As the gas velocity decreases after the throat, the particle velocity continues to increase until the velocities become equal, after which the particle velocity remains higher than the gas velocity due to inertia of the particles but will begin to decrease.

The variation of gas velocity in the nozzle including and excluding thermal coupling effects is shown in Figure 8.18. For no thermal coupling effects, the gas velocity increases in the expected way toward a maximum value at the throat and returns to the initial velocity at the end of the nozzle. With thermal coupling effects, the gas is heated, the density is decreased and the velocity is increased to satisfy continuity. Once through the nozzle, the gas velocity is higher than the entrance velocity because the gas density has been increased. The same analysis could be used to model heat release due to chemical reaction in the nozzle.

The same procedures outlined above apply to an unsteady flow as well. As before, the time is split to provide a better estimate of the source terms over the time step. The extension to unsteady flows will not be addressed here. Multidimensional flows follow directly from the procedures developed for quasi-one-dimensional flows.

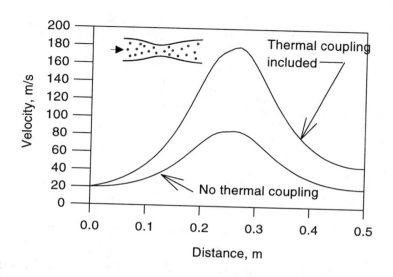

Figure 8.18: Gas velocity distribution with and without thermal coupling.

8.2.2   Multidimensional flows

Single-phase flow

The same approach used for numerical models of quasi-one-dimensional flows carries over directly to multidimensional flows. Consider the steady flow in a sudden expansion section shown in Figure 8.19.

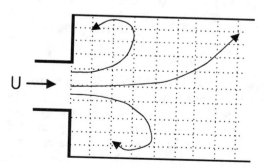

Figure 8.19: Sudden expansion section.

In this case the flow field is subdivided into a series of computational cells. The configuration of the cells around a grid point is shown in Figure 8.20. The continuity cell encompasses a grid note and the mass flux is defined on

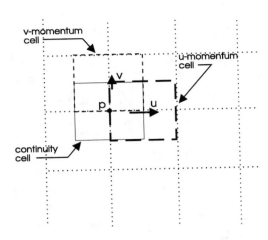

Figure 8.20: Staggered cell arrangement for two-dimensional flow.

the cell boundaries. In this case, for a steady flow, the continuity equation is obtained by setting the sum of the mass flux through all the faces equal to zero.

$$\dot{M}_{c,n} + \dot{M}_{c,e} - \dot{M}_{c,s} - \dot{M}_{c,w} = 0 \qquad (8.81)$$

The control volume used to solve the momentum equation for the velocity components is displaced from the continuity cell to yield a staggered grid system. The velocity at the center of the momentum cells lies on the boundary of the continuity cell. The pressures are defined at the centers of the continuity cells and these pressures are at the edge of the momentum cells thereby providing the correct pressure force. The general form for the momentum equation in the x-direction (1-direction) is

$$\dot{M}_{c,n} u_n + \dot{M}_{c,e} u_e - \dot{M}_{c,s} u_s - \dot{M}_{c,w} u_w =$$

$$A_1(p_w - p_e) + \sum_k A_k \tau_{1k} + \rho_c V g_1 \qquad (8.82)$$

where A_1 is the cross-sectional area of the cell normal to the x-direction, τ_{1k} is the shear stress on the k-face in the x-direction and g_1 is the acceleration due to gravity in the 1-direction. The summation over k means the sum over all four faces (n, e, s, w) with the appropriate signs. Upwind differencing is used to ensure stability of the solution. For example, if the velocity on the east face is positive, then the flux term on the east face is expressed as $\dot{M}_{c,e} u_P$ whereas if the velocity is negative, the momentum flux term becomes $\dot{M}_{c,e} u_E$ where u_E is the velocity on the east velocity node. Upwind differencing on the east face can be expressed more compactly as

$$\dot{M}_{c,e} u_e = \dot{M}_{c,e} [F_e u_P + (1 - F_e) u_E] \qquad (8.83)$$

where

$$F_e = 1 \qquad \text{if } \dot{M}_{c,e} > 0$$
$$F_e = 0 \qquad \text{if } \dot{M}_{c,e} < 0$$

Similarly for the momentum flux on the west face,

$$\dot{M}_{c,w} u_w = \dot{M}_{c,w} \left[F_w u_W + (1 - F_w) u_P \right] \qquad (8.84)$$

where F_w is defined in the same fashion as F_e. The same formulations can be used for the south and north faces. Substituting the momentum flux terms into Equation 8.82 and using the continuity equation yields the following equation for u_P.

$$u_P \sum_k F_k \dot{M}_{c,k} = \sum_k F_k \dot{M}_{c,k} u_1 + A_1 (p_w - p_e) + \sum_k A_k \tau_{1k} + \rho_c V g_1 \qquad (8.85)$$

Typically two additional terms, Sp_u and Su_u, are added to the momentum equation to account for boundary conditions and other effects.

$$u_P \left[\sum_k F_k \dot{M}_{c,k} - Sp_u \right] = \sum_k F_k \dot{M}_{c,k} u_k$$
$$+ A_1 (p_w - p_e) + \sum_k A_k \tau_{1k} + \rho_c V g_1 + Su_u \qquad (8.86)$$

These terms can also be used to incorporate the source terms for multiphase flows. This equation can be easily extended to three dimensions by including the other two faces of a six-sided computational cell.

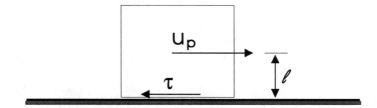

Figure 8.21: Computational cell adjacent to wall.

The application of the boundary conditions is illustrated in Figure 8.21 for a computational cell adjacent to a wall. There is no momentum flux through the wall so $\dot{M}_{cs} = 0$ but the force acting on the fluid due to shear stress on the wall, assuming laminar flow, is

$$F_\tau = -\mu \frac{u_p}{\ell} A_w \qquad (8.87)$$

where ℓ is the distance from the point to the wall and A_w is the area of the computational cell adjacent to the wall. For this cell, $S_{pu} = -\mu A_w / \ell$ and

$Su_u = 0$. Ordinarily, the values of Sp_u and Su_u are zero for computational cells not adjacent to the walls.

The pressure formulation of the continuity equation can now be set up in the same fashion as done for the quasi-one-dimensional model to provide an equation for pressure correction which goes toward satisfying the continuity equation. The continuity equation becomes

$$C_p \Delta p_p = \sum_k C_k \Delta p_k + \sum_k M_{c,k}^* \qquad (8.88)$$

where summation of the mass flows is the net efflux of mass from the continuity cell from the previous iteration. When continuity is satisfied, the term is zero.

The Rhie and Chow scheme also can be applied to multidimensional flows to obviate the need for a staggered grid.

For turbulent flow, additional equations are needed for the turbulent shear stresses. These depend on the models chosen such as the turbulence-dissipation model or Reynolds stress models. With the turbulence-dissipation model two additional equations are needed for the turbulence energy equation (k) and the turbulence dissipation equation (ε). These are also formulated using the continuity cell. The effective viscosity is calculated from the turbulence energy, k, and the turbulence energy dissipation, ε,.

$$\mu_{eff} \propto \rho \frac{k^2}{\varepsilon} \qquad (8.89)$$

The temperature field is determined by setting up the thermal energy equation for the same computational cell as used for continuity. Other equations, such as swirling velocity in an axisymmetric flow or mass fraction of a multicomponent mixture, are formulated in the same way using the continuity cell.

The solution procedure parallels that for the quasi-one-dimensional flow. First, the initial velocity and pressure distribution are assumed. Then, the momentum equations are solved for the velocity components. Next the pressure corrections are found from continuity and the velocity and pressure fields are corrected. If continuity is satisfied, the solution is complete, but this will generally not occur on the first iteration. The other equations for temperature, turbulence energy, dissipation and other variables (depending on the problem) will be solved, and the calculation returns to the momentum equations. Once continuity is satisfied to within a specified criterion, the calculation is terminated. One technique for solving the dependent variable equations is the application of the TDMA line-by-line throughout the field.

The above is a rudimentary description of the numerical scheme for multidimensional flows. There has been considerable work done to improve the accuracy and reduce numerical diffusion, to accelerate convergence and so on. Body fitted, nonorthogonal coordinate systems have been developed as well as nonstructured grids which consist of adjacent triangles. The reader is referred to more recent literature to become cognizant of the current state-of-the-art.

Multiphase flow

The same numerical technique for quasi-one-dimensional gas-particle flows carries over to multidimensional flows. Consider the two-dimensional computational cells with particle trajectories through the cell shown in Figure 8.22.

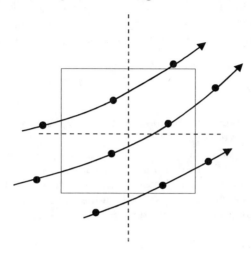

Figure 8.22: Particles passing through a two-dimensional computational cell.

The continuity equation for the carrier phase is now expressed as

$$\dot{M}_{c,n} + \dot{M}_{c,e} - \dot{M}_{c,s} - \dot{M}_{c,w} = S_{mass} \qquad (8.90)$$

The mass flux must include the volume fraction of carrier phase, α_c. For example, the mass flux on the east face is

$$\dot{M}_{c,e} = \alpha_{c,e} u_e \rho_e A_e \qquad (8.91)$$

The volume fraction of the particulate phase in each cell is found from (see
Chapter 7),

$$\alpha_d = \sum_p N_p V_{d,p}/V \qquad (8.92)$$

where the summation is carried out over all parcels in the cell, N_p is the number
of particles associated with each parcel and the $V_{d,p}$ is the volume of an individual particle in the parcel. The volume fraction of the carrier phase on the
east face would then be calculated from

$$\alpha_{c,e} = 1 - \frac{\alpha_{d,P} + \alpha_{d,E}}{2} \qquad (8.93)$$

Obviously for particle volume fractions the order of 10^{-3}, the volume fraction
of the carrier phase can be taken as unity.

The mass source term, given by Equation 7.16, is

$$S_{mass} = -\sum_p N_p \dot{m}_p \qquad (8.94)$$

where the summation is carried out over all parcels in the cell and $\dot{m}_p$ is the rate of change of mass of each droplet or particle associated with parcel p. For steady flows for which the trajectory approach is applicable, the mass source term is given by

$$S_{mass} = -\sum_{traj} \dot{n}_j \Delta t_j \overline{\dot{m}} \qquad (8.95)$$

where $\dot{n}_j$ is the number flow rate along trajectory j and Δt_j is the time it takes for the droplet to traverse the cell.

The momentum equation is the same as the single-phase flow equations except now the mass flux terms contain a continuous phase volume fraction (same as the continuity equation), the pressure gradient term is multiplied by a carrier phase volume fraction and a term must be added to account for the momentum coupling. The momentum coupling term for the u-momentum equation is

$$S_{mom,x} = V \sum_n N_n \left[\beta_{V,n}(v_{x,n} - u_p) - \dot{m}_p v_{x,n} \right] \qquad (8.96)$$

where $v_{x,n}$ is the velocity component of droplet in parcel n in the x-direction. The carrier flow momentum equation now assumes the form,

$$u_p \left[\sum_k F_k \dot{M}_{c,k} - Sp_u \right] = \sum_{n,e,s,w} f_i \dot{M}_{ci} u_i$$

$$+ \alpha_{c,p} A_1 (p_w - p_e) + \alpha_c \sum_k A_k \Delta_k \tau_{1k} + \alpha_c \rho_c V g_1 + Su_u \qquad (8.97)$$

where the Sp and Su terms are

$$Sp_u = -V \sum_n N_n \beta_{V,n} + Sp'_u$$

$$Su_u = V \sum_n (\beta_{V,n} v_{x,n} - \dot{m}_d v_{x,n}) + Su'_u \qquad (8.98)$$

and Sp'_u and Su'_u are the values before the addition of the second phase to account for boundary conditions and other flow features.

The pressure formulation of the continuity equation is

$$C_p \Delta p_p = \sum_{n,e,s,w} C_i \Delta p_i + \sum_{n,e,s,w} M^*_{c,i} + S_{mass} \qquad (8.99)$$

and continuity is satisfied when $\sum_k M^*_{c,k} + S_{mass} = 0$.

Equivalent equations can be developed for the thermal energy of the carrier phase. Equations are also formulated for the turbulence kinetic energy and dissipation. All the equations, except the momentum equations, use the continuity cell as the computational volume.

The calculations for the dispersed phase are initiated by identifying the starting positions of many particles or parcels of particles. The position of the particle parcels is found by integrating the equation of motion for the dispersed phase. These equations can incorporate a turbulence dispersion through a stochastic process as discussed in Section 7.1.4. Also, if the particles impact the wall the equations for particle-wall interaction are implemented to find the rebound velocity and angle, as well as the rotational rate. In some circumstances, the particles may impact and remain on the wall. Also the particle temperature and size (if important) are calculated for each step. Finding the number of parcels in a cell at a given time or the residence time of a particle in a cell is used to evaluate the source terms.

The procedure for solution is the same as for the single-phase flow except now a subroutine has to be included to calculate parcel position and evaluate the source terms for every computational cell. As before, convergence is achieved when the continuity equation is satisfied.

Although the above equations were developed for steady flows, unsteady flows can be easily included by including an accumulation term in the conservation equations. However, it is likely advisable to use a time-splitting procedure to have a more accurate representation of the effect of the source terms.

The Lagrangian method for multiphase flow has been used in a wide diversity of applications. A few of the applications are

- Coal fired furnaces - Hill and Smoot (1993), Kaufman and Fiveland (1995), Fischer et al. (1996)

- Spray dryers - Dilber and Mereu (1996), Oakley and Bahu (1991), Crowe et al. (1985)

- Air pollution control - Griffiths and Boyson (1996), Kim and Crowe (1984), Ushimaru and Butler (1984)

- Fire suppression - Nam (1995)

- Materials processing - Muoio et al. (1996)

- Material erosion - McLaury et al. (1996), Tu et al. (1996), Hamed and Tabakoff (1996), Founti and Kipfel (1996), Tu et al. (1996).

- Aircraft icing - Valentine and Decker (1994)

- Propulsion system - Giridharan et al. (1992)

The Lagrangian method is attractive to industrial applications because of its simplicity and robustness. The major problem is the number of trajectories needed to represent the particle field and the corresponding computer time. Various attempts are being made to alleviate these shortcomings by using parallel processing (Frank and Wassen, 1996) and the use of parcel packets which expand to model the dispersion process (Chen and Pereira, 1996). Some work is also being done to extend the approach to near dense flows (Founti and Klipfel, 1996).

The Lagrangian approach has also found application to bubbly flows (Murai and Matsumoto, 1996).

8.2.3 Direct numerical simulation models

Flow models based on direct numerical simulation are direct solutions of the Navier-Stokes equations with no assumptions concerning Reynolds stress. This means that the grid structure or analytic scheme must be capable of resolving the smallest scale of turbulence, the Kolmolgorov length scale. The Kolmolgorov length scale can be the order of 100 microns which means that the numerical scheme must be capable of resolving a wide range of length scales. For this reason DNS methods are limited to low Reynolds numbers where the range of length scales is manageable. Frequently spectral methods are used in which the flow parameters are described by a Fourier series. Spectral methods provide the desirable accuracy. Modes with sufficiently high frequency have to be included to resolve the Kolmolgorov length scale. Those interested in direct numerical simulations should refer to Hussaini and Zhang (1987).

Multiphase flow analysis using the DNS models utilizes the discrete element approach. The velocity used to integrate the equation of particle motion is obtained by interpolation. Squires and Eaton (1991) have used DNS to track particles in homogenous, stationary turbulence fields. Elghobashi and Truesdell (1992) have addressed particle dispersion in decaying turbulence fields downstream of a grid using DNS. More recently Tong and Wang (1997) and Ling et al. (1997) have used DNS to model particle dispersion in three dimensional mixing layers. The incorporation of particle-particle collisions in a DNS formulation has been reported in a series of two papers by Wang et al. (1997) and Zhou et al. (1997).

In order to include two-way coupling effects the particles are considered to be point forces acting on the fluid. Squires and Eaton (1990) predict that the presence of particles is responsible for an attenuation of turbulence energy. They also demonstrated a tendency for particles to concentrate in regions of high strain and low vorticity. Elghobashi and Truesdell (1993) also addressed two-way coupling effects and predicted that turbulence energy would increase at the high wave numbers. There is some controversy (Liljegren, 1996) concerning the validity of using a point force to simulate the effects of a particle on turbulence. The concern is that a point force does not represent the fluid condition at the particle surface and does not generate a wake.

8.2.4 Large eddy simulation (LES)

Large eddy simulation circumvents the problem of solving the turbulence field over the spectrum of length scales by solving a filtered form of the continuity and momentum equations for the larger scales of turbulence and using an eddy viscosity model for the smaller scale. The flow field generated by this scheme is used to calculate particle trajectories and dispersion patterns. Wang and Squires (1996a, 1996b) have applied it to particle dispersion in a three dimensional mixing layer and flow in a channel. There appears to be no attempt to the present time to apply LES to predict turbulence modulation.

8.2.5 Discrete vortex methods

In the discrete vortex method, the flow is represented by an array of finite vortices which mutually interact as the flow develops. The velocity at each vortex is the cumulative sum of the velocities induced by all the other vortices. In order to avoid the singularities associated with the vortex centers each vortex is given a finite diameter core with a specified velocity distribution. The method works well for simulating the large-scale structures where Reynolds number effects are not important. Particle motion calculations can be carried out by using the velocity induced by the vortices at the particle location. This scheme has been used for plane mixing layers by Chien and Chung (1988), Martin and Meiburg (1994) and Ory and Perkins (1997), for axisymmetric jets by Chung and Troutt (1988) and for bluff body wakes by Tang et al. (1992). These studies show the tendency for the particles to concentrate on the peripheries of the vortices when the Stokes number is the order of unity.

Two-way coupling effects have also been incorporated into the discrete vortex method. For this application, the cloud-in-cell method (Christiansen, 1973) is used in which the vorticity from the discrete vortices is distributed onto a grid and the stream function-vorticity equations are solved for the flow field. The velocities at the discrete vortices are interpolated from the calculated values at the grid nodes and used to translate the vortices over the next time step. The vorticity generated by the particle drag is also accumulated on the grid nodes and used in the stream function calculation. Tang et al. (1989) show that the presence of the particles slows the development of the shear layer. The same approach to include mass and energy coupling effects has been used by Bakkom et al. (1996).

8.3 Dilute flow - two-fluid model

The two-fluid model has also been applied to dilute flows. In this case the equation for the dispersed phase are the differential equations for a continuum and summarized in Table 7.3. A numerical scheme for a quasi-one-dimensional flow will be presented first and then a more general multidimensional model.

8.3.1 Quasi-one-dimensional flow

Once again, consider the steady flow of a fluid particle mixture through the nozzle shown in Figure 8.3. The continuity and momentum equations for the carrier fluid are presented in Table 6.1. The continuity equation for steady quasi-one-dimensional flow through a duct is

$$\Delta(\alpha_c \rho_c u A) = S_{mass} \tag{8.100}$$

It is common practice in two-fluid modeling to express the continuity equation in terms of volume flow rate. By differentiating the continuity equation

$$\rho_c \Delta(\alpha_c u A) + \alpha_c u \bar{A} \Delta \rho_c = S_{mass} \tag{8.101}$$

and dividing by the carrier phase material density, one has

$$\Delta(\alpha_c u A) = \frac{S_{mass}}{\rho_c} - u \bar{A} \frac{\alpha_c}{\rho_c} \Delta \rho_c \tag{8.102}$$

Now one notes that the source term consists of the volume added due to mass transfer and the change of volume associated with the change of density of the carrier phase. The latter term must be evaluated using the equation of state and the pressure and temperature variation through the nozzle. If the carrier fluid is incompressible, the last term vanishes. Thus, for incompressible flow through a nozzle with nonreacting particles, the continuity equation for the carrier phase reduces to

$$(\alpha_c u A)_{i+1} - (\alpha_c u A)_i = 0 \tag{8.103}$$

The steady-state formulation for the continuity equation for the dispersed phase (Equation 7.42) is

$$\Delta(\alpha_d \rho_d v A) = -S_{mass} \tag{8.104}$$

where, for now, it is assumed that all the particles have the same velocity. The volume flow rate form of the equation for constant material density is

$$\Delta(\alpha_d v A) = -\frac{S_{mass}}{\rho_d} \tag{8.105}$$

With no mass transfer, the particle phase continuity equation becomes

$$(\alpha_d v A)_{i+1} - (\alpha_d v A)_i = 0 \tag{8.106}$$

The momentum equation for the carrier phase for a steady quasi one-dimensional flow is

$$\dot{M}_c(u_{i+1} - u_i) = (\alpha_c A)_{i+1}(p_i - p_{i+1})$$

$$\Delta x A_{i+1} \beta_{V,i+1}(v_{i+1} - u_{i+1}) - \Delta x(f_c \alpha_c \rho_c u^2 A / 2 D_H)_{i+1} \tag{8.107}$$

where m is the mass of an individual particle. This equation can be solved for u_{i+1} in the form

$$u_{i+1} = \frac{1}{E_{c,i+1}} \left[\dot{M}_c u_i + (\alpha_c A)_{i+1}(p_i - p_{i+1}) + \Delta x A_{i+1} \beta_{V,i+1} v_{i+1} \right] \tag{8.108}$$

where

$$E_{c,i+1} = \left[1 + (f_c \frac{\Delta x}{2D_H})_{i+1}\right] \dot{M}_c + \Delta x A_{i+1}\beta_{V,i+1}$$

For convenience, this equation will be expressed as

$$u_{i+1} = Q_{u,i+1} + \frac{(\alpha_c A)_{i+1}}{E_{c,i+1}}(p_i - p_{i+1}) \tag{8.109}$$

The momentum equation for the particulate phase is

$$\dot{M}_d(v_{i+1} - v_i) = (A\alpha_d)_{i+1}(p_i - p_{i+1})$$

$$+A_{i+1}\Delta x \beta_{V,i+1}(u - v)_{i+1} - \rho_d \left(f_s \alpha_d v^2 A \frac{\Delta x}{2D_H}\right)_{i+1} \tag{8.110}$$

where the body forces and the unsteady forces on the particles are not included. It is also assumed that all the particles move with the same velocity. This equation can be rewritten for v_{i+1} in the same form as Equation 8.108; namely

$$v_{i+1} = \frac{1}{E_{d,i+1}}\left[\dot{m}_d u_i + (\alpha_d A)_{i+1}(p_i - p_{i+1}) + \Delta x A_{i+1}\beta_{V,i+1}u_{i+1}\right] \tag{8.111}$$

where

$$E_{d,i+1} = \left[1 + (f_s \frac{\Delta x}{2D_H})_{i+1}\right]\dot{M}_d + \Delta x A_{i+1}\beta_{V,i+1}$$

and will be expressed here in abbreviated form as

$$v_{i+1} = Q_{v,i+1} + \frac{(\alpha_d A)_{i+1}}{E_{d,i+1}}(p_i - p_{i+1}) \tag{8.112}$$

There are five dependent variables: namely, α_c, α_d, u, v, and p and four equations. The final equation to complete the set is the sum of the volume fractions equals unity,

$$\alpha_c + \alpha_d = 1 \tag{8.113}$$

This equation is used to develop a pressure formulation for the continuity equation. The volume fraction for the carrier phase at station $i+1$ can be expressed from Equation 8.103 as

$$\alpha_{c,i+1} = \frac{(\alpha_c uA)_i}{(uA)_{i+1}} \tag{8.114}$$

The volume fraction can be expanded through a Taylor series to provide a new value as a function of the velocity change by

$$\alpha_{c,i+1} = \alpha_{c,i+1}^* + \frac{\partial \alpha_{c,i+1}}{\partial u_i}\Delta u_i + \frac{\partial \alpha_{c,i+1}}{\partial u_{i+1}}\Delta u_{i+1} \tag{8.115}$$

which, using Equation 8.114, becomes

$$\alpha_{c,i+1} = \alpha^*_{c,i+1} + \frac{(\alpha_c A)_i}{(uA)_{i+1}}\Delta u_i - \frac{(\alpha_c uA)_i}{(u^2 A)_{i+1}}\Delta u_{i+1} \tag{8.116}$$

The changes in the velocity u as a function of pressure can be obtained from the momentum equation. From Equation 8.109, one can write the change in u_{i+1} as

$$\Delta u_{i+1} = \frac{(\alpha_c A)_{i+1}}{E_{c,i+1}}(\Delta p_i - \Delta p_{i+1}) \tag{8.117}$$

and, correspondingly the change in u_i is

$$\Delta u_i = \frac{(\alpha_c A)_i}{E_{c,i}}(\Delta p_{i-1} - \Delta p_i) \tag{8.118}$$

Substituting Equations 8.117 and 8.118 into Equation 8.116 yields the following expression for the carrier phase volume fraction as a function of pressure change

$$\alpha_{c,i+1} = \alpha^*_{c,i+1} + \frac{(\alpha_c A)_i}{(uA)_{i+1}}\frac{(\alpha_c A)_i}{E_{c,i}}(\Delta p_{i-1} - \Delta p_i)$$
$$-\frac{(\alpha_c uA)_i}{(u^2 A)_{i+1}}\frac{(\alpha_c A)_{i+1}}{E_{c,i+1}}(\Delta p_i - \Delta p_{i+1}) \tag{8.119}$$

or in a more compact form

$$\alpha_{c,i+1} = \alpha^*_{c,i+1} + A_{c,i+1}\Delta p_{i-1} + B_{c,i+1}\Delta p_i + C_{c,i+1}\Delta p_{i+1} \tag{8.120}$$

The identical procedure can be carried out for the dispersed phase velocity change

$$\Delta v_{i+1} = \frac{(\alpha_d A)_{i+1}}{E_{d,i+1}}(\Delta p_i - \Delta p_{i+1}) \tag{8.121}$$

and the equation for the dispersed phase volume fraction becomes

$$\alpha_{dc,i+1} = \alpha^*_{d,i+1} + \frac{(\alpha_d A)_i}{(vA)_{i+1}}\frac{(\alpha_d A)_i}{E_{d,i}}(\Delta p_{i-1} - \Delta p_i)$$
$$-\frac{(\alpha_d vA)_i}{(v^2 A)_{i+1}}\frac{(\alpha_d A)_{i+1}}{E_{d,i+1}}(\Delta p_i - \Delta p_{i+1}) \tag{8.122}$$

Finally, adding Equations 8.119 and 8.122 and setting the sum equal to unity gives

$$1 - \alpha^*_{c,i+1} - \alpha^*_{d,i+1} = L_i\Delta p_{i-1} - (L_i + U_i)\Delta p_i + U_i\Delta p_{i+1} \tag{8.123}$$

where

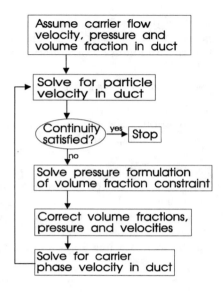

Figure 8.23: Flow diagram for two-fluid model.

$$L_i = \frac{(\alpha_c A)_i}{(uA)_{i+1}} \frac{(\alpha_c A)_i}{E_{c,i}} + \frac{(\alpha_d A)_i}{(vA)_{i+1}} \frac{(\alpha_d A)_i}{E_{d,i}}$$

and

$$U_i = \frac{(\alpha_c uA)_i}{(u^2 A)_{i+1}} \frac{(\alpha_c A)_{i+1}}{E_{c,i+1}} + \frac{(\alpha_d vA)_i}{(v^2 A)_{i+1}} \frac{(\alpha_d A)_{i+1}}{E_{d,i+1}}$$

This equation is the pressure formulation of the continuity (or volume fraction) equation. One notes that if the volume fractions add to unity, then the error is zero and the pressure changes are also zero. Even though the equations are long and cumbersome, the implementation is straightforward.

The solution procedure is shown by the flow diagram in Figure 8.23. First, values for the pressure, carrier phase velocity and volume fraction are assumed through the duct. It is best to assume a velocity and volume fraction distribution which satisfy continuity. Next, the dispersed phase momentum equation is solved for the velocity of the particles. The pressure formulation of the volume fraction equation, Equation 8.123, is used to solve for the pressure. The pressures are corrected as well as the phase velocities from Equations 8.118 and 8.121. The volume fraction of both phases are corrected using Equations 8.116 and 8.122. The momentum equations from both phases are then solved for new velocities, and the procedure returns to the pressure formulation of the volume fraction constraint. When the error in volume fraction is reduced to a specified level, the calculation is terminated.

If there is a distribution of particle sizes, then a volume fraction has to be assigned to each size category and a separate momentum equation must be solved for each size. Then the volume fraction equation becomes

$$\alpha_c + \sum_k \alpha_{d,k} = 1 \tag{8.124}$$

where $\alpha_{d,k}$ is the volume fraction associated with size class k. This may introduce significant numerical errors in calculating the volume fraction in the continuity equation.

The above method has to be modified for dispersed phase flows in which the volume fraction of the dispersed phase is of the order of 10^{-3}. In this case, the scheme is inoperative since there would be significant error in subtracting the sum of the volume fractions from unity. In order to resolve this problem, the mixture volume fraction equation is used. The error in the volume flow rate (excluding mass coupling) between stations i and $i+1$ would be

$$\Delta Q = (uA\alpha_c)_{i+1} + (vA\alpha_d)_{i+1} - (uA\alpha_c)_i - (vA\alpha_d)_i \tag{8.125}$$

The volume fraction of the continuous phase is set equal to unity so the equation is rewritten as

$$\Delta Q = (uA)_{i+1} + (vA\alpha_d)_{i+1} - (uA)_i - (vA\alpha_d)_i \tag{8.126}$$

Using Equations 8.118 and 8.121 to relate the velocities to the change in pressure, this equation becomes

$$\begin{aligned}
\Delta Q = A_{i+1} &\left[u^*_{i+1} + \tfrac{A_{i+1}}{E_{c,i+1}} (\Delta p_i - \Delta p_{i+1}) \right] \\
- A_i &\left[u^*_i + \tfrac{A_i}{E_{c,i}} (\Delta p_{i-1} - \Delta p_i) \right] \\
+ (A\alpha_d)_{i+1} &\left[v^*_{i+1} + \tfrac{(\alpha_d A)_{i+1}}{E_{d,i+1}} (\Delta p_i - \Delta p_{i+1}) \right] \\
- (A\alpha_d)_i &\left[v^*_i + \tfrac{(\alpha_d A)_i}{E_{d,i}} (\Delta p_{i-1} - \Delta p_i) \right]
\end{aligned} \tag{8.127}$$

Setting $\Delta Q = 0$, this equation can be reformulated in the form of Equation 8.123 and solved for the pressure correction using the values from the last iteration. The dispersed phase volume fraction can be determined from Equation 8.122 or from the continuity equation for the dispersed phase. Once again, if there are several particle sizes, a separate volume fraction has to be assigned to each size.

Obviously, as the volume fraction of the dispersed phase becomes very small, the coefficients on the pressure terms become small. The equation reduces to

$$\begin{aligned}
\Delta Q - (Au^*)_{i+1} + (Au^*)_i - (A\alpha_d v^*)_{i+1} + (A\alpha_d v^*)_i = \\
\tfrac{A^2{}_i}{E_{c,i}} \Delta p_{i-1} - \left(\tfrac{A^2{}_i}{E_{c,i}} + \tfrac{A^2{}_{i+1}}{E_{c,i+1}} \right) \Delta p_i + \tfrac{A^2{}_{i+1}}{E_{c,i+1}} \Delta p_{i+1}
\end{aligned} \tag{8.128}$$

The coefficients of the pressure terms can be very small, which can introduce some difficulties with the numerical scheme.

8.3.2 Multidimensional flow

The multidimensional Eulerian-Eulerian approach carries over directly from the quasi-one-dimensional model presented above. The continuity equation is written in terms of a volume flow rate instead of a mass flow rate and the pressure formulation of the volume conservation equation is used. Also, as indicated in the quasi-one-dimensional model, the scheme has to be modified for dispersed phase flows in which the volume fraction of the dispersed phase is of the order of 10^{-3}.

A well-known numerical scheme for the two-fluid formulation was developed by Harlow and Amsden (1975). This scheme has been the basis for the subsequent development of several other codes. The code was developed for two gaseous phases and two condensed phases and included the continuity, momentum and energy equations. For purposes of simplicity, only the continuity and momentum equations are considered here. Also mass coupling will not be included.

The continuity and momentum equations for the continuous and dispersed phases are given in Chapters 6 and 7. They are

$$\frac{\partial \bar{\rho}_d}{\partial t} + \frac{\partial}{\partial x_i}(\bar{\rho}_d v_i) = 0 \tag{8.129}$$

$$\frac{\partial \bar{\rho}_c}{\partial t} + \frac{\partial}{\partial x_i}(\bar{\rho}_c u_i) = 0 \tag{8.130}$$

$$\frac{\partial}{\partial t}(\bar{\rho}_d v_i) = -\frac{\partial}{\partial x_j}(\bar{\rho}_d v_i v_j) - \alpha_d \frac{\partial p}{\partial x_i} + \beta_V(u_i - v_i) + T_{d,i} \tag{8.131}$$

$$\frac{\partial}{\partial t}(\bar{\rho}_c u_i) = -\frac{\partial}{\partial x_j}(\bar{\rho}_c u_i u_j) - \alpha_c \frac{\partial p}{\partial x_i} + \beta_V(v_i - u_i) + T_{c,i} \tag{8.132}$$

where the body forces have been neglected and T_i are the forces due to viscous stress. A staggered grid coordinate system is used where the velocities are defined on the cell boundaries and the other properties at the cell center. The procedure is as follows. First, the bulk phase density is calculated at a new time level using Equation 8.129,

$$\bar{\rho}_d^{n+1} = \bar{\rho}_d^n - \left[\frac{\Delta}{\Delta x_i}(\bar{\rho}_d v_i)\right]^n \delta t = 0 \tag{8.133}$$

where Δ signifies the finite difference formulation. Then the solids phase volume fraction can be found from $\alpha_d = \bar{\rho}_d/\rho_d$ and the continuous phase volume fraction from $\alpha_c = 1 - \alpha_d$. The momentum equations are integrated in two steps. First an intermediate value of mass flux is obtained from

$$(\widehat{\bar{\rho}_c u_i})^{n+1} = (\bar{\rho}_c u_i)^n - \left[\frac{\Delta}{\Delta x_j}(\bar{\rho}_c u_i u_j) + T_{c,i}\right]^{n+1} \delta t \tag{8.134}$$

The same procedure is repeated for the solids momentum equation, Equation 8.131. For the second step, the mass flux becomes

$$(\bar{\rho}_c u_i)^{n+1} = (\widehat{\bar{\rho}_c u_i})^{n+1} - \left[\alpha_c \frac{\Delta p}{\Delta x_i} - \beta_V (v_i - u_i) \right]^n \delta t \qquad (8.135)$$

This equation can be rewritten to give an explicit formulation for u_i^{n+1}. The same procedure is used to find v_i^{n+1}. The pressure is updated by returning to Equation 8.130 and expressing it in finite difference form as

$$D^{n+1} = \frac{\bar{\rho}_c^{n+1} - \bar{\rho}_c^n}{\delta t} + \left[\frac{\Delta}{\Delta x_i} (\bar{\rho}_c u_i) \right]^{n+1} \qquad (8.136)$$

The correction in pressure must be such that $D^{n+1} \to 0$. This is carried out by forming the derivatives $\partial D / \partial p$ and using the Newton-Raphson method to evaluate δp such that this condition is achieved. This is analogous to the SIMPLE procedure. Finally $\bar{\rho}_c$ is calculated using Equation 8.130. Harlow and Amsden applied these scheme to several examples of dispersed phase flows.

A major problem with the Eulerian-Eulerian model is the need to select an effective viscosity for the dispersed phase and to correctly model the boundary conditions. For a dilute flow, some provision must be made for the dispersed phase Reynolds stress arising from the distribution of particle velocities. Also, there is no straightforward way to represent the case where the particles, or droplets, impinge on the wall and stay. The question also arises as to the value of the velocity at the wall. A fluid requires that the normal and tangential components of velocity be zero at the wall. However, the case of particles impacting and sticking on a wall would not give rise to a zero normal velocity. Also, particles bouncing from a wall cannot be modeled as a zero tangential velocity. The tangential velocity at the wall is sometimes regarded as a *slip velocity* as used in modeling the flow of rarefied gases. The most common approach to select an effective viscosity for the dispersed phase is to relate it by a factor to the viscosity of the carrier phase. Equivalent problems are encountered in establishing boundary conditions for the Eulerian particle energy equation.

The two-fluid formulation has been applied to the design of an engine inlet to protect against damage due to dust (Thomas & Dionne, 1994). The two-fluid model has also been used for analysis of two-phase rocket nozzle flows and nozzle plumes (Thorpe et al., 1979). In this case the continuous phase equations are expressed in terms of conservative variables to avoid the discontinuities at shock waves[6] and the particle flow equations are written in Eulerian form. Both the continuous phase and dispersed phase equations can be integrated using the same algorithm.

The two-fluid model has also been used extensively in bubbly flows. An example application is the gas-liquid flow in a tee junction (Issa and Oliveira, 1994).

Although there are several issues involving the constitutive equations for the two-fluid model, it does have the advantage that clouds of particles can be modeled without requiring excessive computational time. Also the method is more adaptable to body-fitted coordinate systems and unstructured grids.

---

[6] This is referred to as a *shock-capture model*.

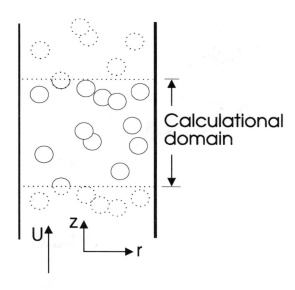

Figure 8.24: Particle transport in a vertical pipe.

8.4   Dense flows - Lagrangian approach

Dense flows describe the situation where the particle motion is controlled by particle-particle collision. The discrete element formulation of the Lagrangian approach is applicable in this situation. Basically there are two categories describing the application of discrete elements for modeling dense phase flows depending on the nature of the particle-particle interaction. If the particle interaction is only by collision the hard sphere model can be used. This condition may occur in a fast fluidized bed or riser flow. This case will be identified as *collision dominated* dense phase flow. However, if the particle motion is primarily controlled by the particle-particle contact, such as in a fluidized bed where several particles are in contact simultaneously, the flow will be identified as *contact dominated* dense phase flow. In this case, the soft sphere model must be used to model the particle dynamics.

8.4.1   Collision-dominated flows

One of the earlier numerical models for pneumatic transport in vertical tubes was reported by Tanaka and Tsuji (1991). The gas-particle flow field they considered is shown in Figure 8.24. In order to maintain a steady flow, they assumed that when one particle left the flow field another particle would enter at the same position in the inlet plane and with the same properties. In this work the number density of the particles was such that the time step was small compared to the average time between particle-particle collisions. It was also assumed that the presence of the particles did not affect the velocity of the carrier fluid.

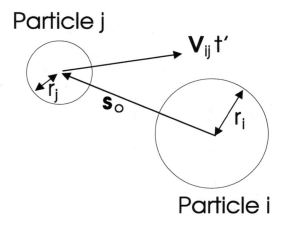

Figure 8.25: Particle collision model.

The calculation was carried out first calculating the particle velocity of every particle in the field and assuming that this velocity persisted over the time step Δt. If any two particles collided during this time step, new velocities and rotational vectors would be calculated and for each pair and would be used in the next time step. The collision model is shown in Figure 8.25. The relative velocity between two particles, i and j, would be $\mathbf{V}_{ij}$ so the path that particle i would follow with respect to particle j would be

$$\mathbf{s} = \mathbf{s}_0 + \mathbf{V}_{ij}t' \tag{8.137}$$

where $\mathbf{s}_0$ is the initial displacement vector between the two particles and t' is elapsed time from the beginning of the time step. A collision will occur if

$$|\mathbf{s}|^2 = (r_i + r_j)^2 \tag{8.138}$$

where r_i and r_j are the radii of the two particles. Substituting Equation 8.137 into Equation 8.138, one can solve for t'. This is a quadratic equation. If there are no real roots, there is no collision. If there are two real roots, the smaller of the roots is taken as the solution. If the value of t' exceeds Δt, there is no collision during the time step. However if there is a solution, then the position of the two particles at contact can be determined and the equation for new velocities and rotational rates can be found to begin the next time step (see Section 5.2). Provision is also made for collision with the wall.

In the work by Tanaka and Tsuji (1991), the motion of every particle was considered. This becomes computationally intractable as the number of particles increases to model actual systems. Another approach is the *direct simulation Monte Carlo method* (DSMC) which was originally proposed by Bird (1976) for simulation of rarefied flows. In this method all the particles in the field are simulated by N computational particles which is the same concept described earlier as a parcel of physical particles. As before, the time interval for computation is chosen to be small compared to the average time between collisions. The probability of collision for particle i is given by

$$P_i = \sum_{j=1}^{N} P_{ij} \qquad (8.139)$$

where P_{ij} is the probability of collision between particle i and j in time step Δt and is given by

$$P_{ij} = \frac{n}{N}\pi d^2 V_{ij} \Delta t \qquad (8.140)$$

where n is the number density of the physical particles and V_{ij} is the relative speed between the particles. This analysis is based on uniform size particles. The occurrence of collision is established by the modified Nanbu scheme (Iller and Neunzert, 1987). A uniformly distributed random number between zero and unity, R, is generated to select a collision partner, k, by

$$k = [RN] + 1 \qquad (8.141)$$

where [ ] means the integer part of the number. Particle i collides with particle k if

$$R > \frac{k}{N} - P_{ik} \qquad (8.142)$$

The velocity of particle i is changed according to the hard sphere equations developed in Chapter 5 while the velocity of particle k is unchanged. The geometry of the collision is also selected using random numbers.

Tanaka et al. (1995) used the above method to calculate dispersed gas-solid flows in a vertical channel. The predictions for the particle concentration, particle and gas velocities are shown in Figure 8.26. One notes the formation of clusters which are the result of repeated particle-particle collisions. One also notes the recirculatory, nonuniform nature of the conveying and particle flow fields. A similar more recent study has also been reported by Tanaka et al. (1996).

8.4.2 Contact-dominated flows

For dense flows such as fluidized beds, there may be many interactions so a more detailed approach is needed (Tsuji et al., 1993). The force acting on a particle is the sum of the particle contact forces and the fluid dynamic forces,

$$\mathbf{F} = \mathbf{F}_f + \mathbf{F}_c \qquad (8.143)$$

The translational acceleration is

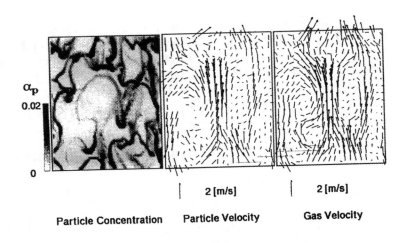

Figure 8.26: Particle concentration, particle velocity and gas velocity in a flow field with clusters. (Tanaka et al., *ASME FED* Vol. 228, 301, 1995. With permission.)

$$\mathbf{v} = \frac{\mathbf{F}}{m} + \mathbf{g} \qquad (8.144)$$

and the rotational acceleration is

$$\boldsymbol{\omega} = \frac{\mathbf{T}}{I} \qquad (8.145)$$

where $\mathbf{T}$ is the torque acting on the particle due to particle-particle or particle-wall contact.

The fluid forces acting on the particle are due to the pressure gradient and the relative velocity between the fluid and the particle. Typically the Ergun equation as discussed in Chapter 4 is used for the force on the particle for fluid volume fractions less than 0.8 and Wen's equation for the rest of the regime.

The carrier flow equations are the same as presented earlier in this chapter. Since the density of the fluid phase is assume constant and there is no mass transfer, the equation of continuity is

$$\frac{\partial \alpha_c}{\partial t} + \frac{\partial}{\partial x_i}(\alpha_c u_i) = 0 \qquad (8.146)$$

and the momentum equation

$$\alpha_c \frac{\partial u_i}{\partial t} + \alpha_c u_j \frac{\partial u_i}{\partial x_j} = -\alpha_c \frac{\partial p}{\partial x_i} + f_{d,i} \qquad (8.147)$$

where $f_{d,i}$ is the force of the particles acting on the fluid. The viscous forces are not included since it is assumed that the particle forces predominate.

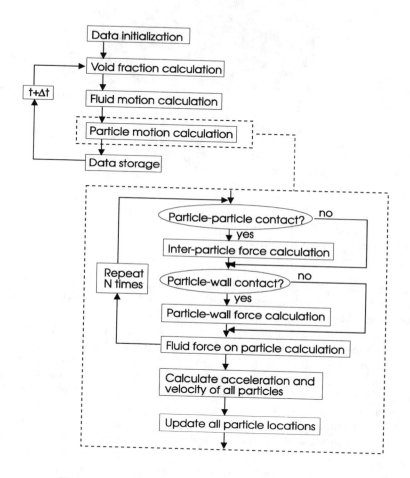

Figure 8.27: Flow diagram for dense phase flow.

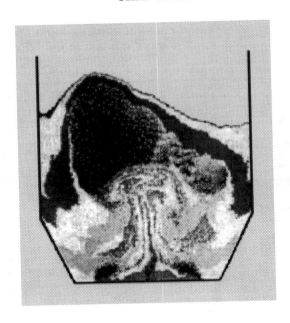

Figure 8.28: Particle field distribution in a two-dimensional fluidized bed using the Eulerian-Lagrangian approach.

The flow field is divided into a series of cells such that the cell size is smaller than the bubbles which appear in a fluidized bed yet larger than the particle dimensions. The flow chart for the numerical scheme is shown in Figure 8.27. First, the volume fraction of the carrier phase is determined by summing the volume of particles in a computational cell, dividing by the cell volume to obtain the particle volume fraction and subtracting from unity to find α_c. The fluid flow equations are then solved to find the velocity in each computational cell at a new time level $t + \Delta t$. The particle motion is then calculated as shown. For particles in contact, the soft sphere model developed in Chapter 5 is used to calculate the force. Basically, the force on every particle is calculated and all the particles are allowed to move simultaneously to a new position. After the data are saved, a new cycle for a new time level is implemented. In order to adequately model the particle motion during contact, small time intervals are needed which leads to long computational times.

A snapshot of a particle flow pattern in a two-dimensional fluidized bed obtained using the collision-dominated formulation for a dense flow is shown in Figure 8.28. One can see the formation of the bubbles on the bottom of the bed and the recirculatory pattern in the bed.

This approach has been used for several other fluidized bed simulations: Tsuji et al. (1993) and Kawaguchi et al. (1995). The model has also been applied to dense phase transport in a horizontal pipe (Tsuji et al., 1992).

This approach is attractive because it models particle-particle contact at the fundamental level. It also includes particle rotation which is, no doubt, important in horizontal flows. The primary shortcoming is the time required to execute a simulation.

8.5   Dense flows - Eulerian approach

The equations for the Eulerian-Eulerian approach have been presented in Chapter 7. In order to use the two-fluid model, relations for the viscosity of the dispersed are needed. These values are usually taken from data which have been used to correlate experimental observations. Also a solids stress modulus is needed which yields the force due to expansion or contraction of the solids mixture. This has also been based on some empirical evaluation.

Numerical solutions to the one-dimensional dense flow equations can be obtained by integrating the equations presented in Chapter 7 using empirical relations for the solids bulk modulus. Solutions are difficult, however, for a distribution of particle sizes. Recently Andrews and O'Rourke (1996) have proposed a scheme in which different sizes are represented by a parcel. Each parcel moves separately but the properties are mapped onto the grid for the continuum flow where the derivatives can be evaluated. These derivatives are then mapped back to the individual parcels to continue the calculation for particle motion.

Several applications of the two-fluid model for multidimensional dense phase flows have appeared in the literature. Tsuo & Gidaspow (1990) have adapted a version of Harlow and Amsden's scheme developed by Rivard and Torry (1977) to model flow in a fluidized bed. Their numerical simulations showed the formation of the bubbles and propagation through the bed. They used empirical form of the bulk solids modulus introduced in Chapter 7.

He and Simonin (1993) have applied the two-fluid model to vertical pneumatic conveying. They used transport equations for the particle phase Reynolds stress[7] which had been developed previously (Simonin, 1991). They compared their results with the simulations of Tanaka and Tsuji (1991) and observed that high level of anisotropy in the particle velocity fluctuations was adequately predicted with their model.

A more recent innovation is the use of kinetic theory to obtain the solids viscosity and stress modulus. The principle idea is the introduction of a granular temperature which is the kinetic energy associated with particle oscillation. An equation is developed based on kinetic theory for the granular temperature which relates the change in particle temperature to diffusion, dissipation and production. The transport properties are then related to the granular temperature. The Rivard and Torrey code was updated to incorporate the kinetic theory formulation and applied to fluidized beds and flow in risers (Gidaspow, 1994). A prediction of the bubble in fluidized bed based on the kinetic theory two-fluid model is shown in Figure 8.29. In this case the bubble is just approaching the surface. This calculation was done using the Gidaspow model in the FLUENT package.

---

[7] They refer to the particle phase Reynolds stress as the "particle phase kinetic stress".

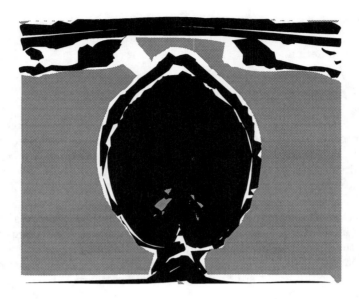

Figure 8.29: Bubble rising in a fluidized bed predicted using two-fluid model.

Balzer et al. (1995) have also reported the results of a two-fluid model for a dense fluidized bed utilizing results from kinetic theory for granular media. They found that the results were sensitive to the value for the coefficient of restitution; for values less than 0.9 the granular temperature decayed.

The introduction of kinetic theory removes some of the empiricism from the model but parameters such as the coefficient of restitution are needed which can have a non-negligible effect on the predictions. The approach does not include the detailed fluid mechanics associated with particle-particle and particle wall collisions. This may not be a serious shortcoming in gas-particle flows but will play a role in liquid-solid flows where the stopping distance of a particle is the order of the interparticle spacing. The kinetic theory developed to this point does not include torque and particle rotation resulting from collisions which should be important in horizontal transport applications.

8.6   Summary

The numerical simulation of dense phase flows is extremely complex because of the fluid-particle and particle-particle interactions. The advantages of the two-fluid approach are that the computations do not depend on the number of particles and can be executed efficiently. This enables the scheme to be applied to complex geometries. The problem with the scheme is the lack of detail on the motion of the particles. For example, no information is available on the particle rotation and particle-particle friction is not a parameter in the problem. Also, the kinetic analysis is based on uniform size spheres and does not currently allow for a size distribution or irregularly-shaped particles.

The discrete element model, on the other hand, provides details on the particle properties. Fundamental information such as Young's modulus, Poisson's ratio and particle-particle friction can be incorporated into the scheme. The major problem is the number of particles needed to have a viable model of a physical system. Even though calculations are currently preformed with sample, or computational particles, there is no criterion for the number of computational particles needed.

The current dense phase two-fluid models have been developed for those cases where particle-particle interaction established the granular temperature and the constitutive relations. As the flow becomes less dense and the fluid-particle interaction becomes an important factor, the current two-fluid models will encounter the same problems as those used for dilute phase flows. The discrete element model, on the other hand, will provide a smooth transition between the two regimes if the carrier-phase turbulence can be modeled correctly.

At the present time models are appearing in the literature in which the details of both the fluid and particle motion are included (Hu, 1996; Johnson and Tezduyar, 1997; Dasgupta et al., 1994). These models, which are truly direct numerical simulations, will provide important information on dense phase flows. Obviously, the modeling of dense phase systems is in the early stages of development. There are many opportunities to make significant contributions to this field of numerical model development.

Exercises

The exercises in this chapter are divided into two parts. This first part suggests areas for development of numerical programs. The second part are applications of the program provided in Appendix D.

Program development

8.1. Modify the subroutine VENTURI to represent an ASME venturi configuration.

8.2. Extend the conservative variable program to model the flow through a de Laval nozzle. First, extend the program excluding two-phase flow effects ($Z = 0$). This is done by using a scheme which iterates on the initial velocity to produce sonic velocity at the throat. When this is achieved, the calculation is repeated with the correct velocity and when the sonic condition is reached the calculation moves one step forward, changes the sign on the discriminant in Equation 8.58 and the calculation is continued into the expansion section. Once the single-phase solution is working, include the two-phase flow effects. You will find that the solution becomes very sensitive when the particle size is small.

8.3. Extend the conservative variable program for flow of a two-component mixture (water + air) of the gaseous phase. This will involve setting up and solving for the fraction of water vapor in the carrier flow by introducing a continuity equation for the water vapor.

8.4. Using the two-fluid equations for the quasi-one-dimensional flow presented in Section 8.2.2, develop a computer solution for the flow nozzle with an applied pressure.

Applications

8.5. Ice tunnels are used to study the effect of icing on engine performance. An ice tunnel is typically a long duct supplied with cold air as shown in the figure below. Droplets are introduced through an atomizer and cooled in a cold air stream. The resulting droplets and air enter the engine and engine performance is tested.

You are to use the program to design an ice tunnel. Assume the tunnel is 1 meter in diameter and 30 m long. Air at 0°C enters the tunnel at 20 m/s and standard pressure (1 bar). The atomizers are capable of generating water droplets of 50, 100 and 150 microns. The anticipated flow rate of the water into the tunnel is 1 kg/s and enters a 20°C. Assume the initial velocity of the droplets is 2 m/s. Do three runs for the initial droplet sizes with zero loading and plot the droplet and gas temperature distribution along the duct. Now do the same runs with the loading corresponding to the water flow rate of 1 kg/s. Compare the temperature of the droplets at the end of the duct with the one-way coupled case. Do some additional runs to investigate the effect of loading on final droplet temperature.

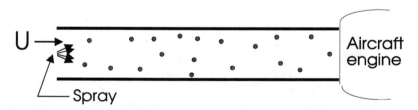

Diagram of icing tunnel.

8.6. Some consideration has been given to using a venturi configuration to meter gas/solids flows. You are to use the program to determine the performance of the venturi as a function of loading and particle size. The program currently has a subroutine which provides a cubic contour. You are to rewrite the subroutine for the geometry of an ASME standard design. The ASME configuration shown in the figure consists of an entrance section with a 21° included angle and a diffuser section of an 8° included angle. The radius of curvature at the throat is equal to the throat diameter. The venturi has a β-ratio (throat/pipe diameter ratio) of 2. The throat diameter is 2 cm.

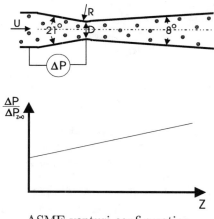

ASME venturi configuration.

Coal particles, 50 microns and 200 microns in diameter, will be used in the venturi. The coal has a material density of 1300 kg/m^3. Air flows through the venturi at an initial pressure of 2 bars and with a velocity of 30 m/s. Make some runs with the two particle sizes for loadings from zero to 10 and make a plot of the pressure ratio versus loading as shown in the figure. Also make a plot (only three points) of the slope versus $Z/(St+1)$ to establish the validity of the scaling parameters.

8.7. A vertical vacuum system is to be designed for removal of particles from a well as shown in the figure. The pipe is 10 cm in diameter and 10 meters long as shown in the diagram below. The pipe is steel. The particles have a material density of 1200 kg/m^3 and two possible diameters, 100 and 500 microns. The gas entering the pipe is air at 20°C and 1 bar. Information is needed on the pressure (vacuum) necessary to operate the system.

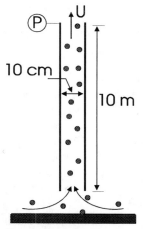

Pneumatic lift system.

First of all, develop a small program to predict the terminal velocity of a particle dropping in air. The terminal velocity is

$$u_T = \frac{g\tau R}{f}$$

but f is a function of Reynolds number. Use chart in Chapter 4 to find terminal velocity.

The terminal velocity represents the minimum velocity for the air in the pipe. Then do your calculations for initial velocity which is 20 percent higher than the terminal velocity. Do several runs with the two particle sizes over a range of velocities up to 150 m/s. For each particle size make runs for a loading of 0, 1 and 5.

8.8. A process to make spherical metal powders is to first melt particles to have the surface tension form spherical particles and then allow them to cool. Assume that aluminum powder is melted by hot nitrogen in a tube as shown in the figure.

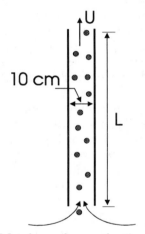

Metal powder production.

The tube has a diameter of 10 cm and the entering nitrogen velocity is 10 m/s at a pressure of 2 bars. The tube is made of a ceramic material with a relative roughness of 0.01. The aluminum powder enters the tube at 20°C with an effective velocity of 1 m/s. The effective diameters of the aluminum powder are 50 and 100 microns. The nitrogen temperature entering the tube is 1000°C.

The program currently does not take into account the variation in temperature across the boundary layer of the particle which means that the particle Reynolds number is different. As an estimate, assume that the effective temperature is the average temperature between the particle and the air for calculating the particle Reynolds number.

Determine how long the tube has to be to completely melt the aluminum powder at the end of the tube for loadings of 0, 2, 5 and 10.

8.9. A sand blaster consists of a tube in which sand is accelerated by a gas stream and impacts a target as shown in the figure. The effectiveness of the system depends on the product of the square of the velocity of the "grit" as it emerges from the duct and the mass of the particle.

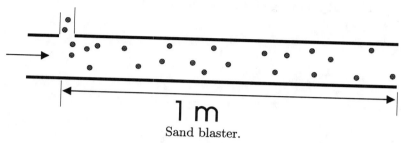

Sand blaster.

Assume the pipe for the sand blaster is 1 cm in diameter and 1 meter long and made of hardened steel. The grit to be used is sand (quartz) with a material density of 2400 kg/m^3. The velocity at the entrance to the tube is 100 m/s and the grit velocity is 2 m/s. The pressure at the exit of the tube is 1 bar. The system is to operate at a loading of 10. Find the velocity at the end of the duct. Plot the particle velocity distribution in the duct to determine if the duct is too long. (Can it be shorter and still achieve nearly the same grit velocity?)

The program is based on a spherical particle. Modify the program to incorporate the drag coefficient for a particle with a cubical shape. Start by assuming some particle size, say 50 microns. Choose different starting pressures until the final pressure is 1 bar.

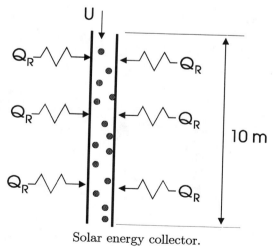

Solar energy collector.

8.10. One technique for solar energy conversion is to heat particles by solar energy in a tube or free falling stream and then utilize the hot particles as the heat supply for a thermal engine. Consider the particles in the tube shown in the figure.

The tube is vertical and particles are released from the top and fall through the tube. Radiant energy from the sun is concentrated on the tube by a series of mirrors. The tube is 20 cm in diameter and 10 m high. The particles have a density of 800 kg/m^3. The air enters the tube at 1 bar and 20°C with a velocity of 1 m/s. The radiant energy acting on the particles is 500 kW/m^2.

Chapter 9

Experimental Methods

Detailed measurements in two-phase flow processes and equipment are necessary for process and quality control, to characterize the global flow behavior and flow regimes, and to obtain local information about the characteristic properties in such flows. Moreover, experimental techniques are required to analyze the behavior and motion of individual particles in order to assess microprocesses occurring on the scale of the particles.

Numerous measuring techniques are available for such experimental studies on two-phase flows. These measuring techniques are categorized in certain groups according to their capabilities as shown in Figure 9.1. The first classification is based on the way the information is extracted from the two-phase system. Sampling methods are still quite frequently applied in the process industry and powder technology industries to provide a detailed characterization of bulk solids at the different stages of the process for quality and process control (e.g., during milling or classification processes). Using mechanical sampling equipment, a number of characteristic samples of the bulk solids are collected and analyzed. The analysis of the samples may be performed using a microscope, mechanical methods such as sieving, fluid classification or sedimentation. Optical methods such as light scattering or light attenuation may also be applied. The result of these analyses could be, for example, characteristic dimensions of the particles, particle shape factors, equivalent particle diameters or particle surface area.

On-line measuring techniques are those methods which can be directly applied within the process or in a bypass line to analyze the properties of both the dispersed and continuous phases. Generally, these methods can be only applied in dispersed two-phase systems with relatively low concentrations of the particle phase. Moreover, on-line measuring techniques are widely used in research programs to characterize the development and the properties of a two-phase flow system. A further classification of on-line measurement techniques may be based on the spatial resolution of the measurement. Integral methods provide time-resolved, spatially-averaged properties of two-phase systems over an entire cross-section of the flow or along a light beam passing through the flow, depending on the method applied.

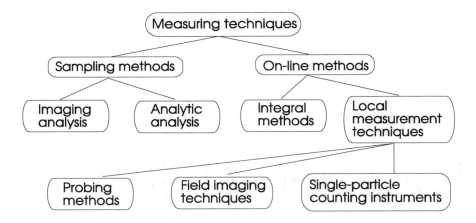

Figure 9.1: Categorization of measurement techniques for analysis of two-phase systems.

On the other hand, local measurement techniques allow the determination of local properties in a two-phase system with a spatial resolution which depends on the measurement technique applied. Those techniques which allow the determination of the properties of a two-phase system with a high spatial resolution (i.e., probe volume dimensions of the order of 100 μm) are mainly based on optical methods. Such optical methods provide information either about the particle properties at a given instant of time or during a time sequence within a finite but relatively large probe volume (e.g., field imaging techniques, such as particle image velocimetry or holography), or are based on measuring the properties of the particles moving through a small finite probe volume within a certain measurement time interval (e.g., light scattering or laser-Doppler and phase-Doppler anemometry). Methods of the first kind are based on visualizing the particles by illuminating the probe volume and recording the images of the particles at a given instant of time or during a relatively short time sequence or at multiple times. Distribution functions of particle phase properties are therefore obtained for all particles that happen to be in the probe volume at a given instant of time. Hence the result is weighted by the particle number concentration.

Instruments based on the second method are also called single particle counting techniques since they require that only one particle be in the probe volume at a given time. Therefore, the size of the probe volume quite often needs to be very small in order to fulfill this requirement, depending on the particle concentration. Since the particle phase properties and the respective distribution functions are obtained for all particles that pass through the probe volume during the measurement time, the result obtained by such methods is weighted by the particle number flux.

For the characterization of a two-phase flow system, the following properties of the dispersed phase are of primary interest:

- particle size (i.e., diameter of spherical particles, equivalent diameters of nonspherical particles),
- particle shape and surface area,
- particle concentration or mass flux,
- particle translational and rotational velocities,
- correlation between particle size and velocity.

In many cases the velocity of the continuous phase (i.e., fluid) in the presence of the dispersed phase is also of interest in order to assess the influence of the particles on the fluid flow. All these properties are generally not obtained as single values, but as a distribution function which is characterized by a mean value, a standard deviation and higher moments of the distribution function.

A summary of the various measurement techniques and the dispersed phase properties which can be obtained by the above described measurement techniques is given in Table 9.1. In the following sections several of the most common measurement techniques and their capabilities are introduced and the limitations of their applications are summarized.

Particle phase property	Sampling methods	On-line methods	
		Integral methods	Local measurement techniques
Size	Sieving Coulter principle Sedimentation	Laser diffraction	Light scattering Phase-Doppler anemometry
Concentration	Isokinetic sampling	Laser diffraction Light absorption	Isokinetic sampling Fiber optic probes Light scattering Phase-Doppler anemometry
Velocity	Direct velocity measurements are not possible with sampling methods	Correlation technique	Fiber optic probes Laser-Doppler anemometry Phase-Doppler anemometry Particle tracking velocimetry Particle image velocimetry

Table 9.1. Categorization of measurement techniques with respect to the measurable properties of a two-phase system.

9.1 Sampling methods

Sampling methods are mainly applied to characterize the shape, equivalent diameter, and surface areas of the solid particles. This implies that a number of representative samples of particles have to be collected from the flowing two-phase system or from the bulk solids conveyed on belts or stored in bins or containers. For this purpose a number of sampling devices and methods are available which enable sampling from stationary or moving powders (Allen, 1990).

The sampling should be performed in such a way that the collected samples closely represent the properties of the two-phase system under consideration. Usually a number of samples have to be taken so the characteristics of the sample will vary and the expected variation may be analyzed by statistical methods. A further important aspect in sampling is the reduction of the sample size to obtain the analysis sample which typically is in the order of a few grams, depending on the method of analysis. The reduction of the sample size is illustrated in Figure 9.2.

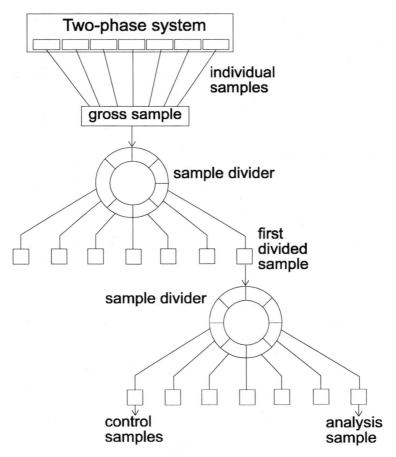

Figure 9.2: Process of sample dividing. (*Mechanische Verfahrenstechnik 1*, Steiss, M., p. 86, Figure 3.1.1, 1995, copyright notice of Springer-Verlag.)

First, a number of individual representative samples have to be taken from the two-phase system under consideration. This process can be continuous, such as sampling from a moving belt or a pneumatic conveying system, or periodic, such as sampling from bins or containers. These gross samples usually are too large for laboratory analysis and are therefore collected, mixed and then reduced to smaller size samples using a sample divider. This dividing process needs to be performed one or more times in order to obtain a sample size which can be analyzed. The required sample size depends on the method of analysis and usually is in the order of a few grams. The entire sampling process must be done very carefully in order to ensure that the sample for analysis is representative of the two-phase system under consideration. More details about sampling equipment and sampling procedures can be found in Allen (1990).

Various methods such as imaging methods, sieving, sedimentation, fluid classification or electrical sensing methods are available for the analysis of the particle samples with regard to shape, dimension and particle size. Some of the most common methods are summarized here. Others are described in detail by Allen (1990).

9.1.1 Imaging methods, microscopy

Microscopy is a direct imaging method to determine particle diameters, characteristic dimensions and shape factors of individual particles. This method can be used for the range of particle sizes between 1 and 200 μm. The lower limit is determined by diffraction (see Section 9.3.3 for more details) which may result in erroneous images of small particles. In order to determine number distributions of linear dimensions, a relatively large number of individual particles have to be analyzed, which is time consuming. In combination with digital image processing, however, such a direct imaging method becomes a powerful technique for particle characterization. The principle of this method is illustrated in Figure 9.3.

First, the sample has to be prepared in such a way that individual particles can be observed under a microscope, which implies a uniform dispersion of the particles on the sample slide. Then, images of a number of particles are taken using a CCD-camera (CCD stands for charged coupled device) which transforms the optical image into a digital image with a certain resolution of gray values for each pixel. Typically, 256 grey values are assigned to one pixel which is one of the photo-sensitive elements of the CCD-array. The number of pixels determines the spatial resolution of the CCD-camera and is, for example, 512 x 1012 for a standard CCD-camera. After enhancement of the image, i.e., thresholding and removing noise objects, the image can be processed to determine the characteristic linear dimensions of the particle. Such linear dimensions are the diameter of spherical particles or certain dimensions of nonspherical particles, which are determined from its projection since the image seen by a microscope is two-dimensional. Therefore, the analysis of nonspherical particles by an imaging method is biased by the fact that these particles will have a preferential orientation on the sample slide, i.e., plate-like particles will lay flat on the slide.

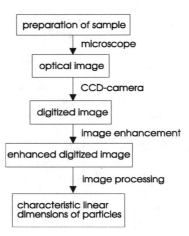

Figure 9.3: Principle of direct imaging technique using digital imaging processing.

Typical statistical linear dimensions of nonspherical particles are shown in Figure 9.4. The so-called *Feret-diameter*, D_f, is the distance between two tangents on opposite sides of the particle image, parallel to some fixed direction. The *Martin-diameter*, D_m, is the length of the line which bisects the particle image and should be obtained in a fixed direction for one analysis. Other linear dimensions of the particle image are the *maximum chord* $D_{c,max}$, the largest dimension and the perimeter diameter, which is the diameter of a circle having the same circumference as the perimeter of the image. Moreover, a *projected area diameter* may be determined which is the diameter of a circle having the same area as the projected image of the particle.

9.1.2   Sieving analysis

Sieving is one of the oldest, simplest and most widely used methods for particle size classification. It yields the mass fractions of particles in specific size intervals from which the resulting particle size distribution can be obtained. Sieving analysis can be performed with a wide range of mesh sizes, from approximately 5 μm to about 125 mm. In order to cover this size range, different sieves and sieving methods have to be used. Also, the properties of the bulk solids determine the method of sieving to be used. Woven sieves are quite often characterized on the basis of mesh size, which is the number of wires per linear inch. For example, the opening for a 400-mesh sieve is 37 μm and the wire thickness is 26.5 μm (see, for example, American ASTM). This results in 34% open area of the sieve.

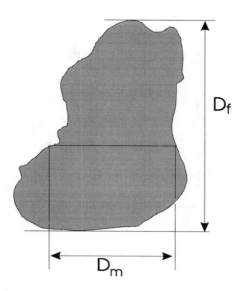

Figure 9.4: Typical linear dimensions of a nonspherical particle or of images of nonspherical particles.

Sieving method	Particle size range	Particle properties
hand or machine sieving	63μm - 125μm	dry powder
air-jet sieving	10μm - 500μm	cohesive and wet powder
wet sieving with micro-mesh sieves	5μm - 50μm	wet powder

Table 9.2. Sieving methods with respect to particle size range and particle properties.

A classification of the sieving methods to be applied based on the size of the particles and the particle properties is given in Table 9.2. The sieves used for hand and machine sieving are usually woven from metal wires which are soldered together and clamped to the bottom of a cylindrical container. For sieving analysis, a series of sieves are mounted on top of one another with decreasing mesh size from top to bottom as shown in Figure 9.5.

The powder to be analyzed is put into the upper sieve bin. By shaking the apparatus for a certain period of time the particles which are smaller than the mesh size fall through the sieves and accumulate in the sieve bins with a mesh size smaller than the particle size. This suggests that for a series of sieves with mesh sizes of 1.6, 0.8, 0.4, 0.2 and 0.1 mm, particles in the size intervals >1.6 mm, 0.8 - 1.6 mm, 0.4 - 0.8 mm, 0.2 - 0.4 mm, 0.1 - 0.2 mm and <0.1 mm are collected in the different sieve bins. Then the content in the different sieve bins is weighed and a histogram of the particle size versus the mass fraction or probability density in each size interval is obtained. The mass

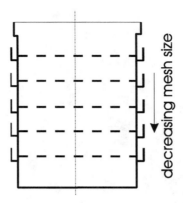

Figure 9.5: Arrangement of sieves for particle size analysis.

fraction in each size interval is determined from $\Delta m_i/m$ where m_i is the mass of particles collected in the different sieve bins and m is the total mass of the analysis sample. The mass probability density (frequency) distribution $f_m(D_i)$ is obtained from[1]

$$f_m(D_i) \simeq \frac{\Delta m_i}{m\Delta D_i} \tag{9.1}$$

where ΔD_i is the size interval $D_{i+1}-D_i$. A typical histogram of such an analysis is shown in Figure 9.6. The sample size for sieve analysis with sieves of 200-mm mesh opening usually range between 100 and 200 grams. The sieving time for dry powder is dependent on the mesh size and should be in the range specified in Table 9.3.

Mesh size	Sieving times
$> 160\mu m$	5 - 10 min
71-60 μm	10 - 20 min
40-71 μm	20 - 30 min

Table 9.3. Approximate sieving times for different mesh sizes.

Machine sieving is usually performed by placing the set of sieves on a vibrating support. Hand or machine sieving of powders which are cohesive or wet may be enhanced by putting additional granular materials on the sieves. This leads to deagglomeration of the powder and avoids the blockage of the apertures, however the sieving time is increased. Other sieving methods, such as air-jet sieving (Table 9.2), should be used for powders which are difficult to sieve. For very

---

[1] The continuous mass frequency distribution is defined as the limit

$$f_m(D) = \lim_{\Delta D \to 0} \frac{\Delta m}{m\Delta D}.$$

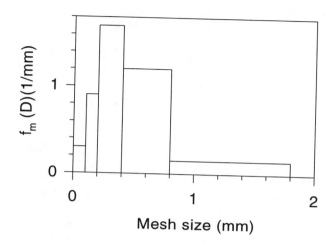

Figure 9.6: Histogram of particle size mass distribution.

fine and wet powders, wet sieving is recommended. Electroformed micromesh sieves are generally used for this method.

9.1.3   Sedimentation methods

The principle of the sedimentation method is based on recording the sedimentation of powders dispersed in a liquid and deriving the particle size distribution from the measured temporal change of the particle concentration. Since the sedimentation process is governed by the free-fall velocity of the particles, a free-fall or Stokes diameter will be determined. Usually small sedimentation columns are used to perform the measurement. Two methods may be applied to introduce the powder into the liquid. In the two-layer method, a thin layer of powder is introduced at the top of the liquid column; while in the second method, the powder is uniformly dispersed in the liquid by shaking the sedimentation column or by ultrasonic agitation before measurement. In order to maintain a stabilized dispersion, wetting agents are quite often introduced into the liquid to reduce the attractive forces and to increase the repulsive forces between the particles in the suspension.

   Two approaches may be used to determine the particle size distribution using the sedimentation method. The first approach is the incremental method in which the rate of change of particle density or concentration is measured at a given location in the sedimentation column. In the second approach, the rate at which the particles accumulate at the bottom of the sedimentation column is measured. This method is called the cumulative method.

Incremental method

As stated above, the incremental method is based on the measurement of the temporal change of the particle concentration close to the bottom of the sedimentation column. The powder has to be homogeneously dispersed throughout the column at the beginning of the experiment so that the initial concentration at the measurement location corresponds to the initial particle concentration, i.e., the mass of particles introduced into the column divided by the volume of the mixture. Various methods such as light attenuation or x-ray attenuation can be applied for the measurement of the density or the concentration of the particles at the measurement location. The *light attenuation method* is based on transmitting a narrow horizontal beam of parallel light through the sedimentation cell at a known depth H and recording the attenuated light on the opposite side using a photodetector. Such instruments are called *photo-sedimentometers*. According to the *Lambert-Beer law*, the attenuation of light is proportional to the cross-sectional area of all particles in the light beam. Since the sedimentation analysis may take a very long time for small particles, the sedimentation cell is usually moved relative to the optical detection system so that the depth H of the measurement section is reduced with time.

The light attenuation method will be described in more detail in Section 9.2.1.

X- rays can also be used as a light source. In this case the X-ray attenuation is directly proportional to the mass of the particles in the X-ray beam and according to the law of Lambert-Beer, one obtains

$$\frac{I}{I_0} = \exp(-kc_m) \tag{9.2}$$

where I and I_0 are the intensities of the attenuated and incident X-rays and k is a constant of proportionality. Instruments based on this principle are called *X-ray sedimentometers*.

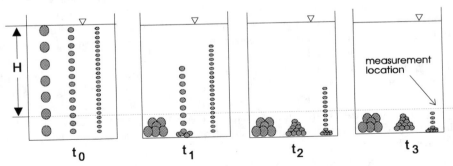

Figure 9.7: Sedimentation model to illustrate incremental method.

The principle of the incremental method is illustrated in Figure 9.7 for a powder with three distinct size classes. At the beginning of the sedimentation

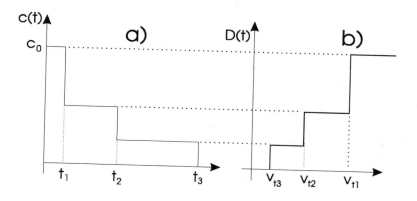

Figure 9.8: Sedimentation model for incremental method; a) time-dependent concentration at measurement location, b) cumulative mass distribution versus sedimentation (terminal) velocity.

process, i.e., at $t_0 = 0$, the particle mass concentration at the measurement location, which is located at a distance H below the surface, corresponds to the concentration c_0 and is given by

$$c_0 = c_{t=0} = \frac{m_s}{V_s + V_l} \qquad (9.3)$$

where m_s is the mass of solid particles and V_s and V_l are the respective volumes of the particles and the liquid in the sedimentation cell. After time $t_1 = H/v_{t1}$, all the largest particles with the highest settling (terminal) velocity v_{t1} will have passed the measurement location. Hence, the concentration has reduced to that of the other two remaining particle size classes as shown in Figure 9.8. Particles of these two size classes have sedimented a shorter distance from the surface due to their smaller free-fall velocity. At time $t_2 = H/v_{t2}$ the second largest particle size class will also have passed the measurement location and the concentration is again reduced. Finally, the smallest size class will have passed the measurement location and the particle concentration drops to zero. Hence one obtains at the measurement location the concentration of the particles for which the sedimentation velocity is smaller than H/t. The cumulative mass fraction is determined by normalizing the measured concentration value by the initial concentration c_0.

$$F_m(t) = 1 - \frac{c(t)}{c_0} \qquad (9.4)$$

In order to determine the cumulative mass fraction as a function of particle size it is necessary to correlate the sedimentation velocity with the particle size (i.e., Stokes diameter). By considering only the drag, the buoyancy and gravity forces, the equation of motion for the particle in a quiescent fluid becomes (see Chapter 4)

$$\frac{dv}{dt} = -\frac{18\mu}{\rho_d D^2} v + \left(1 - \frac{\rho_c}{\rho_d}\right) g \qquad (9.5)$$

At steady state

$$\frac{dv}{dt} = 0 \qquad (9.6)$$

Solving for D yields the Stokes diameter D_{St}:

$$D_{St} = \sqrt{\frac{18\mu v_t}{(\rho_d - \rho_c)g}} \qquad (9.7)$$

This equation is valid for the Stokes regime, i.e., $Re_p < 1$, which is generally applicable for sedimentation analysis since small particles usually are considered.

In actuality, the measurement of particle concentration will be a continuous process so the cumulative percentage of undersize distribution becomes a continuous curve. The continuous curve can be represented by a series of steps and the same process for data reduction can be applied.

Cumulative method

In the cumulative method the rate at which the powder is settling on the bottom of the sedimentation column is measured. One of the most commonly used instruments is the sedimentation balance where the particles collect on a balance pan and the weight of the particles accumulated is continuously measured. Again, the description of this method is based on three distinct particle size classes which are initially homogeneously dispersed in the sedimentation column. During the initial period, the weight of the particles increases at a constant rate until all particles of the large diameter fraction have accumulated on the balance pan as shown in Figure 9.9. Within the time interval $t_1 < t < t_2$ the rate of particle weight accumulation is reduced since all the large particles have already settled out. In the case of the simple three-size-class sedimentation model the accumulated particle mass increases, again at a constant rate. Similarly, a linear increase of the particle mass is observed for $t_2 < t < t_3$. Since only a portion of the smallest particles is in suspension, the rate of weight increase is again reduced.

The mass accumulated on the balance pan at a given time t can be expressed as

$$m(t) = m_o[1 - F_m(D)] + m_o \int_0^D \frac{h}{H} f_m(\lambda) d\lambda \qquad (9.8)$$

where D is the particle diameter just settled out in time t, m_o is the total mass of the sample and h is the distance that particle size λ has settled during time t. Obviously, the first term represents the mass of all particles with diameters larger than D and the second term represents the mass accumulated on the

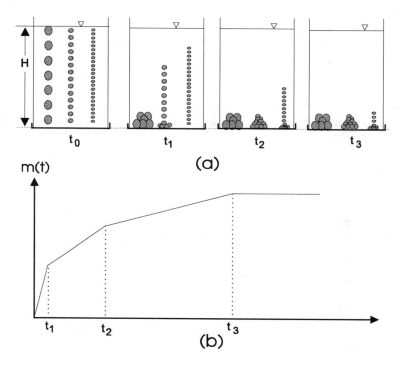

Figure 9.9: Sedimentation model to illustrate the incremental method, a) sedimentation model, b) temporal change of accumulated mass.

balance for all particles with diameters smaller than D. The distance h is related to the settling (terminal) velocity, $v_t(\lambda)$ by

$$h = v_t(\lambda)t. \tag{9.9}$$

Substituting Equation 9.9 into Equation 9.8 and dividing by m_o results in

$$\frac{m(t)}{m_o} = 1 - F_m(D) + \frac{t}{H}\int_0^D v_t(\lambda)f_m(\lambda)d\lambda \tag{9.10}$$

Taking the derivative with respect to time yields

$$\frac{d}{dt}(\frac{m}{m_o}) = -\frac{d}{dt}F_m(D) + \frac{1}{H}\int_0^D v_t(\lambda)f_m(\lambda)d\lambda + \frac{t}{H}v_t(D)f_m(D)\frac{dD}{dt} \tag{9.11}$$

However, at time t, $H = v_t(D)t$ and, by definition, $F_m(D) = f_m(D)dD$, so the above equation reduces to

$$\frac{d}{dt}(\frac{m}{m_o}) = \frac{1}{H}\int_0^D v_t(\lambda)f_m(\lambda)d\lambda \tag{9.12}$$

Using Equation 9.10 to relate the integral to the mass accumulation and cumulative mass fraction, one can write the cumulative mass fraction as

$$F_m(D) = t\frac{d}{dt}\left(\frac{m}{m_o}\right) - \frac{m(t)}{m_o} + 1 \tag{9.13}$$

This equation can be used to determine the cumulative mass fraction from the slope of the mass accumulation curve as illustrated in Figure 9.10. This is known as Oden's equation.

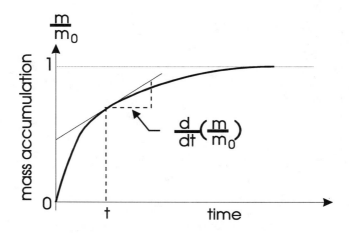

Figure 9.10: Graphical determination of mass fraction and slope from data using a sedimation balance.

Example: A particle size analysis using a sedimentation balance is to be performed. At the beginning of the measurement, 100 grams of particles with a density of 2500 kg/m^3 are homogeneously dispersed in a liquid with a density of 1000 kg/m^3 and a dynamic viscosity of 0.001 Ns/m^2. The balance pan is located 0.3 m below the liquid surface. The measured mass accumulated on the balance pan and the rate of mass increase are provided in the following table.

Sedimentation time (s)	9	20	45	100	200	450	900
Accumulated mass (g)	32.0	55.0	76.0	89.4	95.3	98.7	99.9
Rate of mass increase (g/s)	3.111	1.550	0.4889	0.1140	0.0265	6.0 ×10^{-3}	2.11 ×10^{-3}

1) Determine the cumulative mass distribution from the sedimented mass and rate of mass change using the Oden equation, Equation 9.13. Use the Stokes diameter to transform time to particle size.

2) Determine the particle size for Reynolds number effects using the formulation given in Chapter 4.

3) Approximate the measured particle size distribution by a Rosin-Rammler distribution function and determine the empirical constant n (see Section 3.4.2). Comment on the quality of the approximation.

4) Determine the integral parameters of the size distribution such as the mass, median and Sauter mean diameter.

5) Determine the required measurement time for a particle of 10 μm. Comment on the result.

Solution:

1) From the Oden equation

$$F_m(D) = t\frac{d}{dt}(\frac{m}{m_o}) - \frac{m(t)}{m_o} + 1$$

$$D_{St} = \sqrt{\frac{18\mu H}{(\rho_d - \rho_c)gt}}$$

with $v_t = H/t$. Substituting the values into the equation for Stokes diameter gives

$$D_{St} = 605/\sqrt{t} \text{ (microns)}$$

The results are shown in the table.

t (sec)	m/m_o	$d/dt(m/m_o)$ (1/sec)	$F_m(D)$	D_{St} microns	D microns
9	0.32	3.11×10^{-2}	0.96	202	255
20	0.55	1.55×10^{-2}	0.76	135	150
45	0.76	4.889×10^{-3}	0.46	90.2	95
100	0.894	1.14×10^{-3}	0.22	60.5	65
200	0.953	2.65×10^{-4}	0.11	42.8	45
450	0.987	6.0×10^{-5}	0.04	28.5	30
900	0.999	2.11×10^{-5}	0.02	20.2	21

2) The effect for non-Stokesian flow is obtained by solving, iteratively, the following equation for diameter, D.

$$f v_t = g \tau_V (1 - \frac{\rho_c}{\rho_d})$$

$$\frac{H}{t}[1 + 0.15(\frac{H}{t\nu_c})^{0.67}D^{0.67}] = \frac{g\rho_d D^2}{18\mu_c}(1 - \frac{\rho_c}{\rho_d})$$

The results are given in the table. One notes that the corrected diameter is larger than the Stokes diameter but approaches the Stokes diameter for long sedimentation time (low velocity).

3) The Rosin-Rammler distribution is given by

$$F_m(D) = 1 - \exp\left[-\left(\frac{D}{\delta}\right)^n\right]$$

as discussed in Chapter 3. Plotting $\ln\{-\ln[1 - F_m(D)]\}$ versus $\ln D$ yields $n = 2.1$ and $\delta = 159.5$ microns.

4) From the Rosin-Rammler distribution

$$D_{32} = \frac{\delta}{\Gamma(1 - \frac{1}{n})} = 94.2 \ \mu m$$

$$\mu_m = \delta\Gamma(1 + \frac{1}{n}) = 141.3 \ \mu m$$

$$D_{mM} = \delta\left[\Gamma(1 - \frac{1}{n})\right]^{1/2} = 207 \ \mu m$$

5) For Stokes flow

$$v_t = \frac{(\rho_d - \rho_c)D^2}{18\mu_c}g = 0.082 \text{ mm/s}$$

However, $t = H/v_t = 3658s$. This measurement time is too long, so the use of a sedimentation balance is limited to larger particles.

Pipette method

In the pipette method according to Andreasen (see Allen, 1990), the concentration change near the bottom of the sedimentation cell is determined by drawing off known volumes (typically 10 ml) of the suspension at fixed time intervals by means of a pipette. The content of solids in the samples is generally determined by drying and weighing the remaining solids. This method provides the mass concentration of particles directly.

9.1.4 Electrical sensing zone method (Coulter principle)

The principle of electric sensing zone methods is based on the disturbance of an electrical field by a particle passing a probe volume. The disturbance can be related to the particle size by calibrating with mono-dispersed particles. A well-known instrument based on this principle is the Coulter counter, illustrated in Figure 9.11, which originally was used in medicine to count blood cells and subsequently modified to allow measurements of particle size and number.

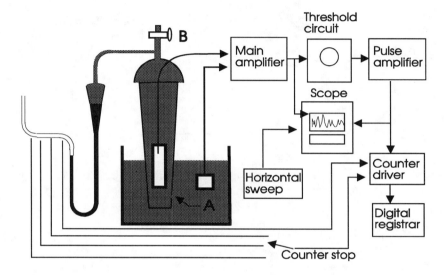

Figure 9.11: Principle of electrical sensing zone method (i.e., Coulter counter).(*Particle Size Measurement*, Allen, T., p. 456, Figure 13.1, Chapman and Hall, 1990. With permission.)

The particles are suspended in an electrolyte and forced to pass through a small orifice, A. Electrodes are immersed in the fluid on both sides of the orifice which produces an electrical field. A particle passing the orifice will change the electrical impedance and generate a voltage pulse. The amplitude of the pulse is proportional to the volume of the particle as shown in Figure 9.12. The Coulter counter principle also belongs to the class of single particle counting techniques along with other optical methods described in Section 9.3. Therefore, the pulse amplitude can be uniquely related to the particle size when only one particle is in the probe volume at one time. To avoid plugging the orifice the diameter of the orifice needs to be larger than the largest particle in the sample.

The probe volume is the volume of the orifice and a region outside the orifice where the signal amplitude is above the noise level as shown in Figure 9.12. The measurable particle size range is between 0.02 and 0.6 of the orifice diameter. This results in a dynamic range of 1:30. By using multiple measurement cells with different orifice diameters a particle size range of 0.5 to 1200 μm can be

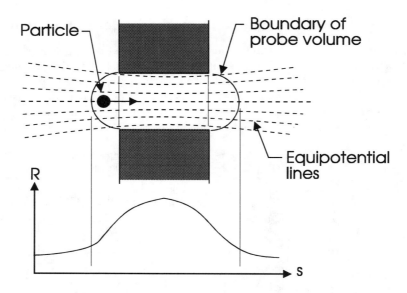

Figure 9.12: Probe volume and modification of resistance by particle in volume for the Coulter principle.

resolved. The upper limit of measurable particle size corresponds to the problem of keeping the particles in suspension. Large particles with density larger than the density of the electrolyte will tend to settle out. A solution to this problem is to increase the viscosity of the electrolyte or to apply stirrers in the reservoirs. The lower limit of detectable particle size is determined by the electronic noise of the system.

The operating principle of a Coulter counter is illustrated in Figure 9.11. Before measurement a vacuum is applied to the glass tube to produce an imbalance in the mercury siphon. After closing the valve (B), the flow through the orifice (A) is created as the mercury siphon is returned to equilibrium. The measurement is initiated and terminated by means of start and stop switches on the siphon. The resistance across the orifice is recorded by means of the electrodes mounted in the glass tube and in the reservoir where the glass tube is immersed. The voltage pulses generated by the modification of the resistance across the orifice as particles pass through are amplified, sized and counted. By setting an appropriate threshold level, only pulses above the noise level are detected and counted. This threshold level also determines the lower limit of the effective probe volume (see Figure 9.12).

The counts are collected in a number of pre-defined amplitude classes. By using a calibration curve which is determined, for example, with monosized Latex particles, the measured amplitude can be related to a particle size. Hence the measurement yields an equivalent diameter of a particle which gives the same voltage pulse as the calibration particle.

More details about the theory of the Coulter principle and signal processing can be found in Allen (1990).

9.1.5 Optical analysis

Different methods are available for the optical analysis of particle samples based on the following physical principles:

- light attenuation

- scattering intensity

- light diffraction.

All these methods require careful preparation of the sample by dispersing the particles either in air or liquid. The dispersion must ensure that the agglomerates breakup into the primary particles in order to allow the measurement of their properties. Some of the most common particle characterization methods based on optical analysis are described in more detail by Allen (1990). The physical principles of these techniques are identical with instrumentation used for on-line measurements and will be described in the following sections.

9.2 Integral methods

Integral methods are characterized as those measurement techniques which provide time-resolved but spatially integrated properties of a flowing two-phase system. Such methods are based on the distortion or attenuation of some energy source such as light, sound, or atomic radiation passing through a two-phase flow system. Hence, these methods are nonintrusive, but provide only information integrated along the beam of the energy source. This is frequently an advantage for process control in industry. In the following sections the most common integral methods for applications to dispersed two-phase flows are summarized.

9.2.1 Light attenuation

The intensity of a light beam passing through a fluid-particle mixture will be attenuated due to scattering and absorption by the particles. The intensity of the transmitted light is sensed by a photodetector and the change of the photodetector resistance can be monitored by appropriate electronic circuitry. According to the Lambert-Beer law the light attenuation by a suspension of monodisperse particles is given by:

$$\frac{I}{I_o} = \exp(-k\frac{\pi}{4}D^2 nL) \tag{9.14}$$

where k is a constant of proportionality known as the extinction coefficient, D is the particle diameter, n is the particle number density and L is the optical path through the suspension as shown in Figure 9.13. Equation 9.14 shows that the light extinction depends on two properties of the particle phase, the particle concentration and the particle diameter. Hence, one of these properties has to be known to enable the measurement of the other property. Introducing $\bar{\rho}_d = nm$ into Equation 9.14 gives:

$$\ln\left(\frac{I}{I_0}\right) = -k'\left(\frac{L}{D}\right)\left(\frac{\bar{\rho}_d}{\rho_d}\right) \tag{9.15}$$

The change in light intensity I for a change in particle bulk density $\bar{\rho}_d$ is given for constant values of L and D by

$$\frac{dI}{I} = -k'\left(\frac{L}{D}\right)\frac{1}{\rho_d}d\bar{\rho}_d \tag{9.16}$$

Hence, for a given geometry, particle composition and particle diameter (i.e., particle diameter distribution) the change in light intensity is proportional to the change in particle cloud density so one has:

$$\frac{dI}{I} = -Kd\bar{\rho}_d \tag{9.17}$$

The constant of proportionality, K, depends on the geometry of the flow system and properties of the particle material, such as material density and diameter distribution. Therefore, the application of the light attenuation method requires the determination of the calibration constant K for each particle composition considered. For polydispersed particles the dependence of the extinction coefficient on the particle size may result in larger measurement errors for the determination of the particle bulk density. Moreover, since the method requires optical access to the flow system, the build-up of particles at the windows must be either negligible or nonvarying during the measurement. As mentioned before, the particle concentration obtained by the light attenuation method provides an integral value along the optical path through the flow system. In many situations this concentration is not representative for the entire cross-section, since there will always be a concentration distribution. Consider, for example, a particle-laden jet. In this case, the particle concentration distribution along the optical path of a light beam through the centerline of the jet follows a normal distribution function. Hence, the value for concentration obtained by light attenuation is some average value of this distribution.

For the determination of the local particle concentration based on the principle of light attenuation, fiber optical probes may be used as will be described in Section 9.3.2.

9.2.2 Laser-diffraction method

As described in Section 9.3.3 the scattering of particles considerably larger than the wavelength may be described by *geometrical optics* and *Fraunhofer diffraction theory*. It has been demonstrated for this regime that most of the light

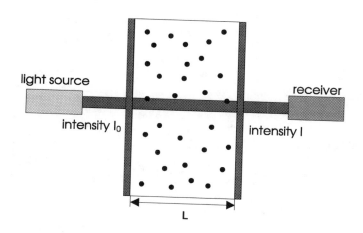

Figure 9.13: Principle of light attentuation method.

is scattered in the forward direction as a result of diffraction. The resulting forward scattering intensity pattern depends on particle size (see Figure 9.26), so the analysis of the scattering pattern may be used to infer information about particle size distribution. This principle, i.e., laser diffraction, has been used for many years to analyze particle samples and two-phase flows (Swithenbank et al., 1977, Hirleman et al., 1984) and a number of commercial instruments based on this principle are available.

Fraunhofer diffraction is one limit of the basic *Fresnel-Kirchhoff theory* of diffraction which describes the interaction of a monochromatic light beam with an aperture. In the *Fraunhofer limit* the diffraction pattern of an aperture is the same for an opaque object with identical cross-section area and shape except for the shadow produced by the object. In the far-field, however, the diffraction pattern is much larger than the geometrical image. Two requirements have to be met in order to obtain the Fraunhofer limit (Weiner, 1984): 1) the area of the object must be smaller than the product of the wavelength of light and the distance from the point source of the light to the diffracting object and 2) the area of the object must be smaller than the product of the wavelength and the distance between the object and the observation plane. The first requirement is easily met for a parallel light beam since the point source can be considered to be at infinity. In order to fulfill the second requirement the detection plane must be positioned far away from the object. Therefore, Fraunhofer diffraction is also known as far-field diffraction. Alternatively, a lens may be used to focus the diffracted light onto a photo-detector positioned in the focal plane of the lens. For such an optical configuration the undeflected light is brought to a point focus on the axis (i.e., center of the detector) and the diffracted light is focussed around this central spot as shown in Figure 9.14. Therefore, the diffraction pattern will also be stationary for a moving object and hence, the size measurement is not biased by the particle velocity, which is an important fact for in-process two-phase flow analysis.

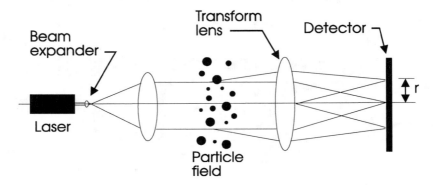

Figure 9.14: Optical arrangement of instruments for particle sizing by laser diffraction.

The intensity distribution of the diffraction pattern in the detector plane for a particle with diameter D may be obtained from

$$I(x) = I_o \left[\frac{2J_1(x)}{x} \right]^2 \tag{9.18}$$

where I_o is the intensity at the center of the pattern and J_1 is the first-order spherical Bessel function. The parameter x is a function of the particle diameter and the optical configuration

$$x = \frac{\pi D r}{\lambda f} \tag{9.19}$$

where r is the radial distance from the center of the detection plane, f is the focal length of the transform lens and λ is the wavelength of the incident light. By introducing

$$\sin \theta = \frac{r}{f} \tag{9.20}$$

and using the fact that θ is usually very small, i.e., $\sin \theta = \theta$, one obtains

$$x = \frac{\pi D \theta}{\lambda}. \tag{9.21}$$

The intensity at the center of the diffraction pattern is given by

$$I_o = c I_{inc} \frac{\pi^2 D^4}{16 \lambda^2} \tag{9.22}$$

where c is a constant of proportionality and I_{inc} is the intensity of the incident light. This results in the following equation for the intensity pattern

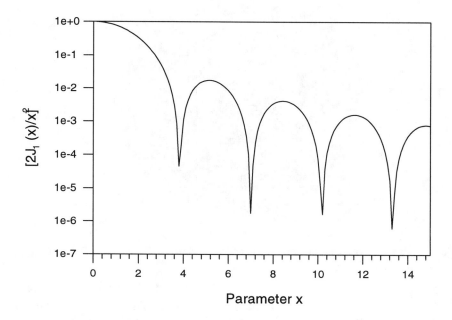

Figure 9.15: Fraunhofer diffraction pattern for a circular aperature or an opaque disk according to Eqn. 9.23.

$$I(D_p, \theta) = cI_{inc} \frac{\pi^2 D^4}{16\lambda^2} \left[\frac{2J_1\left(\frac{\pi D\theta}{\lambda}\right)}{\frac{\pi D\theta}{\lambda}} \right]^2 \tag{9.23}$$

This functional relation is known as the *Airy function* and is plotted in Figure 9.15. The result shows that the diffraction pattern consists of a series of bright and dark concentric rings surrounding the central spot of undiffracted light with a radius which depends only on the dimensionless parameter x.

However, for on-line measurements of particle size distributions, the measurement of the radial intensity distribution is very cumbersome. Therefore, the photodetector consists of a number of concentric ring elements as shown in Figure 9.16 and the light energy over only a finite area between r_i and r_{i+1} needs to be measured and analyzed. By integrating Equation 9.23 from r_i and r_{i+1}, one obtains the light energy received by the detector element

$$E_{i,i+1} = c\pi D^2 \left\{ \left[J_o^2(x) + J_1^2(x) \right]_i - \left[J_o^2(x) + J_1^2(x) \right]_{i+1} \right\} \tag{9.24}$$

The constant c depends on the power of the light source and the sensitivity of the photodetector.

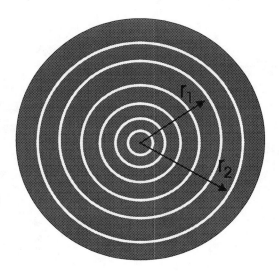

Figure 9.16: Illustration of ring detector element for measurement of the diffraction pattern.

Since the diffraction analysis is done not only for one particle being in the laser beam, but for a suspension (i.e., particles dispersed in air or liquid) with a certain particle size distribution, the light energy falling on one ring element has to be multiplied by the number of particles. For a particle size distribution with M size classes, each with N_j particles, one obtains by neglecting multiple scattering

$$E_{i,i+12} = c\pi \sum_{j=1}^{M} N_j D_j^2 \left\{ \left[J_o^2(x_j) + J_1^2(x_j)\right]_i - \left[J_o^2(x_j) - J_1^2(x_j)\right]_{i+1} \right\} \quad (9.25)$$

where x_j corresponds to the value of x for the particle size D_j. This equation allows one to convert the measured light energy to the particle number distribution represented by M size classes each with N_j particles. The number of possible particle size classes, M, is equal to the number of ring elements, L. Therefore, Equation 9.25 has to be solved L-times, for each ring element. The lower and upper limits of the size classes depend on the magnitude of the smallest and largest ring radii.

The set of L equations is usually solved using the least-squares criteria. Initial values of N_i are either estimated from the raw data or calculated from an assumed functional form of the particle size distribution, e.g., Rosin-Rammler distribution function. Using Equation. 9.25, L light energy values are calculated and compared with the measured light energy values. The assumed N_j-values are then corrected and the final result is obtained iteratively by minimizing the least-square error.

9.2.3 Cross-correlation techniques

The cross-correlation technique allows one to determine the mean transit time of a flowing medium passing two sensors located a given distance apart. Hence, an average velocity can be obtained as the ratio of the sensor separation L to the transit time T_t

$$V_o = \frac{L}{T_t} \tag{9.26}$$

This principle has been used for the design of flow meters (Beck and Plasowski, 1987). The cross-correlation technique may also be used to measure an average velocity in a flowing two-phase mixture, such as in pneumatic conveying (Kipphan, 1977; Williams et al., 1991) as illustrated in Figure 9.17a. The technique requires monitoring the fluctuations of any property of the dispersed phase with two sensors mounted a given distance apart. The type of sensor may be based on a variety of properties, such as ultrasound attenuation, light attenuation or scattering, electrostatic charge variation and conductance or capacitance variations. In most of these situations the resulting signals are proportional to the temporal variation of particle concentration. If the sensors are not positioned too far apart, the two signals generated by the fluctuation of particle concentration are, to some extent, identical but time-shifted as shown in Figure 9.17b. The time delay between the two signals $x(t)$ and $y(t)$ can be determined efficiently by computing the *cross-correlation function* of both signals over a certain measurement time period T_m. The cross-correlation function for the time delay τ is obtained from

$$R_{xy}(\tau) = \frac{1}{T_m} \int_0^{T_m} x(t - \tau) y(t) dt \tag{9.27}$$

For an ideal situation, i.e., a frozen concentration pattern, the signal $y(t)$ is exactly shifted by the transit time T_t, i.e., $y(t) = x(t - T_t)$. In this case the shape of the cross-correlation function is identical to the auto-correlation function (Kipphan, 1977)

$$R_{xx}(\tau) = \frac{1}{T_m} \int_0^{T_m} x(t - \tau) x(t) dt \tag{9.28}$$

and shifted by the transit time T_t as illustrated in Figure 9.17c. In real situations, and especially in turbulent two-phase flows, the concentration signals are stochastic in nature which results in a broadening of the cross-correlation function as shown in Figure 9.17c. For such a condition the transit time is obtained from the location of the maximum in the cross-correlation function $T_{\max}$. This was demonstrated by Kipphan (1977) using different models for the transport process. Moreover, Kipphan demonstrated that the broadening of the cross-correlation function in comparison with the *auto-correlation function* may be used to estimate the average particle velocity fluctuations. It is also obvious that the degree of correlation of the signals depends on the separation of the

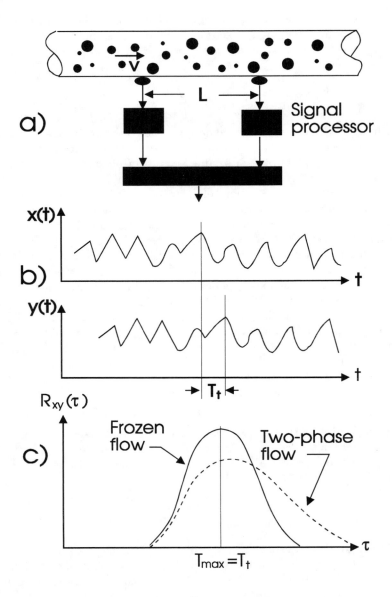

Figure 9.17: Cross-correlation method for particle velocity measurement; a) pneumatic conveying line and sensor locations, b) signals from both sensors and c) cross-correlation functions.

sensors and the intensity of the concentration fluctuation or particle fluctuating motion.

In order to achieve a maximum degree of correlation with a minimum scatter of the data, Kipphan (1977) suggests the following optimum sensor separation

$$L_{opt} \approx 0.35 \frac{b}{\sigma_v/V_0} \tag{9.29}$$

where b is the linear dimension of the sensor in the mainstream direction and σ_v is the mean intensity of the particle velocity fluctuation.

Since most of the sensor principles mentioned above enable the estimation of the average particle concentration, it is also possible to determine the particle mass flow rate from

$$\dot{m}_p = \bar{\rho}_d V_o A \tag{9.30}$$

where A is the cross-section of the sensing region perpendicular to the mainstream flow direction. To what extent the measured integral properties represent the real mean values in the flow system strongly depends on the type of sensor and the homogeneity of the dispersed phase. Considering, for example, horizontal pneumatic or hydraulic conveying, segregation effects due to gravitational settling may yield false measurements. This is illustrated in Figure 9.18. A horizontal arrangement of the two opposed sensors (e.g., based on the light attenuation method) will completely underestimate the particle mass flow rate, while a vertical arrangement results in an overestimation. Hence, accurate measurements are only possible when the condition of the dispersed phase in the instrument probe volume is representative of the entire cross-section of the pipe. Such a condition can be only realized in vertical conveying and when the sensors are placed far enough away from pipe bends or feeding systems to ensure a homogeneous mixture.

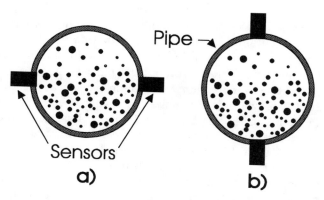

Figure 9.18: Possible sensor arrangements in horizontal pneumatic conveying; a) horizontal arrangement, b) vertical arrangement.

9.3   Local measurement techniques

Local measurement techniques enable the determination of properties in a two-phase flow with a relatively high spatial resolution, depending on the method applied. According to Table 9.1, local measurement techniques may be divided into three groups: probing methods (e.g., isokinetic sampling), field imaging techniques (e.g., particle image velocimetry) and single particle counting methods (e.g., light scattering, laser-Doppler and phase-Doppler anemometry). Probing methods are intrusive and may disturb the flow considerably depending on the application. They are, however, quite robust and hence widely used in industry for process control. The other methods described in this chapter are nonintrusive optical methods. These optical methods may be also divided according to the following two measurement principles.

- Single particle counting methods: The particle phase properties and the respective distribution functions are obtained for those particles passing through a small finite probe volume within a given measurement time. The result is therefore weighted by the particle number flux.

- Field imaging techniques: The distribution functions of the particle phase properties are obtained for those particles which happen to be in the probe volume at a give instant of time. Hence, the result is weighted by the particle number concentration.

Therefore, these two measurement principles yield different results which can be related, as demonstrated by Umhauer et al. (1990).

9.3.1   Isokinetic sampling

Isokinetic sampling can be used to measure particle mass flux, concentration, and density in a flowing suspension. Moreover, the collected particles may be further analyzed to obtain the particle size distribution or other properties such as surface area and shape factor using methods described in Section 9.1.1.

The basic requirement for measuring local particle flux is that the sample extracted must be totally representative of the suspension at the sample point in the two-phase stream, which is not an easy task when using isokinetic sampling. The principle of this method is based on inserting a sampling probe into the flowing two-phase system and gathering a representative sample of particles by a suction fan as shown in Figure 9.19. The sampled particles are collected in a bag filter for a certain time interval and then the particle mass is obtained by weighing the bag filter with and without particles. Hence, a particle mass flux is obtained which, under certain conditions, may be related to the local particle flux, concentration or density of the flowing suspension. The first problem which has to be considered is that the suction velocity u_s established in the sampling probe should be identical with the local gas velocity (i.e., isokinetic sampling condition). This may be achieved by adjusting the suction velocity in such a

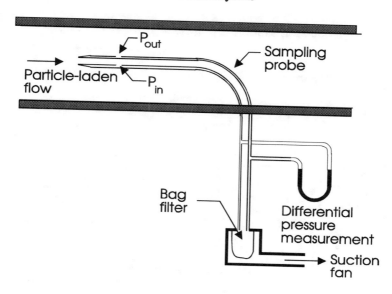

Figure 9.19: Operational principle of isokinetic sampling method.

way that the static pressures outside and inside the sampling probe become identical as shown in Figure 9.19.

In the event isokinetic conditions are not properly achieved the region from which particles are sampled is not identical with the cross-section of the sampling probe as illustrated in Figure 9.20 (Soo et al., 1969). When the suction velocity is smaller than the local fluid velocity, the region from which the particles are collected is smaller than the cross-section of the probe, as seen in Figure 9.20a, and when the suction velocity is too large the particles are collected from a larger area as illustrated in Figure 9.20b. Moreover, the boundary streamline of the flow and the boundary trajectory of the particle are not the same when the suction velocity does not match the local fluid velocity because of particle inertia. Hence, in this case, it is not possible to accurately determine the particle mass flux or concentration as will be shown below.

Determination of particle mass flux

The analysis of mass flux measurements is considered for the nonisokinetic condition $u_s < u_0$ shown in Figure 9.20a. All the properties used for this analysis are defined in Figure 9.22. The mass of particles which enter the sampling probe from the cross-section A_0, which is the collision cross-section for the fluid stream, is given by

$$M_{d1} = \bar{\rho}_{d0} v_0 A_0 \Delta t \tag{9.31}$$

For the condition $u_s < u_0$ a fraction ε of particles from the region $(A - A_0)$ also enter the sampling probe, i.e. $\varepsilon = (A_d - A_0)/(A - A_0)$. The mass collected from

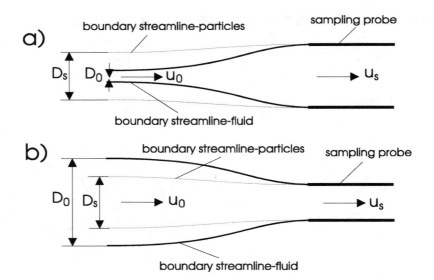

Figure 9.20: Streamlines and particle trajectories for different sampling conditions: a) $u_s < u_0$, b) $u_s > u_0$.

this region is

$$M_{d2} = \varepsilon \bar{\rho}_{d0} v_0 (A - A_0) \Delta t \tag{9.32}$$

Hence, the total mass of collected particles is obtained from

$$M_d = \bar{\rho}_d v A \Delta t = M_{d1} + M_{d2} = \bar{\rho}_{d0} v_0 [A_0 + \varepsilon (A - A_0)] \Delta t \tag{9.33}$$

Introducing the continuity equation for the gas flow,

$$\rho_0 u_0 A_0 = \rho u_s A \tag{9.34}$$

one obtains

$$\frac{\bar{\rho}_d v}{\bar{\rho}_{d0} v_0} = \frac{\rho}{\rho_0} \frac{u_s}{u_0} (1 - \varepsilon) + \varepsilon \tag{9.35}$$

If the fluid is incompressible, i.e., $\rho = \rho_0$, one obtains for the mass flux ratio

$$\frac{\bar{\rho}_d v}{\bar{\rho}_{d0} v_0} = \frac{u_s}{u_0} (1 - \varepsilon) + \varepsilon \tag{9.36}$$

This result shows that a correct measurement of the particle mass flux is only possible either under isokinetic conditions, i.e., $u_s = u_0$, or when the particles have sufficiently large inertia that the particle trajectories in the vicinity of the probe are straight lines and A_p is identical with A, i.e., $\varepsilon \to 1$ and $u_s \neq u_0$. Under these conditions one has:

$$\frac{\bar{\rho}_d v}{\bar{\rho}_{d0} v_0} = 1 \tag{9.37}$$

The above results furthermore reveal that the sampling is unaffected by the slip between the two phases, i.e., the relation between v_0 and u_0.

Definitions for analysis of sampling method.

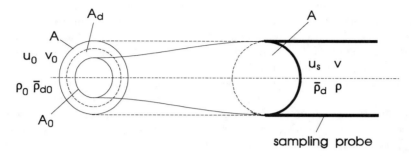

Figure 9.21: Definitions for analysis of sampling method.

Determination of particle bulk density

The particle bulk density is

$$\bar{\rho}_d = \frac{\delta M_d}{\delta V} \tag{9.38}$$

where δV is the volume of the mixture. The particle bulk density can also be expressed as

$$\bar{\rho}_d = \alpha_d \rho_d = \rho_d \frac{\delta V_d}{\delta V_d + \delta V_c} = \frac{\delta M_d}{\delta V_c}\left(\frac{1}{1 + M_d/\rho_d \delta V_c}\right) \tag{9.39}$$

The volume occupied by the fluid is

$$\delta V_c = A u_s \Delta t \tag{9.40}$$

Substituting Equations 9.33, 9.34 and 9.40 into Equation 9.39 gives, after some rearrangement,

$$\frac{\bar{\rho}_d}{\bar{\rho}_{d0}} = \frac{(\rho_c v_0/\rho_{c0} u_0)(1 - \varepsilon) + \varepsilon(v_0/u_s)}{1 + (\bar{\rho}_{d0}/\rho_d)\left[(\rho_c v_0/\rho_{c0} u_0)(1 - \varepsilon) + \varepsilon(v_0/u_s)\right]} \tag{9.41}$$

This equation takes into account nonisokinetic conditions and slip velocity between the carrier phase and the particles.

In the case of an incompressible fluid, i.e., $\rho_c = \rho_{c0}$, and low volume fraction, i.e., $\bar{\rho}_d/\rho_d \ll 1$, Equation 9.61 reduces to

$$\frac{\bar{\rho}_d}{\bar{\rho}_{d0}} = \frac{v_0}{u_0}(1 - \varepsilon) + \varepsilon \frac{v_0}{u_s} \tag{9.42}$$

If, in addition, isokinetic conditions are established, i.e., $u_s = u_0$, Equation 9.42 further reduces to

$$\frac{\bar{\rho}_d}{\bar{\rho}_{d0}} = \frac{v_0}{u_0} \tag{9.43}$$

This shows that the bulk density of the particle cloud can not be determined by isokinetic sampling alone. Additional measurement techniques are required to determine the slip velocity between fluid and particle phase. Only when very small particles are considered, so that $v_0 \approx u_0$, can isokinetic sampling be applied for particle bulk density measurements.

An analysis of the errors associated with nonisokinetic sampling conditions was performed by Bohnet (1973). Since the errors in concentration measurements largely depend on particle inertia, a dimensionless number (i.e., a Stokes number) was defined which characterizes the particle trajectory around the sampling probe. This number depends on the flow velocity u_0, the terminal velocity of the particle v_t, the diameter of the sampling probe D_{Pr}, and the gravitational constant g in the following way:

$$St_{\mathrm{Pr}} = B = \frac{u_0 v_t}{D_{\mathrm{Pr}} g} = \frac{\rho_d D^2 u_0}{18 \mu D_{\mathrm{Pr}}} \tag{9.44}$$

A diagram of the relative error in particle concentration measurements plotted versus the ratio of sampling velocity to flow velocity is shown in Figure 9.22 with the value B as a parameter. It is obvious that the error increases with larger deviations from the isokinetic condition and an increase in parameter B.

Moreover, isokinetic sampling cannot give accurate values for the particle cloud density in a carrier phase with a velocity gradient, i.e., shear flow or wall boundary layer, since the flow around the sampling probe becomes asymmetric and isokinetic conditions can hardly be established.

9.3.2   Optical fiber probes

Measurement systems based on optical fiber probes have been used in various configurations to perform local measurements of particle velocity, size, and concentration in two-phase flows. An optical fiber probe system consists of the probe head, which is inserted into the flow, a light source, a photodetector and the signal processing unit. The probe construction may differ principally in the arrangement of the light emitting fiber, which is connected to the light source, and the light receiving fiber, which transmits the light to the photodetector. Two different measurement principles may be identified.

- light attenuation method

- light reflection method.

The principle of operation also establishes the arrangement of the emitting and receiving fibers. In the light attenuation method, the optical fibers are arranged opposite to each other and the light propagates a given distance through the two-phase mixture before reaching the receiving fiber as shown

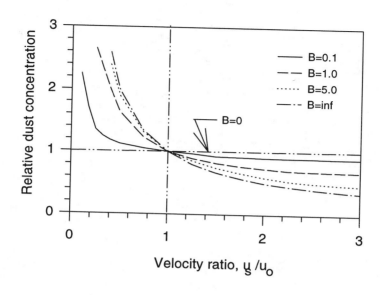

Figure 9.22: Relative error of particle concentration in the sampling probe under nonisokinetic conditions with B as a parameter. (Bohnet, M., *Chemie.-Ing.-Techn.*, 45, 22, 1973. With permission.)

in Figure 9.23a. Particle concentration measurements using this optical fiber arrangement may be based on the Lambert-Beer law for light attenuation (see also Section 9.2.1) or on counting individual particles passing through the probe volume between the two fibers. The method which is applicable depends on the ratio of fiber diameter to particle diameter. The particle counting technique may be applied when the particle diameter is of the same order as the fiber diameter so that the instrument operates as a single particle counter.

Fiber probes operating on the basis of the reflection method are constructed in such a way that the transmitting and receiving fibers are mounted parallel in the probe head as shown in Figure 9.23b. Particles moving in front of the probe head are illuminated by the transmitting fiber and scatter light in the backward direction. This light is received by a second fiber which is usually mounted in line with the emitting fiber.

Instead of separate transmitting and receiving fibers, a single fiber may also be used. In this case, the light from the light source is coupled into the sensor fiber by a fiber optical beam splitter and the light scattered by the particles is received by the same sensor fiber, diverted by the beam splitter and transmitted to the photodetector. Such single fiber reflection probes have been used by Licher and Louge (1991) and Rensner and Werther (1992) for measurements in dense two-phase flows, e.g., fluidized beds. The advantages of the single fiber

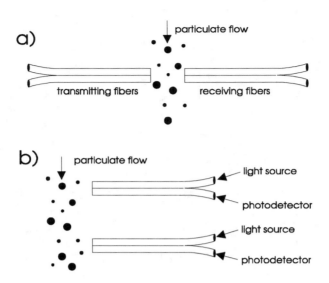

Figure 9.23: Principle arrangement of fiber optical probes, a) light attenuation method, b) light scattering (reflection) method.

reflection probe are its small size (typically smaller than 1 mm in diameter) and the ability to withstand erosion. On the other hand, it is only possible to measure particle concentration. This requires an estimate of the effective probe volume, i.e., the area from which the scattered light is received by the probe. A detailed analysis of the effective measuring volume of single fiber reflection probes was performed by Rensner and Werther (1992).

Particle velocity measurements by fiber optical probes are possible, without calibration, based on the passage time of the particles between two successive sensors or the frequency method which requires a special arrangement of a number of emitting and receiving fibers in the probe head. The *passage-time method* isbased on the cross-correlation of the signals received by two sets of emitting and receiving fibers and can be realized for both the light attenuation and the light scattering method, except for the single fiber reflection probe which only allows for concentration measurements. The cross-correlation method requires that the two sensor pairs are aligned with the main flow direction in order to achieve a high degree of correlation. Furthermore, the required separation of the sensor pairs has to fulfill certain requirements in order to give a high degree of correlation, as recommended by Klipphan (1977). A more detailed description of the cross-correlation method has been given in Section 9.2.3.

The frequency method for measuring particle velocities (Petrak & Hoffmann, 1985) is based on using a number of illuminating fibers combined with a regular arrangement of receiving fibers at a given separation as illustrated in Figure 9.24. All the receiving fibers are connected to one photodetector. Therefore, the line array of receiving fibers acts as a spatial grid whereby the scattered light is modulated so the signal from the photodetector shows a sinusoidal structure with a dominant frequency f_0. This frequency can be related to the particle velocity by:

$$v = f_0 s \qquad (9.45)$$

where s is the spacing between the receiving fibers. From the arrangement of the receiving fibers it is obvious that the frequency method requires a particle motion parallel to the line of fibers, otherwise the signal will have only a low number of zero-crossings which makes an evaluation of the signal frequency difficult.

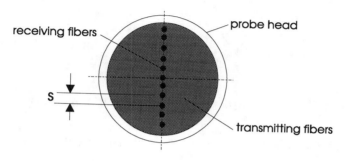

Figure 9.24: Cross-section of fiber optic probe used for the frequency method.

9.3.3 Scattering intensity measurements

In this section the basic theory of light scattering will be briefly summarized and results for the dependence of scattering intensity on particle size will be introduced. The principles of light scattering are also important for laser-Doppler and phase-Doppler anemometry and imaging techniques which will be introduced later.

The interaction of light with particles results in absorption and scattering of the light. The light scattered can be calculated using the theory of Mie (1908) which requires the solution of Maxwell's wave equations for the case of scattering of a plane electromagnetic wave by a homogeneous sphere. For more details about *Mie scattering theory*, the reader is referred to the relevant literature, e.g., van de Hulst (1981).

The absolute intensity of the light scattered by one particle, q, depends on the intensity of the incident light I_0, the wavelength of the light λ, the polarization angle of the light γ, the particle diameter D, the complex index of refraction $\bar{n}$ and the scattering angle φ

$$q = I_0 \frac{\lambda^2}{8\pi^2} i(\lambda, \gamma, \varphi, D, n, \kappa) \qquad (9.46)$$

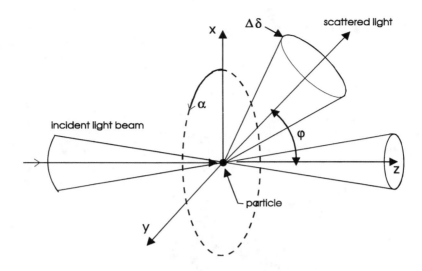

Figure 9.25: Optical arrangement for light scattering measurements.

where I represents the so-called *Mie intensity*. The complex refractive index is given by

$$\bar{n} = n(1 - i\kappa) \tag{9.47}$$

Optical arrangements for scattering intensity measurements usually also involve the integration of the scattered light over the receiving aperture for a selected mean scattering angle. Such a typical optical arrangement is illustrated in Figure 9.25 where $\Delta\delta$ indicates the receiving solid aperture angle. For parallel and monochromatic light as the illuminating source, the resulting scattering intensity is obtained by integration.

$$Q(\lambda, D, n, \kappa, \Delta\delta) = I_0 \frac{\lambda^2}{8\pi^2} \int_0^{2\pi} \int_{\Delta\delta} I(\lambda, \gamma, \varphi, D, n, \kappa) \sin\varphi \, d\varphi \, d\gamma \tag{9.48}$$

In the event white light is used instead of monochromatic light the scattering intensity is obtained by additionally integrating over the wavelength range, which has some advantages with regard to a smooth correlation between scattering intensity and particle size. In the following section the principles of scattering intensity measurements will be introduced by using Mie calculations (DANTEC/Invent 1994), whereby one can get an idea of the range of applicability and design of the instrument.

Typical examples of the scattering intensity around a particle for various particle diameters are shown in Figure 9.26 for parallel polarization. For the smallest particle (i.e., 0.1 μm) the angular intensity distribution has only two lobes for parallel polarization, which is characteristic of the Rayleigh scattering

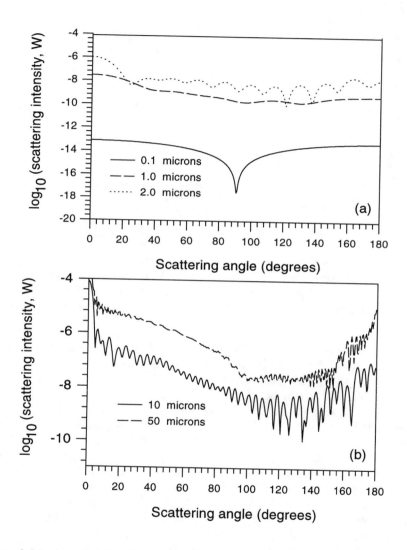

Figure 9.26: Angular distribution of scattering intensity around a particle ($I_0 = 10^7$ W/m^2, $\lambda = 632.8$ nm, $n = 1.5$, parallel polarization with respect to the polarization of the incident light), a) particle diameters 0.1, 1 and 2 μm, b) particle diameter 10 and 50 μm).

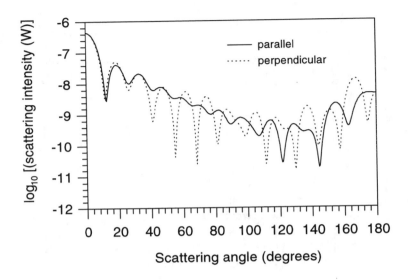

Figure 9.27: Angular distribution of scattering intensity around a particle; comparison of parallel and prependicular polarization with respect to polarization of the incident light ($I_0 = 10^7$ W/m^2, $\lambda = 632.8$ nm, $D = 2.8$ nm, $n = 1.5$).

range. It is obvious that the absolute intensity of the scattered light increases with particle size. Moreover, the number of lobes increases for larger particles. The highest scattering intensity is concentrated in the forward scattering direction and is mainly a result of diffraction. The influence of polarization on the scattering intensity around a 2.8 μm particle is illustrated in Figure 9.27. It is obvious that the intensity distribution for a polarization parallel to the incident light is much smoother compared to aperpendicular polarization.

In the following the scattering intensity as a function of the particle diameter is analyzed in more detail for various properties of the particles and configurations of the optical system with the parameters specified in Figure 9.28. The results presented are again based on the Mie theory and the calculations were performed with the software package "STREU" (DANTEC/Invert, 1994). In the Figure 9.28 a typical dependence of the scattering intensity on the particle diameter is shown. From this correlation three scattering regimes may be identified which are often characterized by the *Mie parameter*, $\alpha = \pi D / \lambda$.

- The so-called *Rayleigh-scattering* applies for particles that are small compared with the wavelength of the incident light, i.e., $\alpha << 1$ or $D < \lambda/10$. This regime is named after Lord Rayleigh, who first derived the basic scattering theory for such small particles (Rayleigh, 1881). Characteristic for this regime is a dependence of the scattering intensity on the fourth to

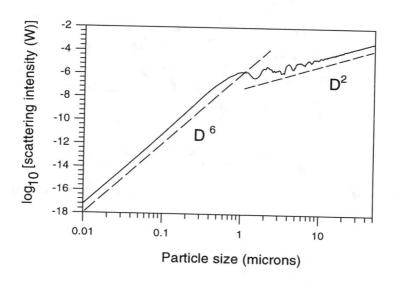

Figure 9.28: Dependence of scattering intensity on particle size ($I_0 = 10^7$ W/m^2, $\varphi = 15^o$, $\Delta\delta = 10^o$, $n = 1.5$).

sixth power of the particle diameter.

- For very large particles, i.e., $\alpha \gg 1$ or $D > 4\lambda$ the laws of *geometrical optics* are applicable under certain conditions (van de Hulst, 1981). The light scattering intensity varies approximately with the square of the particle diameter.

- The intermediate regime (i.e., $D \approx \lambda$) is called the Mie region, which is characterized by large oscillations in the scattering intensity, depending on the observation angle and the particle properties. Hence, the scattering intensity cannot be uniquely related to the particle size.

For large particles, i.e., $D \gg \lambda$, the scattered light is composed of three components: namely, diffracted, externally reflected and internally refracted light as indicated in Figure 9.29. The refracted light may be separated in several modes depending on the number of internal reflections, i.e., $P1, P2, P3, ...Pn$. Light diffraction is concentrated in the forward scattering direction, i.e., the so-called forward lobe, and is the dominant scattering phenomenon. This regime of geometrical optics is called the *Fraunhofer diffraction regime*. The diffraction pattern and the angular range of this forward lobe are dependent on the wavelength of the light and the particle diameter. As already demonstrated in Figure 9.26, the angular range of diffracted light decreases with increasing particle diameter and is given by $\varphi < \pm\varphi_{diff}$ with

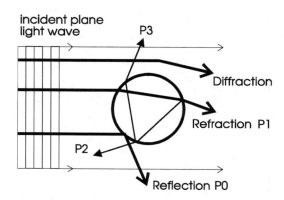

Figure 9.29: Different scattering modes for a spherical particle in the geometric optics regime.

$$\sin \varphi_{diff} = \frac{5}{4} \frac{\lambda}{D} \qquad (9.49)$$

Moreover, it is important to note that the intensity of the diffracted light is independent of the optical constants of the particle material, which might have some advantages in sizing particles of different or unknown refractive indices.

External reflected light is scattered over the entire angular range, i.e. $0° < \varphi < 180°$, whereas refracted light (i.e., $P1$) does not exceed an upper angular limit, $\varphi_{refr.}$, which is given by

$$\cos \left(\frac{\varphi_{refr}}{2} \right) = \frac{n_m}{n_p} \qquad (9.50)$$

for $n_p/n_m > 1.0$. Hence, this upper angular limit is determined by the relative refractive index. For a given fluid and $n_p/n_m > 1.0$, $\varphi_{refr.}$ decreases and more refracted light is concentrated in the forward direction with a decreasing refractive index of the particles. The properties of reflected and refracted light will be addressed in Section 9.3.5. The Mie response functions for three scattering angles, i.e., $\varphi = 0°, 15°$, and $90°$, are shown in Figure 9.30 for different refractive indices in the size range between 0.1 and 20 μm. The calculations were performed for a receiving aperture of $\Delta\delta = 10°$. The results may be summarized as follows:

- In the narrow forward scattering range the scattering intensity is almost independent of the refractive index of the particle. This is a result of the dominance of diffraction in this angular region. As described above, diffraction is independent of the optical properties of the particles. Moreover, the scattering intensity varies approximately with the fourth power of the particle diameter.

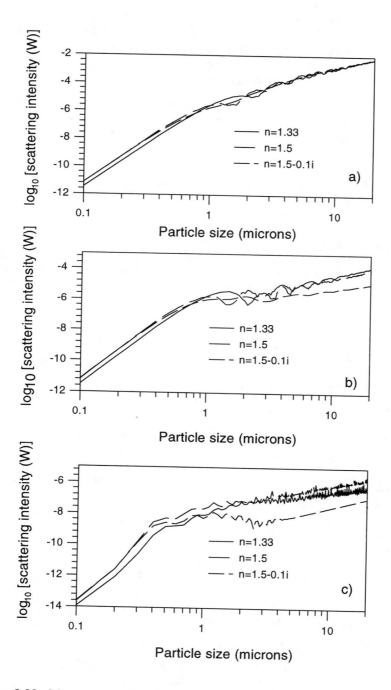

Figure 9.30: Mie response for three scattering angles: a) $\varphi = 0^o$, b) $\varphi = 15^o$, c) $\varphi = 90^o$, for spherical particles with different refractive indices ($I_o = 10^7$ W/m^2, $\lambda = 632.8$ nm, $\Delta\delta = 10^o$).

- For off-axis locations the scattering intensity shows large-scale fluctuations in the Mie range, i.e., the scattering intensity cannot be uniquely related to the particle size. In the Rayleigh range the scattering intensity varies with the 6th power of the particle diameter and in the range of geometrical optics, the square law may be applied.

- At a 90° scattering angle, the largest differences in the scattering intensity are observed for the different refractive indices and the absolute value of the scattering intensity is considerably lower than for the other scattering angles.

To reduce the intensity fluctuations, especially in the Mie range, and to establish a smoother size-intensity correlation, the aperture of the receiving optics may be increased so the scattered light is received from a larger angular region and the oscillations in the angular distribution of the scattering intensity are smoothed out over a larger region as shown in Figure 9.31. Also, the use of white light can improve the smoothness of the response curve, since this is associated with an additional integration over the wavelength spectrum (Broßman, 1966).

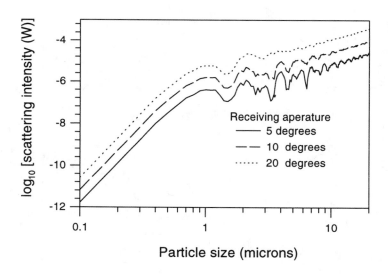

Figure 9.31: Influence of the receiving aperture opening on the Mie response function ($I_o = 10^7$ W/m^2, $\lambda = 632.8$ nm, $\varphi = 15°$).

Particle size measurements by the light intensity method require that only one particle be in the measurement volume at a time in order to enable the determination of the particle size from the measured signal amplitude. The simultaneous presence of more than one particle would result in an erroneous

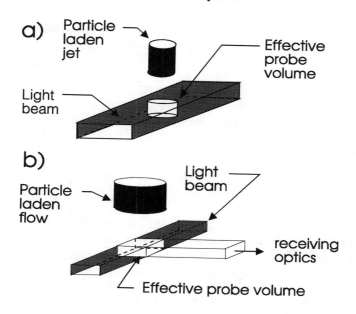

Figure 9.32: Methods of defining the size of the measurement volume: a) aerodynamic or hydrodynamic focusing, b) optical demarcation of the probe volume.

measurement called *coincidence error*. Therefore, sizing methods based on scattering intensity measurements are also called single particle counting techniques as described above. In order to fulfill this requirement it is necessary to limit the measurement volume size.

In principle, the measuring volume can be defined in two ways. The first method is based on directing a narrow particle laden gas or liquid stream through the center portion of the measurement volume as shown in Figure 9.32a. In this case, the diameter of the stream determines the measurement volume size, not the optical arrangement. This method is also called *aerodynamic* or *hydrodynamic focusing* and can be only applied by sampling the particles from the two-phase system under consideration, i.e., by isokinetic sampling.

The second method for defining the measurement volume is based on an appropriate optical design of the transmitting and receiving optics using imaging masks in order to allow a demarcation of the probe volume as shown in Figure 9.32b. In so doing it is also possible to apply the light scattering instrument for an on-line determination of particle size and concentration.

The size of the measurement volume has to be selected in such a way that, for a given maximum particle number concentration, the coincidence error is reduced to a large extent. The probability P_k for the presence of k-particles in the measurement volume follows a Poisson distribution

$$P_k = \frac{N^k}{k!} e^{-N} \tag{9.51}$$

where, $N = nV_m$, is the number of particles in the probe volume V_m. The relative probability for the presence of two particles is

$$P_2' = \frac{P_2}{P_1} = \frac{N}{2} \tag{9.52}$$

For a maximum allowable coincidence error of 5% (i.e., $P_2' = 0.05$), the limiting averaged particle number in the measurement volume must be $N = 0.1$. This results in a maximum particle number concentration of

$$n_{\max} = \frac{0.1}{V_m} \tag{9.53}$$

This equation allows one to estimate the required size of the measurement volume for a given particle concentration. Especially for on-line measurements this criterion is quite often a limiting factor, so rather small measurement volumes have to be realized which are only possible by an appropriate optical configurations. Therefore, quite often, off-axis orientations of the receiving optics have to be selected in order to allow a better demarcation of the measurement volume. This results in lower scattering intensities compared to a forward scattering arrangement (see Figure 9.30) and may limit the lower detectable particle size.

Moreover, the measurement volume must be illuminated uniformly, which can be achieved by properly shaping the light beam using masks in the transmitting optics. A more sophisticated approach is the elimination of the boundary zone error as illustrated in Figure 9.33. Particles passing through the edge of the illuminating light beam (i.e., particle 2) will be only partly illuminated and hence their scattered light intensity is too low and not proportional to their size. If the scattering intensity is still above the detection level (i.e., trigger level), such particles will be detected as smaller particles. For particles passing the edge of the beam which is imaged into the photodetector (i.e., particle 3 in Figure 9.33) only a portion of the scattered light will be received and again their size will be underestimated. The boundary zone error can be eliminated, for example, by extended optical systems with two colocated measurement volumes as introduced by Umhauer (1983).

A typical optical setup for a particle size analyzer operating with a 90° scattering angle together with the signal processing system (Umhauer, 1983) is shown in Figure 9.34. The transmitting optics consists of a white light source, a condenser, an imaging mask and an imaging lens. The scattered light is focused into the photomultiplier using a lens system and an imaging mask which limits the area from which scattered light is collected.

The photomultiplier provides an analog signal which is first filtered to remove signal noise and then digitized to obtain the pulse height of the signal. A counter is used to determine the number of signals detected. The data are then acquired by a computer and further processed to determine the particle size distribution. The number frequency distribution is obtained by grouping the pulse height U into a number of classes of width ΔU_i

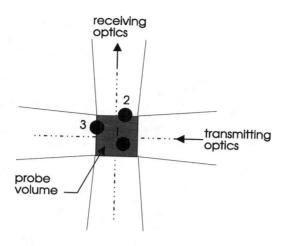

Figure 9.33: Illustration of the boundary zone error.

$$f_n(U_i) = \frac{1}{N} \frac{\Delta N(U_i)}{\Delta U_i} \qquad (9.54)$$

where $\Delta N(U_i)$ is the number of samples acquired in the amplitude interval i and N is the total number of samples acquired during the measuring time. In general, a calibration curve for U as a function of D has to be used to transform the distribution with respect to pulse height into a number frequency size distribution.

Moreover, it is possible to determine the particle number flux in the direction perpendicular to the probe volume cross-section from

$$\dot{n}_{tot} = \sum_{i=1}^{I} \dot{n}_i = \sum_{i=1}^{I} \frac{\Delta N(D_i)}{t_m A_m} = \frac{N}{t_m A_m} \qquad (9.55)$$

where I is the number of size classes, t_m is the measuring time and A_m is the cross-section of the measurement volume. By multiplying the number flux $\dot{n}_i$ by the particle mass $m_{d,i}$ the particle mass flux can also be obtained.

calibration curve has to be used to relate pulse height (i.e., scattering amplitude) to some characteristic particle size, such as the diameter of a spherical particle or an equivalent diameter. A determination of the calibration curve using calculations based on the Mie theory is only possible for the rather limited case of homogeneous spherical particles.

9.3.4 Laser-Doppler anemometry

Laser-Doppler and Phase-Doppler anemometry are the most advanced and accurate nonintrusive measuring techniques to obtain velocities of the fluid and

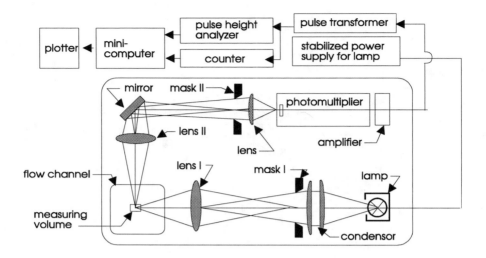

Figure 9.34: Optical configuration of a particle size analyzer operating at $90°$ scattering angle and signal processing. (Reprinted from *J. Aerosol Sci.*, 14, Umbauer, H., Particle size distribution analysis by scattered light measurements using an optically defined measuring volume, p. 765, Copyright 1983, with kind permission from Elsevier Science Ltd., The Boulevard, Langford Lane, Kidlington OX5 1GB, UK.)

particles in a two-phase flow. These measuring techniques enable one to obtain instantaneous (i.e., time series) and time-averaged measurements of the velocities with a high spatial resolution. Using phase-Doppler anemometry (PDA), the size of spherical particles, the refractive index of the particle and the particle concentration can be determined accurately as well.

The physical principle underlying LDA and PDA for velocity measurements is the *Doppler effect*, which relates the interaction of sound or light waves with a moving observer or the modulation of sound or light waves received by a stationary observer from a moving emitter. In LDA this principle is used in such a way that a laser emits plane light waves which are transmitted from a moving emitter, the particle. Hence, the frequency or wavelength of the light received by the particle is already modulated. Since the moving particle scatters the light into space, an additional Doppler shift occurs when the scattered light is received from a stationary observer as shown in Figure 9.35. Hence, the frequency of light received at the photodetector can be determined from

$$f_r = f_e \frac{1 - \frac{\vec{v} \cdot \vec{l}}{c}}{1 - \frac{\vec{v} \cdot \vec{k}}{c}} \tag{9.56}$$

where f_e is the frequency of the laser source (emitter), $\vec{v}$ is the velocity of the moving particle, c is the velocity of the light, and $\vec{k}, \vec{l}$ are unit vectors in the

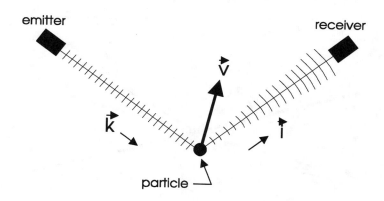

Figure 9.35: Doppler shift of scattered light from a moving particle.

direction as defined in Figure 9.35. The frequency of the scattered light f_r is, however, too high to allow direct detection by a photodetector. Therefore, two different methods are used in LDA such that the frequency of light to be detected is considerably reduced: the *reference beam method* and the *Doppler frequency difference method*.

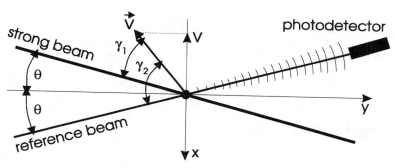

Figure 9.36: Configuration of reference beam LDA system.

This is achieved in the reference beam method by illuminating the particle with a strong light beam and interfering the resulting scattered light with a weak reference beam from the laser light source at the photodetector as shown in Figure 9.36. Subtracting the frequency of the reference beam gives the Doppler frequency:

$$f_D = f_r - f_e \tag{9.57}$$

Using Equation 9.56 and introducing $\vec{v} \cdot \vec{k} = -|\vec{v}|\cos\gamma_1$ and $\vec{v} \cdot \vec{l} = -|\vec{v}|\cos\gamma_2$ gives

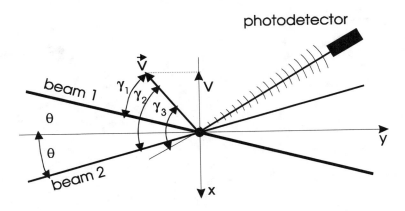

Figure 9.37: Configuration of the Doppler difference frequency method (dual beam LDA system).

$$f_D = f_e \frac{1 + \frac{|\vec{v}|}{c}\cos\gamma_1}{1 + \frac{|\vec{v}|}{c}\cos\gamma_2} - f_e \qquad (9.58)$$

$$f_D = \frac{1}{\lambda_e}\left(\frac{|\vec{v}|\left(\cos\gamma_1 - \cos\gamma_2\right)}{1 + \frac{|\vec{v}|}{c}\cos\gamma_2}\right) \qquad (9.59)$$

The velocity component perpendicular to the bisector of the two incident beams is

$$v = -|\vec{v}|\sin\frac{\gamma_1 + \gamma_2}{2}$$

Using the trigometric relationship

$$\cos\gamma_1 - \cos\gamma_2 = -2\sin\frac{\gamma_1 + \gamma_2}{2}\sin\frac{\gamma_1 - \gamma_2}{2}$$

one obtains

$$f_D = \frac{1}{\lambda_e}\left[\frac{2v\sin\theta}{1 + \frac{v}{c}\sin\theta}\right] \qquad (9.60)$$

where $\theta = \frac{1}{2}(\gamma_1 - \gamma_2)$. Since, in general, $v \ll c$, Equation 9.60 finally becomes

$$f_D = \frac{2v\sin\theta}{\lambda_c} \qquad (9.61)$$

It should be noted that the reference beam mode can be only operated at a fixed observation angle which coincides with the reference beam angle θ. Furthermore, the solid angle for light collection is limited to satisfy coherence requirements, i.e., the amount of scattered light which is collected is restricted.

More frequently, the Doppler frequency difference method is used for LDA measurements. Here the moving particle is illuminated by two laser beams from different directions as shown in Figure 9.37. In this case, the frequency of the scattered light is obtained from the difference of the contributions from two incident beams:

$$f_D = f_{r1} - f_{r2} \tag{9.62}$$

Using, once again, Equation 9.56 and the velocity components in the directions of the two beams, one obtains

$$f_D = f_e \left(\frac{1+(|\vec{v}|/c)\cos\gamma_1}{1+(|\vec{v}|/c)\cos\gamma_3} - \frac{1+(|\vec{v}|/c)\cos\gamma_2}{1+(|\vec{v}|/c)\cos\gamma_3} \right)$$

$$= \frac{|\vec{v}|}{\lambda_e} \frac{(\cos\gamma_1 - \cos\gamma_2)}{1+(|\vec{v}|/c)\cos\gamma_3} \tag{9.63}$$

For the velocity component perpendicular to the bisector of the two incident beams and with $v \ll c$, Equation 9.63 becomes

$$f_D = \frac{2v\sin\theta}{\lambda_c} \tag{9.64}$$

This expression is identical to the one obtained for the reference beam method. However, the observation angle can be arbitrarily selected in the Doppler difference method. This implies that the observation angle and solid angle of scattered light collection may be selected for convenience according to the desired application.

The principle of the LDA may be explained using the *fringe model* in the following way. If two coherent light beams cross, the interference of the light waves results in a fringe pattern parallel to the bisector plane, (i.e., the $y - z$ plane in Figure 9.39) which can be visualized on a screen when a lens of small focal length is placed at the interaction of the beams. As the particle passes through the LDA probe volume, the scattering intensity detected by a photodetector is modulated in such a way that the Gaussian-shaped absolute scattering intensity (which results from the Gaussian intensity distribution in the probe volume) is superimposed with an alternating pattern produced by the particles passing through the bright and dark fringe pattern. As pointed out by Durst (1982), it should be noted, however, that the fringe pattern does not exist for the particle and is the result of integration by the human eye and the photodetector, both of which have a response time much larger than the inverse of the frequency of the light waves. The fringe spacing d_f is basically the conversion factor to determine the particle velocity from the measured Doppler difference frequency.

$$v = \frac{1}{f_D} \frac{\lambda_e}{2\sin\theta} = \frac{d_f}{f_D} \tag{9.65}$$

For particles smaller than the fringe spacing, a completely modulated Doppler signal will be generated as shown in Figure 9.44. As the particle becomes larger than the fringe spacing, the signal modulation is reduced and the scattering intensity received by the photodetector does not reduce to zero in the Doppler signal.

A typical optical setup of an LDA system operated in the forward scattering mode is shown in Figure 9.38. The transmitting optics consist of the laser, a beam splitter, one or two Bragg cells and a transmitting lens. The Bragg cells introduce a frequency difference between the two incident beams whereby it is possible to detect the direction of particle motion in the measurement volume (see, for example, Durst et al., 1981). The receiving optics consists of an imaging lens with a mask in front of it and a photodetector with a pinhole.

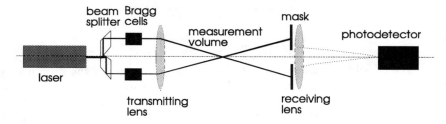

Figure 9.38: Typical optical setup of the dual beam laser-Doppler anemometer.

The spatial resolution of the velocity measurement depends on the dimensions of the LDA probe volume which is determined by the initial laser beam diameter, the beam crossing angle (determined by the initial beam spacing and focal length of the transmitting lens), the focal length of the receiving lens and the observation angle of the receiving optics. Since the incident focused laser beams have a Gaussian intensity distribution as shown in Figure 9.39, their waist diameter at the focal plane is taken to be that value at which the light intensity has diminished to $1/e^2$ of the maximum value at the beam axis. The waist diameter is given by

$$d_m = \frac{4 f_e \lambda_e}{\pi d_0} \qquad (9.66)$$

where d_0 is the $1/e^2$ unfocused laser beam diameter and f_e is the transmitting lens focal length. The probe volume established by the two crossing beams has an ellipsoidal shape as shown in Figure 9.39. The dimensions of the $1/e^2$ ellipsoid, corresponding to Figure 9.39, are given by

$$
\begin{aligned}
\Delta x &= \frac{d_m}{\cos\theta} \qquad (9.67) \\
\Delta y &= d_m \\
\Delta z &= \frac{d_m}{\sin\theta}
\end{aligned}
$$

The number of fringes in the $1/e^2$ measurement volume can be determined from

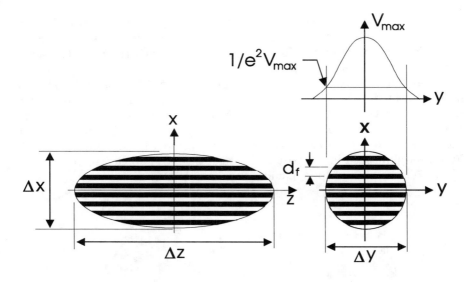

Figure 9.39: Dimensions of LDA probe volume.

$$N_f = \frac{8}{\pi} \frac{f_e}{d_o} \tan \theta = \frac{4}{\pi} \frac{\Delta b}{d_o} \tag{9.68}$$

where Δb is the initial spacing of the transmitting beams.

By using an off-axis orientation of the receiving optics, the length of the portion of the measurement volume imaged into the photodetector can be further reduced and the spatial resolution improved. However, due to the angular dependence of the light scattering intensity, any off-axis orientation of the receiving optics results in reduced scattering intensities as shown in Section 9.3.3.

For more details about the principle of the LDA, frequency shifting, properties of photodetectors and signal processing methods, the reader is referred to the relevant literature, such as Durst et al. (1981, 1987) and a recent review on advances in signal processing by Tropea (1995). Applications of LDA for two-phase flow measurements follow.

The basic ideas for LDA applications for two-phase flows were introduced by Farmer (1972, 1974), Durst and Zaré (1975) and Roberts (1977). They showed that LDA may also be used for velocity measurements of large reflecting and refracting particles. The light waves produced by the two incident laser beams reflect or refract at large particles, interfere, and produce fringes in space as indicated in Figure 9.40. The rate at which the fringes cross any point in space, i.e., at the photodetector, is the same at all points in the surrounding space and is linearly related to the velocity component of large non-deformable particles perpendicular to the symmetry line between the two incident beams. The theoretical derivations of Durst and Zaré (1975) revealed that the relations for the Doppler difference frequency for large reflecting or refracting particles are identical to the universal equation for laser-Doppler anemometry (Equation 9.64) when the intersection angle of the two incident beams is small and the photo-detector is placed at a distance much larger than the particle diameter from the measurement volume.

These findings are the basis for the application of LDA for particle velocity measurements in two-phase flows. Because LDA is a nonintrusive optical technique, it may be used for measurements in two-phase flow systems so long as optical access is possible and the two-phase system is dilute enough to allow the transmission of the laser beam and the scattered light. Numerous studies have been published in the past where LDA has been applied to various types of gas-solid two-phase flows, liquid sprays, and bubbly flows.

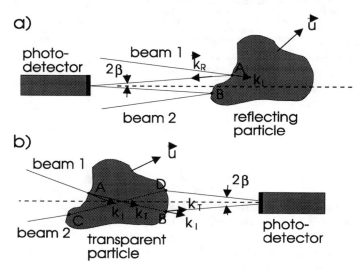

Figure 9.40: Interference of two laser beams for large reflecting (a) and refracting (b) particles. (The intersection points of the incident and refracted beams with the surface are indicated by A, B, C and D.)

There also have been several attempts to apply laser-Doppler anemometry for the simultaneous measurement of particle velocity, size and concentration (Farmer, 1972; Chigier et al., 1979; Durst, 1982; Hess, 1984; Hess and Espinola, 1984; Allano et al., 1984; Negus and Drain, 1982). The sizing of particles by LDA is based on:

- the absolute value of the scattering intensity (i.e., pedestal of the Doppler signal, Figure 9.43), or

- the signal visibility.

The *pedestal* of the Doppler signal is obtained by using the low frequency component of the signal obtained by using a low pass filter unit. As shown in

Section 9.3.3, the intensity of the scattered light depends on the particle size. However, the size-amplitude relation shows strong fluctuations in the Mie region when the particle size is comparable to the laser wavelength. Moreover, particle sizing based on intensity measurements generally require calibration. An additional problem with sizing particles by a standard LDA system is the effect of the non-uniform distribution of intensity within the measurement volume. Laser beams normally have a Gaussian intensity distribution. Particles passing through the edge of the measurement volume have a lower scattering intensity and are detected as smaller particles. This effect, which is also called *trajectory ambiguity*, results in the dimensions of the probe volume being dependent on particle size. A small particle passing the edge of the measurement volume might be not detected by the data acquisition system due to its low scattering intensity, while a large particle at the same location still produces a signal which lies above the detection level as shown in Figure 9.41. This effect also has consequences for the determination of particle concentration, which will be described later. Therefore, measurements of particle size and concentration by LDA require extensions of the optical system or data acquisition procedures in order to reduce errors due to the Gaussian beam effect.

The following methods have been mainly used for particle sizing by LDA:

- Limitation of the probe volume size by additional optical systems (i.e., gate photodetector or two-color systems),

- Modification of the laser beam to produce a "top-hat" intensity distribution,

- Computational deconvolution of signal intensity distributions.

There are several examples of particle size measurements using LDA which were developed before the introduction of PDA. Most of the techniques, based on intensity and visibility measurements described below, were shown not to be very reliable even though, in a few cases, they were used to develop commercial instrumentation. It should be emphasized, however, there is still a market for reliable instruments for local, single point size and velocity measurements in two-phase flows with nonspherical particles common to industrial processes.

In order to limit the region of the probe volume from where signals are received, Chigier et al. (1979) used additional receiving optics placed at 90° off-axis to trigger a second receiving system mounted in forward scattering. For further reduction of the trajectory ambiguity, an inversion routine was used to convolute the signal amplitude distributions obtained from many particles by applying an equation relating the signal peak amplitude to both the particle diameter and the particle location in the probe volume. A comparison of particle size distribution measurements by LDA with results obtained by the slide impaction method gave only fair agreement.

By superimposing two probe volumes of different diameter and color it is possible to trigger the data acquisition system only when the particles pass through the central part of the larger probe volume where the intensity is more

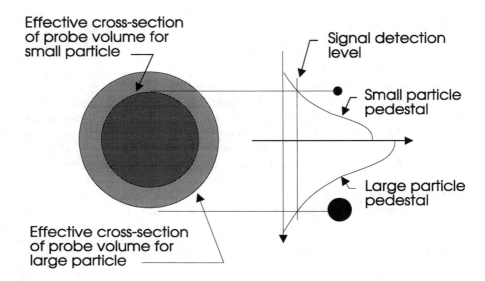

Figure 9.41: Illustration of Gaussian beam effect on intensity measurements by LDA and its influence on the effective cross-section of the measurement volume.

uniform. Such a co-axial arrangement of two probe volumes may be realized by using a two-component LDA system with a different waist diameter for each color (Yeoman et al., 1982; Modarres and Tan, 1983) or by overlapping a large diameter single beam with the LDA probe volume (Hess, 1984). When a particle passes through the LDA probe volume, the light scattering intensity from the larger single beam is measured to determine the particle size as shown in Figure 9.42.

For producing laser light beams with uniform intensity distribution, the "top-hat" technique may be applied. To produce such a top-hat profile, Allano et al. (1984) used a holographic filter and related the scattering intensity with the particle diameter by using the Lorenz-Mie theory. Grehan and Gouesbet (1986) studied this system for simultaneous measurements of particle size and velocity. Also, a combination of LDA with light scattering instruments has been applied for simultaneous particle size and velocity measurements (Durst, 1982).

In addition the signal visibility or signal modulation may be used for particle sizing by LDA (Farmer, 1972). Compared to the scattering intensity measurements the method has a number of advantages since the visibility is independent of scattering intensity and, hence, is neither biased by laser power nor detector sensitivity. The visibility is determined from the maximum and minimum amplitudes of the low-pass filtered Doppler signal as indicated in Figure 9.43.

$$V = \frac{I_{\max} - I_{\min}}{I_{\max} + I_{\min}} \tag{9.69}$$

The visibility of the Doppler signal decreases with increasing particle size as

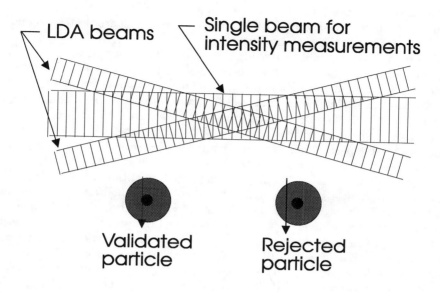

Figure 9.42: Co-axial arrangement of two probe volumes of different color.

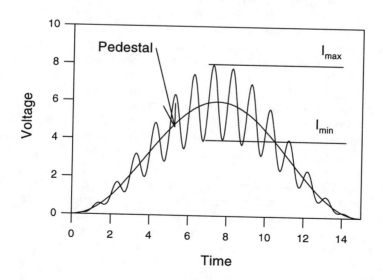

Figure 9.43: Doppler signal and definition of visibility and modulation.

illustrated in Figure 9.44. The first lobe in the visibility curve covers the measurable particle size range. With a further increase in particle size secondary maxima appear in the visibility curve.

Moreover, the visibility curve strongly depends on the optical configuration of the receiving optics, i.e., the off-axis angle and the size and shape of the imaging mask in the receiving optics. The latter effect was evaluated in detail by Negus and Drain (1982). Example Mie calculations for the visibility curves for different optical configurations are shown in Figure 9.45. It is obvious that the shape of the imaging mask influences the measurable size range. However, the measurable particle size range is significantly increased by using an off-axis arrangement of the receiving optics.

Extensive research has been performed on the suitability of the visibility method for particle sizing. It was found that this method appears to be very sensitive to the positioning of the aperture mask, the accuracy of the mask dimensions and the particle trajectory through the LDA probe volume. The latter effect may be minimized by using a two-color probe volume with an appropriate validation scheme to ensure that only particles passing through the center of the probe volume are validated as done, for example, by Yoeman et al. (1982). A detailed review on the visibility method has recently been published by Tayali and Bates (1990) in which a number of other LDA-based methods are also described. More recently an LDA combined with a direct imaging technique, referred to as the *shadow-Doppler technique*, was developed for combined size and velocity measurements in two-phase flows with arbitrary shaped particles as found, for example, in coal combustion systems (Hardalupas et al., 1994).

Because of the particle size-dependent dimensions of the probe volume and the difficulties associated with particle size measurements using LDA, the measurement of particle concentration is generally based on a calibration procedure using the information for a global mass balance. In practice, accurate particle concentration measurements by laser-Doppler anemometry are only possible for simple one-dimensional flows with mono-sized particles. In this case, the probe volume size may be determined by calibration. This, however, does not remove the problem related to a spatial distribution of particle concentration in the flow which affects the scattering intensity received by the photodetector from the measurement location due to different optical path lengths through the particle-laden flow and the associated different rates of light absorption (Kliafas et al., 1987).

For a simultaneous determination of fluid and particle velocity by LDA, the fluid flow has to be additionally seeded by small tracer particles which are able to follow the turbulent fluctuations. The remaining task is the separation of the Doppler signals resulting from tracer particles and the dispersed phase particles. In most cases, this discrimination is based on the scattering intensity combined with some other method to reduce the error due to the Gaussian beam effect. The discrimination procedure introduced by Durst (1982), for example, was based on the use of two receiving optical systems and two photodiodes which detect the blockage of the incident beams by large particles. Together with sophisticated signal processing it was possible to successfully separate signals

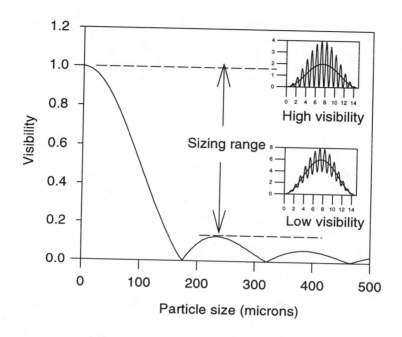

Figure 9.44: Variation of signal visibility with particle size (off-axis light collection).

from large and tracer particles.

An improved amplitude discrimination procedure using two superimposed measurement volumes of a different size and color was developed by Modarres and Tan (1983). The smaller or pointer probe volume was used only to trigger the measurements from the larger control volume. By this scheme it was ensured that the sampled signals were received only from the center part of the larger probe volume where the spatial intensity distribution is relatively constant.

A combined amplitude-visibility discrimination method which did not rely on additional optical components was proposed by Börner et al. (1986). After first separating the signals based on the signal amplitude, the visibility of all signals was determined to ensure that no samples from large particles passing the edge of the measurement volume were collected as tracer particles. This method required additional electronic equipment and sophisticated software for signal processing.

A much simpler amplitude discrimination method was introduced by Hishida and Maeda (1990). To ensure that only particles traversing the center of the measurement volume were sampled, a minimum number of zero crossings in the

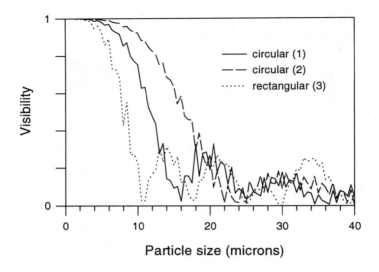

Figure 9.45: Visibility curves for different optical configurations of the receiving optics ($\lambda = 632.8$ nm, $\varphi = 0^\circ$; [1] $d_f = 10.2$ μm circular mask, $\Delta\delta = 4^\circ$; [2] $d_f = 18$ μm circular mask, $\Delta\delta = 4^\circ$; [3] $d_f = 6.55$ μm rectangular mask, receiving aperature angle in horizontal and vertical direction, $\Delta\delta_h = 11^\circ$, $\Delta\delta_v = 4^\circ$).

Doppler signal were required for validation.

The discrimination procedures described here can be successfully applied only when the size distribution of the dispersed phase particles is well separated from the size distribution of the tracer particles.

9.3.5 Phase-Doppler anemometry

The principle of phase-Doppler anemometry (PDA) is based on the Doppler difference method used for conventional laser-Doppler anemometry and was first introduced by Durst and Zaré (1975). By using an extended receiving optical system with two or more photodetectors it is possible to measure particle size and velocity simultaneously. The phase shift of the light scattered by refraction or reflection from the two intersecting laser beams is used to obtain the particle size

A typical optical set-up of a two detector PDA-system is shown in Figure 9.48. The transmitting optics is the conventional dual beam LDA optics which, in this case, uses two Bragg cells for frequency shifting. The PDA receiver is positioned at the off-axis angle φ and consists of a collection lens which produces a parallel beam of scattered light. This parallel light beam passes through a mask which defines the elevation angles of the two photo-detectors.. In this case

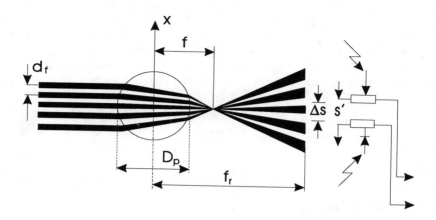

Figure 9.46: Fringe model for phase-Doppler principle for the case of refraction.

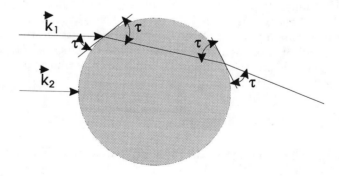

Figure 9.47: Phase difference of a light ray refracted at a spherical particle.

the mask has two rectangular slits in order to allow the light to pass through to the photo-detectors.. The slits are located parallel to the $y - z$ plane at the elevation angle $\pm\Psi$. The light is then focused on a spatial filter, i.e., a vertical slit which defines the effective length of the probe volume from where the scattered light can be received. The effective length of the probe volume is determined by the width of the spatial filter l_s (typically 100 μm) and the magnification of the receiving optics; that is, the focal length of the collecting lens to the second lens, $L_s = l_s f_1 / f_2$. Finally, the scattered light passing the two rectangular slits is focused onto the photo-detectors using two additional lenses.

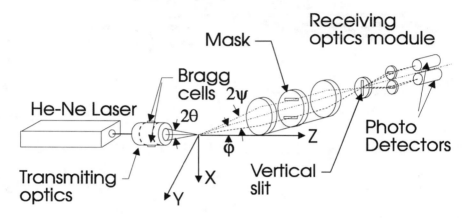

Figure 9.48: Optical configuration of a two-detector phase-Doppler anemometer.

The operational principle of PDA can be explained using the simple fringe-type model assuming that the interference fringes in the intersection region of the two incident light beams of the LDA are parallel light rays (Saffman, 1987). A spherical transparent particle placed into this fringe pattern will act as a kind of lens which will project the light rays into space as indicated in Figure 9.46. The separation of the projected fringes at a distance f_r from the particle is given approximately by

$$\Delta s \approx (f_r - f)\frac{d_f}{f} \tag{9.70}$$

where d_f is the fringe spacing in the measurement volume. The focal length of the particle is given by

$$f = \frac{m}{m-1}\frac{D}{4} \tag{9.71}$$

where $m = n_d/n_m$, the ratio of the refractive index of the particle to that of the surrounding medium. Since small particles are considered and f_r is usually much larger than the particle diameter, one obtains

$$\Delta s \approx f_r \frac{d_f}{f} \tag{9.72}$$

The separation of the projected fringes is obtained from

$$\Delta s \approx \frac{4 f_r d_f}{D} \frac{m-1}{m} \tag{9.73}$$

In general, the particles move through the probe volume and the fringes move as well so it is difficult to measure this spatial separation. However, if two photodetectors are symmetrically placed at f_r with a separation $\Delta s'$, as shown in Figure 9.46, the fringes produced by the moving particle will sweep across the two detectors at the Doppler difference frequency. The signals seen by the two detectors will have a relative phase difference given by

$$\phi = 2\pi \frac{\Delta s'}{\Delta s} = 2\pi \frac{2 f_r}{\Delta s} \sin \psi \tag{9.74}$$

or by using Equation 9.73

$$\Delta \phi = \pi \frac{D}{d_f} \frac{m}{m-1} \sin \psi = \frac{2\pi D}{\lambda} \frac{m}{m-1} \sin \theta \sin \psi \tag{9.75}$$

where ψ is the elevation angle of one photodetector measured from the bisector plane of the two incident beams which is the $y - z$ plane in Figure 9.48 where the optical axis of the PDA receiver is also located. It should be emphasized that Equation 9.75 is an approximation valid only for small scattering angles φ which represents the off-axis angle measured from the forward scattering direction, the z direction defined in Figure 9.48. The equation is very useful, however, to provide a rough estimate of the measurable particle size range for a given system or to perform a preliminary design of the optical configuration for small scattering angles. To determine the particle size from the measured phase difference, the required correlations are derived from geometrical optics which is valid for particles that are large compared to the wavelength of light (van de Hulst, 1981). The phase of the scattered light is given by

$$\phi = \frac{2\pi D n_m}{\lambda} \left(\sin \tau - p \frac{n_d}{n_m} \sin \tau' \right) \tag{9.76}$$

where n_m and n_d are the refractive indices of the medium surrounding the particle and medium of the particle itself. The parameter p indicates the type of scattering, i.e., $p = 0, 1, 2, ...$ for reflection, first-order refraction, second-order refraction, and so on. Moreover, τ and τ' are the angles between the incident ray and the surface tangent and the refracted ray and the surface tangent, respectively, as shown in Figure 9.47. For a dual beam LDA system, the phase difference of the light scattered from each of the two beams is given in a similar way

$$\Delta \phi = \frac{2\pi D n_m}{\lambda} \left[(\sin \tau_1 - \sin \tau_2) - p \frac{n_d}{n_m} (\sin \tau_1' - \sin \tau_2') \right] \tag{9.77}$$

where the subscripts 1 and 2 are used to indicate the contributions from both incident beams. For two photodetectors placed at a certain off-axis angle φ and placed symmetrically with respect to the bisector plane at the elevation angles $\pm\psi$, one obtains the phase difference (see, for example, Bauckhage, 1988)

$$\Delta\phi = (2\pi D n_m / \lambda)\, \Phi \tag{9.78}$$

The parameter Φ depends on the scattering mode. For reflection ($p = 0$)

$$\Phi = \sqrt{2}[(1 + \sin\theta\sin\psi - \cos\theta\cos\psi\cos\varphi)^{1/2} \\ -(1 - \sin\theta\sin\psi - \cos\theta\cos\psi\cos\varphi)^{1/2}] \tag{9.79}$$

and for refraction ($p = 1$)

$$\Phi = 2\left\{ \begin{array}{c} [1 + m^2 - \sqrt{2}m(1 + \sin\theta\sin\psi + \cos\theta\cos\psi\cos\varphi)^{1/2}]^{1/2} \\ -\left[1 + m^2 - \sqrt{2}m(1 - \sin\theta\sin\psi + \cos\theta\cos\psi\cos\varphi)^{1/2}\right]^{1/2} \end{array} \right\} \tag{9.80}$$

where $m = n_d / n_m$ has been used for convenience and 2θ represents the angle between the two incident beams. Since the phase difference is a function of p, one expects a linear relation for the correlation between particle size and phase (see Equation 9.76 and Equation 9.77) for only those scattering angles where one scattering mode is dominant (i.e., reflection or refraction). Therefore, the values for Φ have been given for these two scattering modes only (i.e., Equations 9.79 and 9.80). Other scattering modes, i.e., $p = 2$, may be also used for phase measurements, especially in the region of backscattering as will be shown later. Such a backscatter arrangement might have advantages with regard to optical access since both the incident beams and the scattered light may be transmitted through one window.

By recording the band-pass filtered Doppler signals from the two photodetectors the phase $\Delta\phi$ is determined from the time lag between the two signals as indicated in Figure 9.49.

$$\Delta\phi = 2\pi \frac{\Delta t}{T} \tag{9.81}$$

where T is the time of one cycle of the signal. With Equation 9.81 it is now possible to determine the particle diameter for a given refractive index n_m and wavelength λ.

$$D = \frac{\lambda}{2\pi m_m} \frac{1}{\Phi} \Delta\phi \tag{9.82}$$

From Equation 9.81 and Figure 9.49 it is also obvious that only a phase shift between zero and 2π can be distinguished with a two detector PDA system, which limits the measurable particle size range for a given optical configuration. Therefore, three-detector systems are also used in which two phase differences are obtained from detector pairs having different spacing (Figure 9.50). This method enables one to extend the measurable particle size range while maintaining the resolution of the measurement. Moreover, the ratio of the

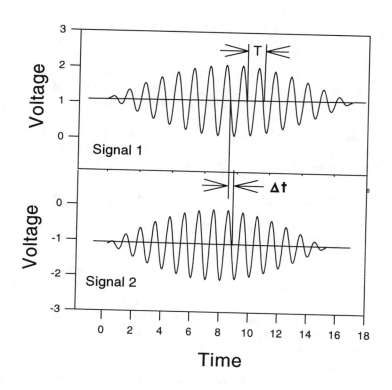

Figure 9.49: Determination of phase shift from two band-pass filtered Doppler signals.

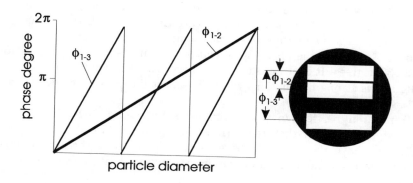

Figure 9.50: Phase size relations for a three-detector phase-Doppler system.

two phase measurements may be used for additional validation, such as checking the sphericity of deformable particles (liquid droplets or bubbles).

The principle of geometrical optics introduced previously to calculate light scattering from particles is limited to particle sizes larger than the wavelength of light and takes into account reflection and refraction. This implies that no interference of the scattering modes is considered (van de Hulst, 1981). In most cases, this theory is sufficient to support the optical design of PDA systems. Also this theory can be easily extended to account for the Gaussian light intensity distribution in the probe volume (Sankar and Bachalo, 1991).

For small particles, diffraction represents an especially important contribution to the light scattering which may affect and disturb the phase measurement. Therefore, the more general Mie theory has to also be applied to determine the scattering characteristics for a particle of any given size. The Mie theory relies on the direct solution of Maxwell's equations for the case of the scattering of a plane light wave by a homogeneous spherical particle for arbitrary size and refractive index. In order to calculate the scattered field of a PDA system, it is necessary to add the contributions of the two incident beams and the average over the aperture of the receiving optics by taking into account the polarization and phase of each beam. Hence, it is possible to determine the intensity, visibility and phase for arbitrary optical configurations. To allow for the influence of the Gaussian beam, the generalized Lorenz-Mie theory (GLMT) has also been recently applied to optimize PDA systems (Grehan et al., 1992).

There are several issues with regard to the optimum selection of the optical systems for different types of particles (i.e., reflecting and transparent particles) based on calculations by geometrical optics and Mie theory (DANTEC/Invent, 1994). The calculations based on geometrical optics are performed for a point-like aperture while the Mie calculations account for the integration over a rectangular aperture with given half angles in the horizontal (δ_h) and vertical (δ_v) directions with respect to the $y - z$ plane as shown in Figure 9.48. It should be noted that the integration of the scattered light over the receiving aperture is important for obtaining a linear phase−size relation.

For totally reflecting or strongly absorbing particles, any scattering angle may be used except the near-forward scattering range where diffraction will destroy the linearity of the phase−size relation. This effect is illustrated in Figure 9.51 showing the phase−size relation for three scattering angles, 10 and 20 and 30 degrees. It is obvious that the phase−size relation for small scattering angles shows strong fluctuations while an almost linear relation is obtained for the larger scattering angle.

Transparent particles may be distinguished between those having a refractive index larger or smaller than the surrounding medium. Liquid droplets or glass beads in air have a relative refractive index which is larger than unity typically in the range 1.3 to 1.5, and bubbles in liquid have a relative refractive index less than unity.

The selection of the optimum optical configuration should be based mainly on the relative importance of the scattering mode considered (i.e., reflection, refraction or second-order refraction) with respect to the other modes and the

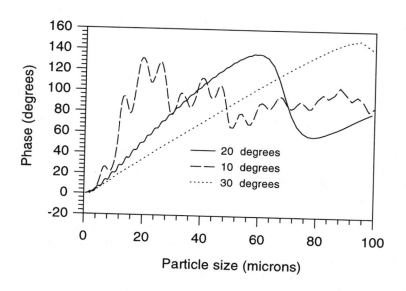

Figure 9.51: Phase–size relation calculated by Mie theory for different off-axis collection angles ($\lambda = 0.6328$ μm, p polarization, $\theta = 2.77^\circ$, $\Psi = 1.85^\circ$, $\delta_h = 5.53^\circ$, $\delta_v = 1.85^\circ$).

resulting linearity of the phase–size relation. The relative intensities of the different scattering modes are determined by using calculations based on geometrical optics where both parallel (p) and perpendicular (s) polarization are considered as shown in Figure 9.52. As pointed out earlier, reflected light covers the entire angular range for refractive index ratios below and above unity. However, a distinct minimum is found for parallel polarization at the so-called *Brewster's angle* which is given by

$$\varphi_B = 2\tan^{-1}(1/m) \tag{9.83}$$

The Brewster's angle decreases with an increasing refractive index ratio. First-order refraction is concentrated in the forward scattering range and extends up to the critical angle which, for different relative refractive indices $m = n_p/n_m$, is given as:

$$\varphi_c = 2\cos^{-1}(m) \qquad m < 1 \tag{9.84}$$

$$\varphi_c = 2\cos^{-1}(1/m) \qquad m > 1 \tag{9.85}$$

The critical angle increases with an increasing relative refractive index and first-order refraction becomes dominant over reflection over a wider angular range.

Second-order refraction again covers the entire angular range for a relative refractive index less than unity. For m larger than unity, second-order refraction is concentrated in the backward scattering range and is limited by the *rainbow angle*.

$$\varphi_R = \cos^{-1}\left[\frac{2}{m^4}\left(\frac{4-m^2}{3}\right)^3 - 1\right] \qquad (9.86)$$

The angular range of second-order refraction is reduced with an increasing relative refractive index and the rainbow angle increases. The characteristic scattering angles given above are summarized in Table 9.4 for different typical refractive indices. A map of the presence of the different scattering modes as a function of scattering angles and relative refractive indices was introduced by Naqwi and Durst (1991) for supporting the layout of the optical configuration of PDA systems. The angular distribution of scattering intensities resulting from different modes and for different refractive indices typical of practical two-phase flow systems are discussed in more detail below and summarized in Table 9.4.

Mie calculations have been performed for the range of the optimum scattering angle suggested by the relative intensity distributions. For bubbles in water, the optimum scattering angle seems to be rather limited, i.e., between 70° and about 85° where reflection is dominant for either polarization as shown in Figure 9.52a. The phase–size relations show reasonable linearity in this range, but a scattering angle of 55° also gives a linear response function as shown in Figure 9.53. Strong interference with refracted light exists in forward scattering and the phase-size relation becomes nonlinear (i.e., at a scattering angle of 30°). Similar observations are made for water droplets or glass particles in oil.

For two-phase systems with relative refractive indices larger than unity, refraction is dominant for parallel polarization in the forward scattering range up to about 70° to 80° depending on the value of the refractive index ratio. Since below approximately 30° diffraction interferes with the refracted light, especially for small particles, the lower limit of the optimum scattering angle is limited by this value. This is also obvious from the angular distribution of the phase for different particle diameters shown in Figure 9.54. The phase–size relations for water droplets in air, illustrated in Figure 9.55, show that a reasonable linearity is obtained in the range between 30° and 80°.

Flow	$m=n_p/n_m$	φ_B	φ_C	φ_R
air bubbles in water	1.0/1.33	106.12	82.49	-
water droplets in oil	1.33/1.50	96.88	55.09	-
oil droplets in water	1.5/1.33	83.12	55.09	94.10
water droplets in air	1.33/1.0	83.12	82.49	137.48
diesel droplets in air	1.46/1.0	68.82	93.54	153.34
glass particles in air	1.52/1.0	66.68	97.72	158.92

Table 9.4. Characteristic scattering angles for different combinations of the dispersed and continuous phase, i.e., different relative refractive indices.

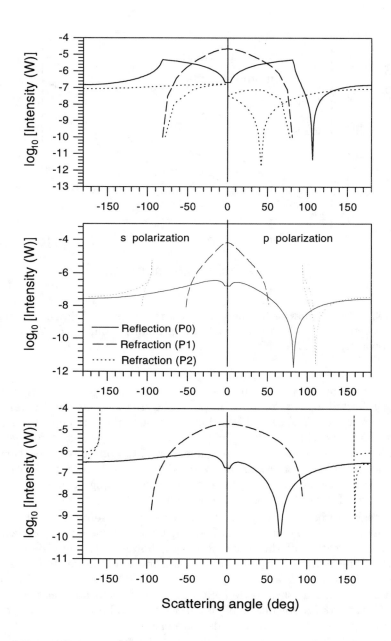

Figure 9.52: Angular intensity distribution of different scattering modes obtained by geometrical optics for a point receiving aperature ($\lambda = 632.8$ nm, $D = 30$ μm; a) $m = 0.75$, b) $m = 1.128$, c) $m = 1.52$).

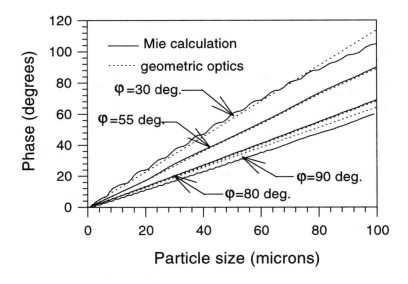

Figure 9.53: Mie calculation of the phase-size relations for different scattering angles between 30° and 80° ($\lambda = 0.6328$ μm, p polarization, $m = 0.75$ (i.e., air-bubble in water), $\theta = 2.77^{\circ}$, $\Psi = 1.85^{\circ}$, $\delta_h = 5.53^{\circ}$, $\delta_v = 1.85^{\circ}$).

The relative intensity distributions shown in Figure 9.52 also suggest that for $m > 1$ reflected light is dominant between the critical angle and the rainbow angle. However, interference with third-order refraction exists here (not shown in Figure 9.52) and this angular range can be only recommended for perpendicular polarization where a reasonable linearity of the phase response curve is obtained for water droplets in air only around 100° (see Figure 9.56). At 120° the Mie calculations do not correspond to the results using geometrical optics and, therefore, this result is not shown in Figure 9.56.

In the region of backscatter the intensity of secondary refraction is dominant only in a narrow range above the rainbow angle for perpendicular polarization. The optimum location of the receiving optics, however, strongly depends on the value of the relative refractive index, as shown in Figure 9.57, which is critical in fuel spray applications where the refractive index varies with droplet temperature and, hence, the location of the rainbow angle is not constant. From Figure 9.57 it becomes obvious that just above the rainbow angle, i.e., for $\varphi = 140^{\circ}$, a linear phase–size relation is also obtained.

As described above, the proper application of PDA requires that one scattering mode is dominant and the appropriate correlation (i.e., Equation 9.78 and Equation 9.79 or Equation 9.80) has to be used to determine the size of the particle from the measured phase. However, on certain trajectories of the

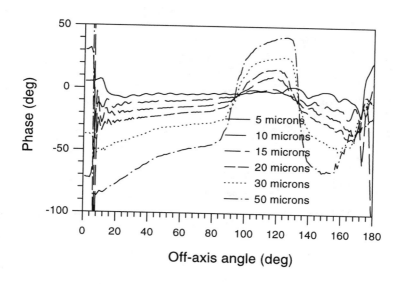

Figure 9.54: Angular distribution of phase for different particle diameters ($\lambda = 0.6328~\mu$m, p polarization, $m = 1.33$ (i.e., water droplets in air), $\theta = 2.77°$, $\Psi = 1.85°$, $\delta_h = 5.53°$, $\delta_v = 1.85°$).

particle through the focused Gaussian beam, the wrong scattering mechanism might become dominant and lead to erroneous size measurements (Sankar and Bachalo, 1991, Greham et al., 1992). This error is called *trajectory ambiguity* and is illustrated in Figure 9.58 where a transparent particle moving in air is considered and the desired scattering mode is refraction, which is dominant for collection angles between 30° and 80°. When the particle passes through the part of the measurement volume located away from the detector (i.e., negative y-axis), it is illuminated inhomogeneously. The refracted light is coming from the outer portion of the measurement volume where the light intensity is relatively low, while reflected light comes from a region close to the center of the probe volume where the illuminating light intensity is considerably higher due to the Gaussian profile. In this situation the reflected light might become dominant resulting in an incorrect size measurement since the particle diameter is determined from the correlation to refraction. It is obvious from Figure 9.58 that the trajectory ambiguity is potentially very important for large particles whose size is comparable to the dimensions of the probe volume.

The phase error, as a function of probe volume diameter to particle diameter, is illustrated in Figure 9.59 for a particle moving along the y-axis through the probe volume. The phase and amplitudes are calculated using GLMT (Grehan et al., 1992). The phase error might become negative or positive depending on

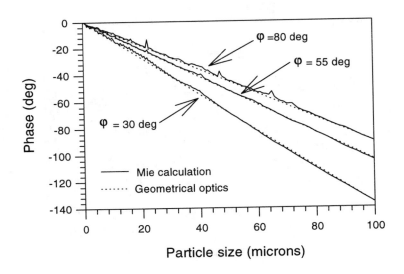

Figure 9.55: Mie calculation of phase-size relations for different scattering angles between $30°$ and $80°$ ($\lambda = 0.6328$ μm, p polarization, $m = 1.33$ (i.e., water droplets in air), $\theta = 2.77°$, $\Psi = 1.85°$, $\delta_h = 5.53°$, $\delta_v = 1.85°$).

the diameter ratio. The smallest errors are observed, however, for small particles which lead to the recommendation that the probe volume diameter should be about 5 times larger than the largest particles in the size spectrum considered. This requirement has restrictions for applications in dense particle-laden flows where the probe volume must be small enough to ensure that the probability of two particles being in the probe volume at the same time is small (see Section 9.3.3).

Therefore, phase-Doppler systems have recently been developed which allow the elimination of the trajectory ambiguity by the use of multiple detectors, such as the dual mode phase-Doppler anemometer (Tropea et al., 1995).

Since PDA allows the measurement of particle size and velocity, it is also possible to estimate the particle number or mass concentration and the particle mass flux. The particle number concentration is defined as the number of particles per unit volume. This quantity, however, cannot be measured directly since the PDA is a single particle counting instrument which requires that, at most, only one particle at a time be in the probe volume. The particle concentration has to be obtained from the number of particles moving through the probe volume during a given measurement period. For each particle one has to determine the volume of fluid which passes through the probe volume cross-section with the particle during the measurement time Δt_s. The volume depends on

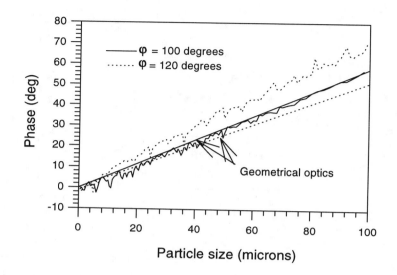

Figure 9.56: Mie calculation of phase–size relations for scattering angles 100° and 120° ($\lambda = 0.6328$ μm, s polarization, $m = 1.33$ (i.e., water droplets in air), $\theta = 2.77°$, $\Psi = 1.85°$, $\delta_h = 5.53°$, $\delta_v = 1.85°$).

the instantaneous particle velocity **v** and the probe volume cross-section perpendicular to the velocity vector, i.e., $Vol = A' |\mathbf{v}| \Delta t_s$ as shown in Figure 9.60. Additionally, the effective cross-section of the probe volume is a function of the particle size and, therefore, $A = A(\alpha_k, D_i)$ where α_k is the particle trajectory angle for each individual sample k and D_i is the particle diameter of size class i. Hence, the concentration associated with one particle is

$$n = \frac{1}{Vol} = \frac{1}{|\mathbf{v}| A'(\alpha_k, D_i)\Delta t_s} \tag{9.87}$$

This implies that for accurate particle concentration measurements one has to know the instantaneous particle velocity and the effective probe volume cross-section. The dependence of the probe volume cross-section on particle size is a result of the Gaussian intensity distribution in the probe volume and the finite signal noise. As illustrated in Figure 9.41, a large particle passing the edge of the probe volume will scatter enough light to produce a signal above the detection level. A small particle will produce a detectable signal for only a small displacement from the probe volume center. Therefore, the probe volume cross-section decreases with particle size and approaches zero for $D_p \rightarrow 0$ as shown in Figure 9.61.

Additionally, the effective probe volume cross-section is determined by the off-axis position of the receiving optics and the width of the spatial filter used

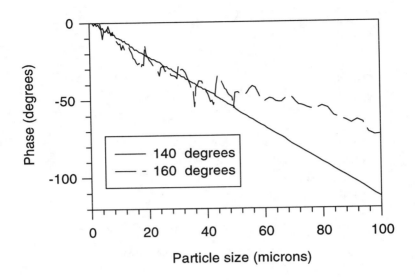

Figure 9.57: Mie calculations of phase—size relations for scattering angles 100° and 120° ($\lambda = 0.6328$ μm, s-polarization, $m = 1.33$ (i.e., water droplets in air), $\theta = 2.77^o$, $\Psi = 1.85^o$, $\delta_h = 5.53^o$, $\delta_v = 1.85^o$).

to limit the length of the probe volume imaged onto the photodetectors. For a one-dimensional flow along the x-axis, as shown in Figure 9.62, the effective size-dependent cross-section of the measuring volume is determined from

$$A(D_i)_x = 2r(D_i)L_s/\sin\varphi \qquad (9.88)$$

where L_s is the width of the image of the spatial filter in the receiving optics (see Figure 9.48) which depends on the slit width and the magnification of the optics, D_i is the diameter of size class i and $r(D_i)$ is the particle size-dependent radius of the probe volume as illustrated in Figure 9.62. For any other particle trajectory through the probe volume the effective cross-section obtained from the particle trajectory angle α_k is

$$A'(\alpha_k, D_i) = \frac{2r(D_i)L_s}{\sin\varphi}\cos\alpha_k \qquad (9.89)$$

The particle trajectory angle can be determined from the different instantaneous particle velocity components,

$$\cos\alpha_k = \frac{1}{\sqrt{1 + (\frac{w}{u})^2 + (\frac{v}{u})^2}} \qquad (9.90)$$

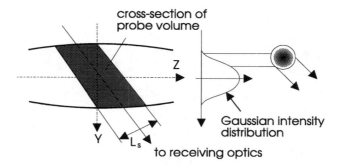

Figure 9.58: Illustration of Gaussian beam effect resulting in a trajectory ambiguity.

The particle size-dependent radius of the probe volume $r(D_i)$ may be determined in situ by using the burst length method (Saffman, 1987) or the so-called logarithmic mean amplitude method (Qiu and Sommerfeld, 1992). The latter is more reliable for noisy signals, i.e., low signal-to-noise ratio as demonstrated by Qiu and Sommerfeld (1992). The reader is referred to the above publications for more details.

These findings suggest that in complex two-phase flows with random particle trajectories through the probe volume, a three component PDA system is required for accurate concentration measurements. For a spectrum of particle sizes the local particle number density is evaluated using

$$n = \frac{1}{\Delta t_s} \sum_{k=1}^{M} \left[\sum_{i=1}^{N_k} \frac{1}{A'(\alpha_k, D_i)} \sum_{j=1}^{N_j} \frac{1}{|\mathbf{v}_{k,j}|} \right] \tag{9.91}$$

The sums in Equation 9.91 involve the summation over individual realizations of particle velocities (index j), in predefined directional class (index k) and size class (index i). The summation over particle size classes includes the appropriate particle-size dependent cross-section of the measurement volume for each size class, Equation 9.89, and finally the summation over all directional classes.

The particle mass concentration can be obtained by multiplying Equation 9.91 with the mass of the particles. Quite often the particle mass flux is a useful quantity to characterize a two-phase flow. The mass flux in direction n is obtained from

$$\dot{M}_d = \frac{1}{\Delta t_s} \sum_{k=1}^{M} \left[\sum_{i=1}^{N_k} \frac{m_p}{A'(\alpha_k, D_i)} \sum_{j=1}^{N_j} \frac{v_n}{|\mathbf{v}_{k,j}|} \right] \tag{9.92}$$

where v_n is the particle velocity component in the direction in which the flux is to be determined and m_p is the mass of one particle. For a directed two-phase flow, i.e., when the temporal variation of the particle trajectory through the

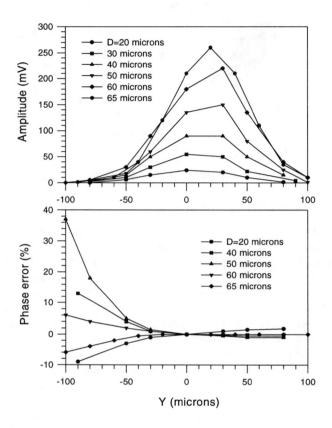

Figure 9.59: Phase error and scattering amplitude along the y-axis for different particle diameters and for a probe volume diameter of 100 μm.

probe volume is relatively small (such as in a spray), the particle trajectory angle may be determined from independent measurements of the individual velocity components (Equation 9.90) as shown by Qiu and Sommerfeld (1992).

An alternative method for determining the particle number concentration is based on the averaged residence time of the particles in the probe volume (Hardalupas and Taylor, 1989).

$$n = \frac{1}{\Delta t_s} \sum_{i=1}^{I} \left[\frac{\sum_{N_i} t_{ri}}{\forall(D_i)} \right] \tag{9.93}$$

where t_{ri} is the particle residence time in the probe volume, $\forall(\dot{D}_i)$ is the particle size-dependent volume and N_i is the number of samples in one particle size class (index i). As mentioned previously, the particle residence time or burst length

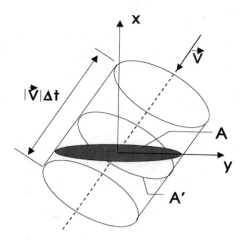

Figure 9.60: Probe volume associated with one particle moving across the detection region during the measurement time Δt.

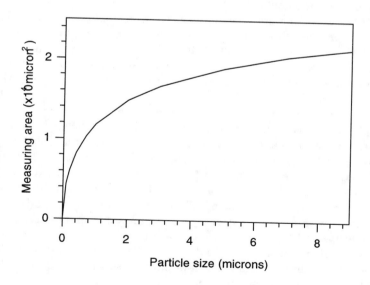

Figure 9.61: Correlation between the radius of the effective probe volume and particle size.

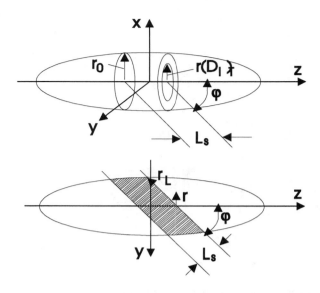

Figure 9.62: Geometry of a PDA measurement volume.

cannot be accurately determined for noisy Doppler signals. For this reason the alternative method is not very reliable and leads to considerable error in particle concentration measurements.

Recently a novel method was introduced which allows accurate particle concentration or mass flux measurements even in complex flows using a one-component PDA system (Sommerfeld and Qiu, 1995). This method uses the integral value under the envelope of the band-pass filtered Doppler signal in order to estimate the instantaneous particle velocity and its trajectory through the probe volume.

Example: The optical configuration of a two-detector PDA system for measurements in a particle-laden channel flow (air with $\rho_c = 1.18$ kg/m^3) with a mass concentration of two is to be designed. The particles (spherical glass beads with $\rho_d = 2500$ kg/m^3, $n_p = 1.5$) have a size range from 40 to 150 μm with a number averaged diameter of 60 μm. Transmitting optics with a focal length of 500 mm and a beam spacing Δb of 30 mm together with a He-Ne laser ($\lambda = 632.8$ nm) with a beam diameter d_o of 1 mm are available. The receiving optics should be positioned at an off-axis angle of 30° where refraction is the dominant scattering mode. The spatial filter in the receiving optics should have a width of 100 μm and a magnification of about 2 so that a length of 200 μm of the probe volume is imaged onto the receiver.

- Determine a suitable detector spacing and focal length of the receiving lens by ensuring that particles up to 200 μm can be measured. Select from available lenses; $f = 250, 500, 1000$ mm. Use Equation 9.75 to estimate the elevation angle and check the measurable particle size range using Equations 9.78 and 9.80.

- Estimate the coincidence error for the presence of two particles in the probe volume. Assume that the gas and particles have the same mean velocity and carry out the estimate assuming that the particles are monodisperse with a size corresponding to the most probable diameter in the size distribution, i.e., 60 μm.

- Redesign your optical system to reduce the coincidence error to 5% by considering changing the transmitting lens.

- Are modifications of the receiver required in order to maintain the measurable particle size range?

Solution:

First of all, find the half angle of the beam intersection θ from Equation 9.63

$$\tan \theta = \frac{\Delta b}{2f_e}, \quad \theta = 1.72°$$

Using Equation 9.75, one can solve for ψ in the form

$$\sin \psi = \Delta \phi_{max} \frac{\lambda_e}{2\pi D_{max}} \frac{m-1}{m} \frac{1}{\sin \theta}$$

With $\Delta \phi_{max} = 2\pi$, $D_{max} = 200$ μm, the value for ψ is 2°. Selecting a focal length of 310 mm the separation of the receiving detectors 21.65 mm according to the relation

$$\tan \psi = \frac{\Delta s'}{2f_r}$$

Thus, select a detector separation of $\Delta s' = 20$ mm. Using Equations 9.78 and 9.80 gives $D_{max} = 250$ μm.

From Equation 9.66, the waist diameter is

$$d_m = \frac{4f_e \lambda_e}{\pi d_o} = 402.85 \ \mu m$$

and the probe volume (see Figure 9.62) is

$$V_m = \frac{\pi}{4} d_m^2 L_s = 0.0225 \ \text{mm}^3$$

The relative probability of having two particles in the probe volume is given by Equation 9.52 as

$$P_2' = \frac{N_m}{2} = \frac{n V_m}{2}$$

Using $z = \frac{\dot{m}_d}{\dot{m}_c} = \frac{\bar{\rho}_d A v}{\bar{\rho}_c A u}$ and assuming $u \simeq v$ and $\bar{\rho}_c \simeq \rho_c$, the number density can be written as $n \simeq z \frac{\rho_c}{\rho_d} \frac{6}{\pi D^3}$. Substituting in the appropriate values gives $n = 8347$ particles/mm^3. The corresponding coincidence error is 0.106 or 10.6%.

The maximum measuring volume is $V_{m,\max} = 0.1/n_{\max} = 0.012$ mm^3. Introducing the measuring volume diameter, d_m, into the equation for V_m gives

$$f_e = \sqrt{\frac{\pi}{4} \frac{V_{m,\max}}{L_s}} \frac{d_o}{\lambda_e} = 343 \text{ mm}$$

Select a focal length of 310 mm for f_e. This gives $d_m = 249.8$ μm, $V_m = 0.0099$ mm^3 and $\theta = 2.77°$.

According to Equations 9.78 and 9.80, the optical configuration gives a maximum measurable size of $D_{\max} = 155.4$ μm. To increase the particle size range, the elevation angle ψ has to be reduced. Take $\Delta s = 15$ mm and $f_r = 310$ mm. This gives a maximum diameter of 207 μm. On the other hand if one selects $f_r = 500$ mm and $\Delta s = 20$ mm, a maximum diameter of 250 μm is measurable.

The first system is preferred because of the larger receiver aperture and resulting higher scattering intensities.

9.3.6 Imaging techniques

Not only may imaging techniques be used to determine the instantaneous spatial distribution of particles over a finite region of interest in the flow, but also properties of individual particles such as particle velocity, particle size and shape. Such instantaneous images can be obtained by either using pulsed light sources which can deliver a high light energy during a very short time period combined with a slow image recording system or by using a continuous light source in combination with high speed photography or other high speed recording systems, e.g., drum cameras. Moreover, a continuous illumination combined with long-exposure photography can yield information about the time-averaged distribution of the dispersed phase within a flow field.

Light sources used for imaging techniques may deliver diffuse illumination or parallel light beams. Some of the most common light sources are:

- continuous white light sources, such as halogen lamps,

- spark discharges or flash lights (e.g., Xenon flash lamp),

- and continuous or pulsed laser light sources (e.g., Argon Ion and Nd-Yag laser).

The required power of the light source is determined by the particle size under consideration and the scattering properties of the particle (i.e., reflecting, transparent or absorbing particles). In the range of geometrical optics, i.e., for particles larger than the wavelength of the light, the scattering intensity is approximately proportional to the square of the particle diameter. Moreover, the sensitivity of the image recording system affects the required laser power.

The image recording system also determines the possible spatial resolution of the images. The highest spatial resolution is obtained by conventional film photography or high speed motion picture cameras which also use photographic film. Such recording systems, however, have the disadvantage that the image evaluation is rather cumbersome and cannot be automated. The evaluation of the photographs requires postprocessing by scanning and digitizing which generally results in a reduction of the spatial resolution. More convenient for image analysis are methods using direct electronic imaging such as CCD (Charged Coupled Device) cameras which, however, have a limited spatial resolution. Typically, CCD-cameras have 256 x 256, 492 x 800 or 756 x 581 detector elements which are called pixels. High resolution CCD-cameras have up to 2048 x 2048 pixels. On the other hand, digitized images can be more easily enhanced by various software-based algorithms and also statistical information on the properties of the dispersed phase can be evaluated more easily.

Imaging techniques for two-phase flows may be categorized according to the applied illuminating and recording systems and the information can be extracted from the images in the following way:[2]

- direct imaging techniques,

- whole field visualization to determine the particle phase distribution,

- particle tracking velocimetry (PTV) or streak line technique,

- particle image velocimetry (PIV)

Direct imaging techniques are mostly based on continuous illumination with high speed recording and are used to visualize the time-dependent motion and behavior of individual bubbles or droplets. Such a technique allows one to study the shape of bubbles or droplets under different flow conditions. Since there are numerous examples for applications of such techniques in two-phase flows, the reader is referred to the book by Van Dyke (1982) to appreciate the potential of this technique.

The visualization of the instantaneous distribution of the dispersed phase particles in a two-phase flow is generally performed by producing a thin laser light sheet which illuminates a specific cross-section of the flow. Images are taken by a recording system oriented perpendicular to the light sheet as shown in Figure 9.63. To obtain an instantaneous image of the dispersed phase distribution, either the light source has to be operated in a single pulse mode or the recording systems must allow for short time exposure. The exposure time

---

[2]Laser-speckle velocimetry and holographic methods will not be considered here. More information about laser-speckle velocimetry can be found in Adrian (1986, 1991). Applications of holographic methods to sprays can be found in the papers of Chigier (1991) and Chavez & Mayinger (1990).

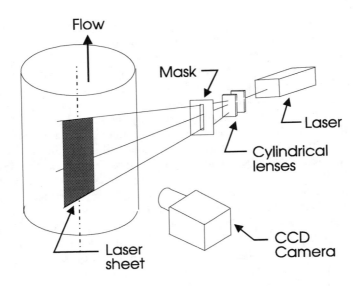

Figure 9.63: Light sheet flow visualization technique.

or the pulse duration of the laser determines the temporal resolution and has to be adjusted according to the flow velocity in order to yield sharp images. Moreover, the image recording system must be well focused.

Such a visualization technique has been applied by Longmire and Eaton (1992) to evaluate the particle concentration distribution and, in turn, the particle response characteristics in a forced particle-laden free jet. A pulsed copper vapor laser was used as a light source and a light sheet was produced by a cylindrical-spherical lens combination. The light sheet was transmitted through the center region of the jet and aligned parallel to the flow direction. Images of the particles were taken with a 35 mm camera activated by an electronic trigger. The single-pulse photographs were printed on a 8 in. x 10 in. photographic paper so that the individual particle images could be identified. A scanner was used to produce a binary image matrix which was transferred to a PC for further processing. The images were enhanced and all individual particle images were identified. The particle concentration field was evaluated by dividing the image into a number of small cells and then counting the individual particles in each of the cells. In order to correlate the particle concentration distribution with the flow structure, independent images were recorded when the jet was only seeded with smoke particles.

A similar visualization technique was applied by Wen et al. (1992) for studies of the particle response to the vortical structures evolving in a plane shear layer. Either a diffuse flash light unit or a continuous Argon-Ion laser light sheet were used as light sources. The images were recorded by a large format camera (6 x 4 cm) using exposure times between 1/250 and 1/10,000 seconds. The photographs were processed by a commercial image processing system. The air flow and the vortical structures in the shear layer were visualized using the smoke wire technique. From the instantaneous images of the particle distribution in the shear layer, the response characteristics of different sized particles were analyzed.

In a recent study of Huber and Sommerfeld (1994), a continuous Argon-Ion laser light sheet was used together with a CCD-camera for recording images of the cross-sectional particle concentration distribution in pneumatic conveying systems. Using digital image processing the images were enhanced and the instantaneous particle concentration distribution in the pipe cross-section was evaluated. The mean particle concentration field was determined by averaging a number of individual images.

Particle tracking and particle image velocimetry enable the determination of the instantaneous particle velocity distribution over a finite domain of the flow field. Both techniques have been used in single-phase flows for more than ten years, especially for the analysis of unsteady and turbulent flows. However, only recently have these methods become increasingly popular due to the enhanced computational capability of personal computers used mostly in laboratories for data processing. In single-phase flows, small tracer particles are added as light scattering centers in order to trace the fluid elements. By observing the locations of the images of these marker particles two or more times, the velocity can be determined from the displacement $\Delta \vec{x}(\vec{x}, t)$ and the time period Δt between two successive light pulses or recordings from

$$\vec{u}(\vec{x}, t) = \frac{\Delta \vec{x}(\vec{x}, t)}{\Delta t} \tag{9.94}$$

The response time of the marker particles must be sufficiently small that the particles follow the turbulent fluctuations in the flow. In two-phase flows the velocity distribution of the dispersed phase can easily be determined by PTV or PIV since the particles are usually rather large and, therefore, ideal scatterers.

Methods for simultaneous velocity measurements of both phases in a two-phase flow are more sophisticated and have only recently been developed. In the following section the basic principles of PTV and PIV are explained and some examples of applications to two-phase flow studies are provided.

According to the classification of Adrian (1991) both methods, i.e., PIV and PTV, may be grouped under the topic, pulse-light velocimetry (PLV), which indicates that pulsed light sources are used for illumination of the flow region of interest. For PLV, powerful laser light sources are used most of the time where the emitted light beam is expanded by a system of lenses to form a narrow light sheet typically up to a few millimeters in thickness. Moreover, it is possible to sweep the laser beam through the flow field using a rotating mirror (e.g., polygon mirror) whereby the effective intensity in the light sheet is increased. The time separation Δt between subsequent laser pulses determines the measurable particle velocity range and the pulse duration establishes the degree to which the image of the particle is frozen on the recording system.

Lasers typically used for PLV are:

- continuous lasers (e.g., Argon-Ion laser) which are continuously chopped using acousto-optical modulators,

- continuously pulsed metal-vapor lasers (e.g., copper-vapor laser),

- Q-switched ruby lasers,

- frequency doubled-pulsed Nd:Yag lasers which can produce trains of double pulses.

Several combinations of laser pulse coding (i.e., single, double or multiple pulse) and image recording (e.g., single frame recording and multiframing) may be used to determine the velocity of particles (see, for example, Adrian 1991).

The differences between PTV and PIV are the way the images are processed and the displacement is evaluated. With PTV, the tracks of individual particles are reconstructed from one or multiple images so this technique is only applicable for relatively dilute two-phase flows where the length of the recorded particle tracks should be smaller than the mean inter-particle spacing. Moreover, a sophisticated point-by-point analysis of the images is generally required. PIV requires higher particle concentrations since the displacement of a group of particles located in an interrogation area (a small subregion of one image) is determined by special processing techniques such as auto- and cross-correlation methods. Hence, with this method, an average displacement and an average velocity of all particles located in the interrogation area is determined. The number of particles in each interrogation area should typically be from 5 to 10 in order to yield reasonable averages of the particle velocity.

The resolution of velocity measurements by PIV and PTV depends on the accuracy of the displacement measurement Δx and the uncertainty in the time difference between the light pulses Δt. Normalized by the maximum velocity in the flow field, u_{max}, the root mean square (rms) uncertainty of the velocity is determined from (Adrian, 1986)

$$\frac{\sigma_u}{u_{max}} = \frac{\sigma_{\Delta x}}{\Delta x} + \frac{\sigma_{\Delta t}}{\Delta t} = \frac{\sigma_{\Delta x}}{u_{max}\Delta t} + \frac{\sigma_{\Delta t}}{\Delta t} \qquad (9.95)$$

where $\sigma_{\Delta x}$ is the root mean square uncertainty in the displacement between two images and $\sigma_{\Delta t}$ is the uncertainty in the time between two successive light pulses or recordings. In the determination of the displacement, the major source of error is related to establishing the centers of the images which is determined by the applied processing algorithm and the image size. With the assumption that this uncertainty is related to the image diameter through a number c which is established by the accuracy of the processing method used for particle location, one has

$$\sigma_{\Delta x} = c d_I \qquad (9.96)$$

The image size depends on the resolution of the recording system, d_r (i.e., photographic film or electronic recording) and the magnification of the imaging system, M. It can be estimated from the following relation (Adrian, 1986)

$$d_I = (M^2 D^2 + d_s^2 + d_r^2)^{1/2} \tag{9.97}$$

where D is the particle diameter and d_s is the diffraction-limited spot diameter of the optical system given by (Adrian, 1986)

$$d_s = 2.44(M+1)f^{\#}\lambda \tag{9.98}$$

where $f^{\#}$ is the f-number of the lens and λ is the wavelength of light. The relative importance of the first two terms in Equation 9.97 depends on particle size. For an optical system with $M = 1$, $f^{\#} = 8$ and $\lambda = 514.5$ nm, the image diameter becomes independent of particle size for diameters much less than 20 μm. On the other hand, for large particles (i.e., much larger than 20 μm which is mostly the case in two-phase flows) the image diameter is given by

$$d_I = (M^2 D^2 + d_r^2)^{1/2} \tag{9.99}$$

In addition the image size may be increased by particle motion during the exposure time or pulse duration Δt_{exp}. Hence the maximum velocity in the flow field times the pulse duration should be smaller than the image diameter to avoid this blur effect.

$$u_{\text{max}}\Delta t_{\text{exp}} \le d_I \tag{9.100}$$

If one also can assume that the pulse frequency of the light source can be adjusted with a high accuracy, i.e., the time jitter may be neglected, the uncertainty of velocity measurement is given by

$$\sigma_u = c\frac{d_I}{\Delta t} \tag{9.101}$$

This equation reveals that high accuracy for velocity determination is achieved with small image diameters.

Two different methodologies may be used for PTV, the single-frame/multi-pulse technique and the multiframe/single-pulse method where the particle trajectories are reconstructed by overlapping several single exposure images. The latter method actually corresponds to conventional cinematography and has the advantage that the particles may be tracked for a long time (i.e., to evaluate Lagrangian trajectories) and that the direction of particle motion is fully determined from the sequence of the recordings. The disadvantage is that the frequency of the image recording system must be equal to the laser pulse frequency or a known integer multiple lower. This limits the measurable velocity range when, for example, CCD-cameras with a standard framing rate of 50 Hz are used. Otherwise high speed videos or motion picture cameras are required. Such a method was used by Perkins and Hunt (1989) to experimentally study particle motion in the wake of a cylinder. In order to reconstruct the particle trajectories, a cross-correlation method was used to avoid a time consuming search algorithm to identify images belonging to one particle track.

In the other mode of PTV, i.e., the single-frame/multipulse method, the time sequence of the particle images are recorded on one frame for a certain time interval. This method has the advantage that slow image recording systems such as CCD-cameras may be used. However, the direction of particle motion cannot be resolved by this method so its applicability is limited to simple flows with a known direction of particle motion. This limitation can be overcome with pulse coding or image shifting which is described later. In combination with low frequency laser light sources this method produces a sequence of particle streak lines where the particle velocity is determined from the displacement of the centers of individual streak lines. For this mode of operation the pulse duration needs to be longer than the image diameter divided by the particle velocity, i.e., $\Delta t > d_I / |\vec{u}|$ and the time between pulses needs to be longer than the pulse duration.

The analysis of PTV images usually consists of the following steps (see Hassan et al., 1992):

- thresholding of the image, i.e., removal of noise objects with intensity below a certain gray level scale,

- analysis of the image to determine the location of the particle images,

- point-by-point matching of the particle images from one frame to the next (if the multiframe technique is used),

- determination of the displacement and particle velocity for each individual particle,

- postprocessing in order to erase unreasonable velocity vectors resulting from errors in the previous analysis steps.

Some examples of the application of PTV techniques to two-phase flows are presented here. The particle streak line technique was used by Ciccone et al. (1990) to study the saltating motion of particles over a deposited particle layer. A section of the flow field was illuminated with a laser light sheet which was chopped by a rotating grid with an equal number of open and closed sections. The images were recorded by conventional film photography. Subsequently the photographs were digitized and enhanced using Fast Fourier Transforms to enable reconstruction of the trajectories of individual particles from a number of imaged streak lines.

A similar technique was used by Sommerfeld et al. (1993) to obtain statistical information on the collision of individual particles with rough walls. The laser light sheet was produced in such a way that it propagated along a horizontal channel and illuminated a thin sheet perpendicular to the wall of the channel. The light beam from a continuous Argon-Ion laser was chopped using a Bragg-cell. The images of particle trajectories in the near-wall region were recorded by a CCD-camera. The CCD camera which was mounted perpendicular to the light sheet was equipped with a macro zoom lens and extension bellows. To enable the processing of a large number of individual wall collision events, the images were recorded on a video tape and processed off-line later using a frame grabber card and a personal computer. A special processing software package was developed to determine the streak lines belonging to one particle which collided with the wall. The processing routines involved the enhancement of the images, the identification of streak lines with a minimum of 50 pixels and the grouping of two streak lines before and after wall impact for each wall collision event. The particle velocity components and the trajectory angles before and after impact were determined from the separation of the geometric centers of the two respective streak lines.

A visualization scheme for solid particle motion in a turbulent pipe flow was reported by Govan et al. (1989) using an axial viewing technique. In these studies, an expanded laser light beam was transmitted axially through a vertical pipe. A high speed motion picture camera was mounted at the opposite end of the pipe. The particle trajectories in the pipe cross-section were reconstructed from a sequence of 30 frames for the smallest to 200 frames for the largest particles. The axial viewing technique for visualizing particle motion in a pipe flow was further improved in subsequent studies (Lee et al., 1989). The particle motion in several cross-sections of the pipe was observed using a number of adjacent pulsed laser light sheets. The pulse frequency was adjusted in such a way that the images of the particles appeared as dots as opposed to streaks. The use of a number of adjacent light sheets pulsed in sequence allowed one to follow the particle trajectories in the cross-section of the pipe and to determine two components of particle velocity.

In PIV two methodologies are most common, the single-frame/double-pulse (or less common multi-pulse) technique and the multiframe (i.e., mostly two-frame)/single-pulse method. The first technique is usually combined with a spatial autocorrelation algorithm in order to evaluate the average displacement of image pairs in each interrogation area. This technique has the advantage of not requiring sophisticated recording systems with high speed shutters. When used with a double pulse laser, it cannot resolve the particle direction. Hence, additional means are required to determine the particle direction in complex or turbulent flows. Two methods, among others, are most common:

- The recording system (e.g., CCD or photographic camera) is placed on a linear rail and moves at a constant and defined speed during the exposure.

- A rotational mirror is mounted between the flow field and the recording system.

Both methods result in an additional known shift of the second image with respect to the first one. By subtracting this shift from the recorded image displacement, the effective displacement is determined which may be either positive or negative, depending on the sense of particles velocity vector.

The second methodology of PIV is the multiframe/single-pulse method. In this case subsequent images are recorded on separate frames and hence, similar to the corresponding PTV method, a high speed recording system is required with a speed identical to the laser pulse frequency or a frequency which is a integer multiple smaller than the laser pulse frequency. Moreover, the shutter of the camera has to be synchronized with the laser pulses. Since the displacement is evaluated from at least two subsequent images, the direction of particle motion is fully determined and no additional equipment is needed. Cross-correlation methods are most favorable for the determination of the average displacement of the particles in the interrogation area.

Examples of the application of PIV in two-phase flows are numerous. In most cases, however, the velocity fields of both phases were not recorded simultaneously in order to assess the effect of the dispersed phase on the fluid flow. A few examples of measuring the dispersed phase velocity field by PIV are described below.

The effect of vortices on the particle motion in a forced jet impinging onto a circular plate was investigated by Longmire and Anderson (1995). The velocity fields of both phases in the vicinity of the wall were evaluated independently using the single-frame/double-pulse PIV method. In order to resolve directional ambiguity, the image shift technique based on a rotating mirror placed between the flow field and the camera was used. For analyzing the vortex structure in the single-phase jet, both the jet and the outer flow were seeded with smoke. For the determination of the particle phase velocity field, it was required that a sufficient number of particles be present in each interrogation area in order to ensure a good autocorrelation. Since the particle loading was very low (i.e., dilute system) a complete velocity vector field could not be obtained for the dispersed phase, although ensemble averages of 10 realizations were evaluated. Moreover, problems in resolving the velocity of particles moving towards the wall and those rebounding from the wall using PIV became obvious. In those interrogation areas where both classes of particles were present, some average velocity depending on the number of particles in each class was determined. This special case represents a disadvantage of the PIV method.

For the simultaneous determination of the velocity fields of both phases using PTV or PIV, the fluid flow has to be seeded with appropriate tracer particles which are able to follow velocity fluctuations of the flow. In order to discriminate the images from tracer and dispersed phase particles, two approaches are most common:

- Discrimination based on the size of the images which requires a considerable difference in the size of the dispersed phase particles and the tracer particles as, for example, in a bubbly flow.

- Introduction of florescent tracer particles which are excited by the incident laser light and emit light with a different wavelength. This method requires two recording systems, each equipped with an appropriate color filter.

Simultaneous measurements of both velocity components in a bubbly two-phase flow were conducted by Hassan et al. (1992) using PTV and size discrimination. Bubbles about 1 mm in size were periodically released at the bottom of a test tank filled with transparent mineral oil which was seeded with 70 μm plastic particles. Particle and bubble trajectories and velocities were reconstructed from a sequence of 10 single-pulse images and a sophisticated point-by-point analysis.

The second discrimination approach is based on laser-induced fluorescence.. In this case either the seed particles or the dispersed phase particles (i.e., solid particles or droplets) are treated with a fluorescent dye solution. Solid particles have to be impregnated with the dye solution, while in the case of liquid droplets, the dye solution may be mixed with the liquid prior to atomization. When the dyed particles are illuminated by laser light with the appropriate wavelength, they emit light at a different wavelength. For example, rhodamine 6G may be excited by green light (i.e., 514.5 nm) and the emitted fluorescent light has a wavelength around 580.5 nm. By using two CCD-cameras (or any other cameras) with appropriate color filters, one camera may receive only the scattered light around the laser wavelength and the other camera responds to the fluorescent light. Hence separate images of the seed particles and the dispersed phase particles are produced.

Such a method was for example applied by Philip et al. (1994) using red fluorescent seeds in order to visualize the flow near a collapsing steam bubble. By using a color filter in front of the CCD-camera, most of the green light reflected by the bubble was blocked and the red light emitted by the seeds was recorded. A similar technique was used by Tokuhiro et al. (1996) to measure the fluid velocity distribution around a stationary bubble. In this case a second CCD-camera was used in order to image the bubble shape using a shadow technique.

The few examples presented on imaging techniques show the wide variety of applications and the potential of imaging techniques for analyzing two-phase flows. Further examples may be found in relevant journals or conference proceedings.

9.4 Summary

In this chapter some of the most common measurement techniques for particle characterization (i.e., sampling methods) and on-line analysis of two-phase systems have been introduced. Most of the techniques for the analysis of particle samples with regard to size, shape and surface area are well established. However, on-line methods such as light scattering, laser-Doppler and phase-Doppler anemometry for particle characterization and for the reliable and accurate measurement of flow properties in industrial processes or in research laboratories are still under development. The main emphases in these developments are related to improved measurement accuracies, more reliable mass flux and concentration measurements and extensions of the optical system to enable determination of additional properties, such as the refractive index and temperature of droplets as reported by van Beek and Riethmuller (1996).

Industrial applications of single particle counting instruments for process control are, at present, minimal. More robust and easy-to-use instruments need to be developed for such applications. Miniaturized optical systems based on semiconductor lasers which have been developed over the past 10 years (Wang et al., 1994) may give rise to compact robust systems in the future.

However, one should keep in mind that optical techniques are limited to applications in dilute two-phase systems only. For on-line measurements and control in dense systems, radiation sources, such as X-rays and gamma-rays which are highly penetrative due to their extremely short wavelengths, have to be used. Tomographic methods based on these radiation sources have received considerable attention during the past 10 years for imaging the phase distribution in dense multiphase systems. One is referred, for example, to the work of Williams and Beck (1995).

Exercises

9.1. The particle concentration in a particle-laden pipe flow is to be measured using a simple sampling probe. A homogeneous particle concentration in the pipe cross-section may be assumed. The suction velocity is adjusted to the average gas velocity in the pipe, i.e., 20 m/s. The mass concentration (particle bulk density) may be determined from the accumulated particle mass m_d during the sampling time Δt and the gas volume flow rate, $\dot{V}_c$, and particle volume flow rate, $\dot{V}_d$, according to:

$$\bar{\rho}_d = \frac{m_d}{(\dot{V}_c + \dot{V}_d)\Delta t}$$

Assume that the gas is incompressible and the gas velocity profile is given by:

$$\frac{u}{U_o} = \left(\frac{y}{R}\right)^{1/n} \qquad \text{(with } n=7\text{)}$$

$$\frac{\bar{u}}{U_o} = \frac{2n^2}{(n+1)(2n+1)}$$

where U_o is the centerline gas velocity, $\bar{u}$ is the average velocity and y is the distance from the wall and R is the pipe radius.

Determine the error in the particle concentration measurement on the pipe centerline and 50 mm away from the wall using the result given in Figure 9.22. Determine the errors in the measurement of the particle mass flux at the same locations and comment on the result.

The following properties are given:

pipe diameter	200 mm
sampling probe diameter	10 mm
sampling time	20s
particle size	200 μm
particle material density	2500 kg/m^3
density of gas	1.18 kg/m^3
kinematic viscosity	18.4×10^{-6} Ns/m^2

9.2. Determine the required dimensions of the probe volume for a light scattering instrument allowing a coincidence error of 5 percent. The measurements will be carried out in a gas-solid flow with a mass loading ratio of 2. Consider two particle sizes, 15 and 60 μm. The particle material density is 2500 kg/m^3.

9.3. Design an LDA system for measurements in a high-speed, gas-particle flow (initial velocity 500 m/s) through a normal shock wave. You have available a signal processor which is able to resolve a signal frequency of 50 MHz. No frequency shifting is to be used. The laser has a beam diameter of 2 mm. Determine the beam crossing angle and select a suitable combination of initial beam spacing (typically: 10, 20, 30, 40, 50 mm) and focal length (typical focal lengths of lenses are: 160, 300, 500, 700 and 1000 mm). Also calculate the diameter of the probe volume and the number of fringes. Ensure that the number of fringes is larger than 10 to allow for reliable signal processing.

9.4. Select an appropriate configuration of a PDA system for measurements in a spray of water droplets (n_p =1.33) with a size spectrum from 50 to 200 μm. The spray is issuing into an air flow. A three-detector system with the following configuration is available:

laser wavelength	514.5 nm
laser beam diameter	1.35 mm
focal length of transmitting lens	500 and 1000 mm
initial beam spacing	variable between 20 and 40 mm
receiving lens focal length	variable between 160 and 400 mm

The receiving optics have interchangeable masks with the smallest detector spacing (i.e., detector 1 - 2 in Figure 9.50) being 5, 10, and 15 mm.

Select a scattering angle of 30 degrees and try to minimize the Gaussian beam effect by choosing the diameter of the probe volume to be larger than the largest drop.

9.5. Estimate the errors for a PTV measurement in a flow with particles of 80 μm and a maximum velocity of 100 m/s. A copper-vapor laser is available with a pulse duration of 30 ns and a pulse-to-pulse jitter of 20 ns. Assume that the image can be located within an accuracy of 20 percent; i.e., $c = 0.2$ in Equations 9.96 and 9.101.

The following properties of the optical system are given:

magnification	1
laser wavelength	510.6 nm
f-number of lens	8
resolution of recording system	20 μm

Determine the following parameters;

a) image diameter,

b) a suitable pulse frequency to allow for a image separation of 5 d_I,

c) error due to blur in relation to the image diameter uncertainly associated with velocity measurement.

Chapter 10

Final Remarks

This book has introduced the basic physical principles of fluid particle flows, provided a background in numerical modeling and reviewed measurement techniques. The purpose of this final chapter is to provide some examples of computational models and measurements of fluid-particle systems and to address the field with respect to current and future needs. The wide diversity of applications of fluid particle flows in industry will continue to drive the need for better predictive models and measurement techniques.

10.1 Applications

There are numerous applications of numerical models and experimental measurements of multiphase flows with droplets and particles. The cyclone separator, fire-sprinkler system, coal-fired furnace, co-current spray dryer, dense phase pneumatic transport and fluidized bed will be addressed here.

10.1.1 Cyclone separator

The wide use of the cyclone separator has led to many experimental studies and empirical models for collection efficiency and pressure loss. Most recently Kim and Lee (1990) report an extensive experimental study of collection efficiency of several designs including the generic (Stairmand) cyclone. There have been numerous predictive models that have been correlated with data from industrial cyclones. These models, however, do not perform well for small cyclones used for sampling systems. The particle loading in a cyclone is typically low so models based on one-way coupling are adequate.

Recently Griffiths and Boysan (1996) have reported on the application of a CFD code (FLUENT) to predict the collection efficiency of the cyclone shown in Figure 10.1 and compared their predictions with the data obtained by Kim and Lee. They modeled the three-dimensional flow in the cyclone by discretizing the volume into approximately 35,000 control volumes. The RNG (Randomized

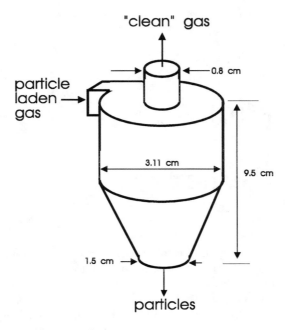

Figure 10.1: Configuration of cyclone separator used for computational model and measurement.

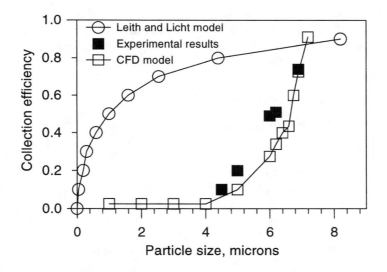

Figure 10.2: Comparison of measured and predicted collection efficiency of a small cyclone separator.

Normal Grouping)-based $k - \epsilon$ model was used for turbulence simulation because of the swirling flow. The trajectory approach with no coupling was used for the particle phase. The trajectories for a range of particle sizes were calculated from 100 starting locations along the inlet plane. By repeating the calculations for different particle sizes, the curve for collection efficiency versus particle size were generated.

The comparison of the experimental data for collection efficiency obtained by Kim and Lee and the predictions of the CFD code used by Griffths and Boysan are shown in Figure 10.2. One notes good agreement between the predicted and measured collection efficiencies. Shown on the same figure is the prediction based on the classic Leith and Licht model (1972) which does not compare favorably with the experimental results. Good agreement between CFD predictions and measurements was also obtained for the pressure loss in the cyclone. The capability of the numerical model to simulate the two-phase flow in a cyclone under normal operating conditions suggests its utility in assessing cyclone performance in other applications such as high-temperature gas clean-up and biotechnology.

10.1.2 Fire sprinkler system

The dynamics of sprays is important in the design and operation of sprinkler systems for fire suppression. A series of experiments and numerical predictions for the penetration of sprays in a fire have been reported by Factory Mutual Research Corporation (Nam, 1995). A schematic diagram of the experimental setup is shown in Figure 10.3. An ESFR (Early Suppression Fast Response) sprinkler is mounted about three meters above a fire produced by nine heptane spray nozzles. The convective heat released by the fire varied from zero (no fire) to 1500 kW. Collection pans for the water were placed below the heptane spray nozzles to collect the droplets that penetrated through the flame. The weight of water collected in the pans was reduced to give the volume flow rate per unit area (l/ min /m^2) and reported as "actual delivered density".

A numerical model for actual delivered density was developed and based on the trajectory model. A total of 275 trajectories were used with 25 trajectories assigned to each size. The size distribution was obtained from experimental data and fit with either the log-normal or Rosin-Rammler distribution. The initial droplet speed was obtained from experimental data and the starting angle was distributed over a range of angles. In this application, the capability of a numerical model to handle thermal coupling is very important

The comparison of the predicted and measured actual delivered density is shown in Figure 10.4. The delivered density distributions for no fire and for a fire with an energy release rate of 1500 kW are shown in the figure as a function of the water supply rate at the nozzle (l/min). The actual delivered density for the case with no fire is higher, as expected, than for the case with a fire. In general, the agreement is good and lends some confidence in using numerical modeling to complement fire sprinkler design and performance.

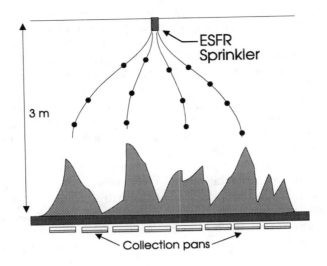

Figure 10.3: Schematic diagram of sprinkler fire suppression system.

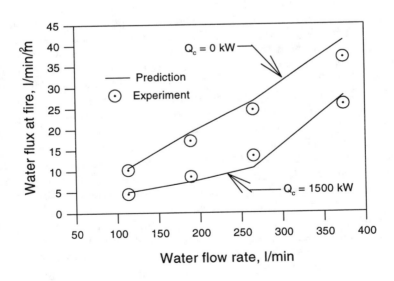

Figure 10.4: Predicted and measured actual delivered density as a function of water flow rate through a fire suppression sprinkler with no fire and with a fire with a heat release of 1500 kW.

10.1.3 Coal-fired furnace

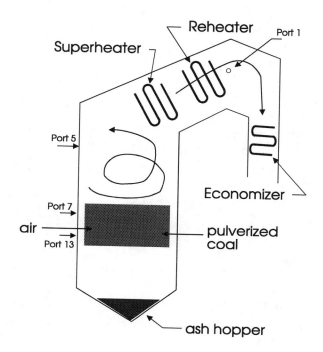

Figure 10.5: Schematic diagram of coal-fired utility burner.

The application of numerical modeling to coal-fired furnaces has been reported by several authors (Hill and Smoot, 1993, Kaufman and Fiveland, 1995). Hill and Smoot report the development of a generalized, three-dimensional combustion model (PCGC-3) to simulate a large-scale furnace fired by pulverized coal. It uses the trajectory approach for the particulate phase and incorporates equilibrium gas-phase chemistry. Convective and radiative heat transfer are included with radiation being modeled by the discrete ordinates method. The chemical reactions and turbulent flow field are coupled by integrating the equations over a probability density function.

The large-scale coal-fired 85 MW_e utility burner modeled by Hill and Smoot is depicted schematically in Figure 10.5. The air and pulverized coal are fed into the burner and the resulting combustion gases rise vertically in a swirling motion toward the heat exchangers. The ash drops into the ash hopper below. Various types of measurements were made for comparison with numerical predictions at the ports indicated in the figure. Gas temperatures were made with a triply-shielded water-cooled pyrometer. Measurements of gas velocity, gas composition and radiant flux were made using probes. A laser-based particle velocimeter yielded information on particle number density, velocity and size.

The numerical model was based on 240,000 computational cells. The particle

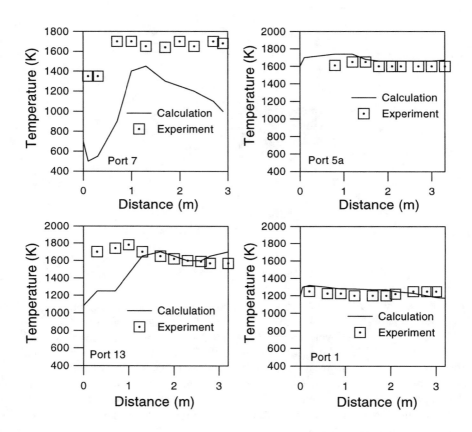

Figure 10.6: Predicted and measured temperature distributions in coal-fired utility burner. (Reprinted with permission from Hill, S.C. and Smoot, L.D., *Energy and Fuels*, v. 7, p. 874. Copyright 1993 American Chemical Society.)

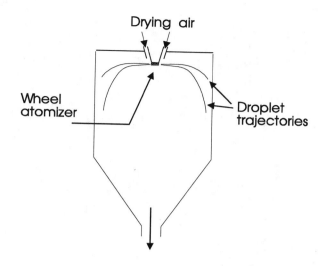

Figure 10.7: Configuration of co-current spray dryer.

size distribution was discretized into 5 particle sizes with 144 starting locations for a total of 720 trajectories. The turbulent particle dispersion was based on an effective particle turbulent diffusivity reduced from experimental observations. The coal devolitization was modeled as a simple, two-step mechanism that produces an off-gas and a solid char.

The comparison of the numerical predictions with temperature measurements is shown in Figure 10.6. At port 7, which in the burner region between the upper two burners, the measured temperatures exceed the predicted values. At port 13, which is also in the burner region between the lower two burners also shows predicted values lower than the measured values. This is attributed to a slower heating and burning of the coal particles than is observed in the furnace. The prediction of correct ignition location is apparently a common problem in coal flames (Hill and Smoot, 1993). Much better agreement is achieved for the ports 1 and 5 which are located beyond the burner region.

This study illustrates the complexity of modeling industrial systems but does show promise in the use of numerical methods to complement the design and operation of coal-fired systems.

10.1.4 Spray drying

Spray drying is used to produce a wide variety of products from detergents to pharmaceuticals. It is carried out in three steps: atomization of the liquid or slurry spray by swirl, pressure or pneumatic atomization; drying of the droplets to a powder by hot gases; and removal of the powder usually by means of a cyclone separator. The advantage of spray drying is that the product is exposed to the hot drying medium for a short period of time.

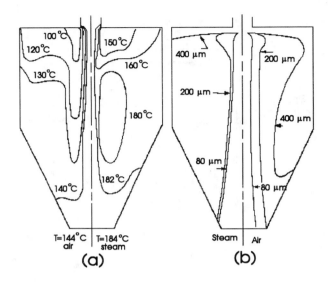

Figure 10.8: Predicted temperature field (a) and droplet trajectories (b) in an air-operated and steam-operated cocurrent spray dryer. (From Crowe et al., *Drying'85*, 261, 1985. With permission.)

A co-current spray dryer is shown in Figure 10.7. Drying air is introduced through the top and into the chamber through swirl vanes to produce a swirling inlet gas flow. A wheel atomizer introduces droplets into the air stream. The system has to be designed such that the droplets become a sufficiently dry powder without overheating. Also, there can be no significant accumulation of powder on the wall to avoid expensive shut-down time and the possibility of fire. In this application thermal coupling is very significant. The heat transfer to the droplets results in a cooling of the drying gases and a reduction in the drying rate.

Several empirical and numerical models have appeared in the literature for spray dryers (Crowe et al., 1985; Oakley and Bahu, 1991). Crowe et al. modeled the co-current spray dryer using the trajectory approach for the dispersed phase. The purpose of the study was to compare the effectiveness of superheated steam and hot air as the drying medium. The dryer was 4 meters in diameter and 6 meters high. The length of the cylindrical section 2.8 meters. Air entered at 200°C with absolute humidity of 1%. Superheated steam entered at the same temperature. No swirl component was added to the drying air although a swirl did develop in the dryer because of momentum coupling with the tangential velocity component of droplets leaving the wheel atomizer. The slurry to be atomized and dried was an aqueous solution of an insoluble material with an initial moisture content (mass of water/mass of bone-dry material) of 1.2. The mass flow rate of the slurry was 5% that of the drying gases. The wheel atomizer produced a distribution with a mass median droplet diameter of 200 μm. The droplet distribution was discretized to five sizes with equal mass fractions.

The predicted isothermal lines for steam and air are shown in Figure 10.8a. There is a very rapid depression of temperature near the atomizer where thermal coupling is intense. The temperature of the steam is uniformly higher due to the higher thermal capacity of steam. The results showed that steam drying was more effective as a drying agent because of its higher thermal capacity and heat transfer coefficient.

The predicted droplet trajectories are shown in Figure 10.8b. For air drying the 400-μm droplet almost penetrates to the wall before being redirected inwardly by the recirculating flow pattern. On the other hand the 400-μm droplet released in the steam dryer quickly impacts the wall. This is attributed to the lower viscosity of steam and resulting smaller drag force on the droplet.

The availability of detailed numerical models for spray dryers should prove useful in complementing the design of drying systems for new applications and improving the performance of existing systems.

10.1.5 Horizontal dense phase flow

A good example of a dense phase flow is the dense phase transport in a horizontal pipe. In this case particle contact dominates the flow. Tsuji et al. (1992) have modeled the developing flow pattern of a slug of particles in a horizontal tube as shown in Figure 10.9. In this application the discrete element model for contact dominated flows was used. The initial slug is shown in the first frame and the particles are tagged by different shades to better illustrate the deformation of the slug. The flow is from right to left. This simulation was carried out for a 50-mm diameter pipe with a superficial gas velocity of 2.4 m/s. The particles were 10 mm in diameter with a density of 1000 kg/m^3. A total of 1000 particles were used in the simulation. The simulation is carried out such that those particles which leave the left boundary are introduced through the right boundary.

The formation of slugs is evident. The results show that the forward part of the slug falls away and the top part of the slug begins to move faster that the bottom portion. One also notes that there is considerable mixing as the top and bottom portions of the particulate flow move at different speeds. With extended time the particle velocity in the slugs becomes nearly uniform and there is clearly a particle velocity near the bottom wall.

Calculations of this type require considerable computational time. The number of particles needed to model a typical dense phase flow is beyond current computational capacity. However the simulation allows a detailed model of the flow in which both particle rotation and translation due to particle-particle contact are included.

10.1.6 Fluidized bed

The fluidized bed is a very important device in the chemical industry. It is characterized as dense phase flow where particle-particle interaction controls

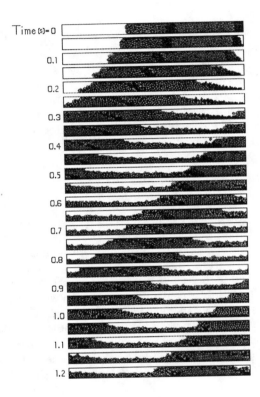

Figure 10.9: Flow pattern development of a slug of particles in a horizontal tube. (Reprinted from *Powder Tech.*, 71, Tsuji, Y. et al., Lagrangian numerical simulation of plug flow of collisionless particles in a horizontal pipe, 239, 1992, with kind permission from Elsevier Science S.A., P.O. Box 564, 1001, Lausanne, Switzerland.)

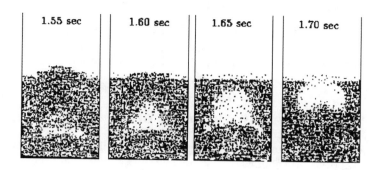

Figure 10.10: Predicted temporal development in a fluidized bed based on the two-fluid model and kinetic theory. (Ding, D. and Gidaspow, D., A bubbling fluidization model using kinetic theory of granular flow, *AIChE J.*, 36, 523. Reproduced with permission of the American Institute of Chemical Engineers. Copyright 1990 AIChE. All rights reserved.)

the dynamics of the particle field. Both the discrete element method (Tsuji et al., 1993) and the two-fluid model (Tsuo and Gidaspow, 1990) have been applied to this problem. The two-fluid model has been traditionally based on with empirical relations for the constitutive equations. More recently, kinetic theory (Gidaspow, 1993) has been used to obtain the terms for particle-particle interaction.

The predicted temporal development of the particle distribution in a fluidized bed using the two-fluid model with kinetic theory (Ding and Gidaspow, 1990) is shown in Figure 10.10. The bed was cylindrical with a diameter of 13.8 cm. The particles had a uniform diameter of 500 μm and a density of 2.5 kg/m^3. The air entered with a superficial velocity of 64.1 m/s. The bubbles tend to move toward the bed center and merge around the centerline. Although not visible in the figure, a ring-like vortex appears in the upper portion which is related to the bubble motion. The overall behavior of the bubble agrees with observations reported by others. The same pattern of bubble development is predicted using the discrete element approach (Tsuji et al., 1993).

Although the predictions of the dynamics of a fluidized bed agree with qualitative observations, considerable work remains to be done. It will be important to include such effects as particle size distribution and shape, particle coalescence and attrition. Also the modeling of the carrier phase in a dense phase flow is still in the primitive stage.

10.2   Current and future needs

The needs can be subdivided into three categories: fundamental parameters, numerical models and measurement capabilities. There are specific needs for improvement of current systems and to support the development of new technologies.

10.3   Fundamental parameters

The fundamental phenomena controlling coupling in fluid particle flows are mass transfer (evaporation, condensation or devolitization), drag and heat transfer.

10.3.1   Particle drag

The drag coefficient of an isolated particle in a low turbulent, steady has been investigated both experimentally and analytically for many years and is well established. The unsteady effects (apparent mass and Basset force) have been derived for Stokes flow but are not well established for high Reynolds numbers. Particle shape effects have been studied experimentally but there are no correlations valid over the entire spectrum of possible shapes. Also the influence of the particle shape on the unsteady effects is unknown.

The effects of free stream turbulence have been quantified by the turbulence intensity based on the particle fluid relative velocity. Turbulence appears to produce a critical Reynolds number effect for Reynolds numbers as low as 10^3 but no critical Reynolds number effects seem to appear for Reynolds numbers less than 10^2. The effect of turbulence on the drag coefficient for the intermediate range is unknown and requires further study.

Particle drag in rarefied flows has been addressed by analytic and numerical models and there is some experimental validation.

The more serious lack of particle drag information is in dense phase flows where the particles are separated by less than five diameters. The correlations available in the literature were derived from packed bed pressure drop and may provide some sort of average effect but should be used cautiously for predicting the motion of individual particles in dense phase flows. The effects of unsteady flow on particle motion in a dense phase mixture have not been addressed.

Another area where more data and modeling are needed is the drag of particles in a high temperature plasma.

10.3.2   Particle heat transfer

As with particle drag, particle heat transfer in a low turbulent, steady flow has been extensively measured and correlated with the Reynolds number and Prandtl number. Analytic solutions are available for unsteady effects in Stokes flow but have not been verified experimentally. The effects of unsteady flow on heat transfer at high Reynolds number are unknown. Also, the effect of particle shape on heat transfer has not been fully addressed.

Some data are available for the effect of free stream turbulence on heat transfer which show an increase in Nusselt number with turbulence intensity but considerably more data are needed to include a more extensive Reynolds number range.

As with particle drag, there is need to have data on heat transfer to individual particles in a dense phase flow. Although heat transfer may be less sensitive to interparticle spacing than drag coefficient, the need still exists to support the numerical models for dense phase flows.

Heat transfer to particles in high temperature plasmas has received some attention in the literature but requires more study before correlations or models can be used with confidence.

10.3.3 Turbulence effects

The effects of turbulence on fluid particle flows and the corresponding effects of particles on the carrier phase turbulence are important. Turbulence is responsible for particulate and fluid phase mixing as well as for the Reynolds stress in the carrier phase.

Numerical and experimental studies show that turbulence in fluid particle flows may not lead to uniform mixing as the gradient transport model would predict. Particles with Stokes number of the order of unity tend to avoid regions of vorticity and concentrate in regions of high strain. This may be an especially important consideration in modeling droplet or particle combustion systems.

Experimental studies also show that the presence of particles can modulate the carrier phase turbulence. The general trend is that large particles augment turbulence while small particles attenuate it. Several empirical models have been proposed to explain these trends. Direct numerical simulations in which the effects of the particles are treated as point forces have been used to predict turbulence modulation. The results of these simulations should be considered carefully because modeling the particle as an applied force on the fluid may not adequately account for the effect of the particle surface. This important area of endeavor requires considerably more work, both experimental and numerical.

10.4 Numerical models

Major developments in numerical modeling of fluid-particle flows have been made during the past decade. Several commercially available CFD codes include options for fluid particle flows. Many of these codes offer both the trajectory and two-fluid approaches.

In general, the trajectory approach can be used with confidence to model dilute flows. Applying the two-fluid approach to dilute flows must be carefully evaluated to determine the sensitivity of the results to the boundary conditions and the values used for particle viscosity. Modeling a flow with a particle size distribution may require discretizing the particle equations to represent each particle size.

Some commercial codes are also available for dense phase flows. The two-fluid codes use either empirical values for the particle bulk modulus and viscosity or values derived from kinetic theory. The discrete element codes treat individual particles (or randomly selected particles) and model the force on each particle. In general, the two-fluid model relies somewhat more on empiricism but does not require excessive CPU time. The discrete element model includes the physics at the particle level but requires extensive CPU time to include a sufficient number of particles.

A point of caution when using the dense phase flow models is that they should not be applied to flow conditions for which they are not designed. For example, the kinetic theory model does not include the tangential forces and particle rotation and so would not be reliable in applications where these effects are important. Also dense flow models may not include the effect of the interstitial fluid on the particle-particle interaction which would imply that models are not valid for liquid-solid (slurry) flows.

The challenge in numerical modeling is to develop a viable model for dense phase flows which has general applicability and can be executed in reasonable time. It is difficult to predict the ultimate form this model will take. It may involve using direct numerical simulation to develop constitutive equations for a two-fluid model. Concurrently the model must be able to predict the turbulence modulation of the carrier phase.

Another challenge in numerical modeling is the atomization process. This is important because in many processes the initial droplet field is the result of atomization and the utility of the particle-fluid model is limited by the information available on atomization.

10.5   Experimental measurements

Significant advances have been made in the last decade in experimental measurements in fluid-particle systems because of the continuing development of laser technology. It is routinely possible to measure the velocity and size of spherical particles in a dilute phase flow. Methods are also becoming available to make *in situ* measurements of particle temperature and index of refraction. Also, PTV methods enable one to measure the velocities of many particles at an instant. Some of these methods are finding commercial applications.

The ultimate challenge will be the noninvasive measurement of particle and carrier flow properties in dense phase mixtures which are inaccessible by optical methods. One method that has been used in liquid-solid flows is the matched index of refraction in which the index of refraction of the liquid and (transparent) particles is the same. Of course, this approach will not function in a gas particle flow. One possibility is to use radioactive tracking, but this is expensive and difficult to use. Another approach may be nuclear magnetic resonance which also is expensive and requires specialized skills.

10.6 Summary

Even though significant advances have been made in understanding, predicting and measuring multiphase flows with droplets and particles, major challenges remain. Because of the importance of droplet and particle flows in current and emerging technologies, the practicing engineer and research scientist will be addressing these challenges for many years to come.

References

Achenbach, E., Experiments on the flow past spheres at very high Reynolds numbers, *J. Fluid Mech.*, 107, 565, 1972.

Adams, M.J. and Edmondson, B., Forces between particles in continuous and discrete liquid media, *Tribology in Particular Technology,* Briscoe, B.J. and Adams, M.J., (Eds.), Adams Hilger, Bristol and Philadelphia, Chap. 2.4, 1987.

Adrian, R.J., Multi-point optical measurements of simultaneous vectors in unsteady flow - A review, *Int. J. Heat and Fluid Flow,* 7, 127, 1986.

Adrian, R.J., Particle-imaging techniques for experimental fluid mechanics. *Ann. Rev. Fluid Mech.,* 23, 261, 1991.

Allano, D., Gouesbet, G., Grehan, G. and Lisiecki, D., Droplet sizing using a top-hat laser beam technique. *J. of Physics D: Applied Physics*, 17, 43, 1984.

Allen, T., *Particle Size Measurements.* Chapman and Hall, London, 4th ed. 1990.

Anderson, D.A., Tannehill, J.C. and Pletcher, R.H., *Computational Fluid Mechanics and Heat Transfer,* Hemisphere Publ. Corp., Bristol, PA, 1984.

Andrews, M.J. and O'Rourke, P.J., The multiphase particle-in-cell method for dense particulate flows, *Int. J. Multiphase Flow,* 22, 379, 1996.

Asanaliev, M.K., Zheenbaev, Zh. and Makesheva, K.K., Measuring the coefficient of aerodynamic drag for a sphere subject to nonisothermal streamlining, *J. Engr. Physics, (trans.)* 57, 1148, 1990.

Bagnold, R.A., Experiments on a gravity-free dispersion of large solid spheres in a Newtonian-fluid under shear, *Proc. Roy. Soc.,* A225, 49, 1954.

Bailey, A.B. and Haitt, J., Free-flight measurements of sphere drag at subsonic, transonic and near-free-molecular flow conditions, *AIAA J.,* 10, 1436, 1972.

Bakkom, A.W., Crowe, C.T., Troutt, T.R. and Xu, C., Particle-fluid coupling effects on bluff bodies, *ASME FED* Vol. 236, 273, 1996.

Balzer, G., Boelle, A. and Simonin, O., Eulerian gas solid flow modeling of dense fluidized bed, Electricité de France, Rpt. no. HE-44/95/026/A, 1995.

Barkla, H.M. and Auchterlonie, L.J., The Magnus or Robins effect on rotating spheres, *J. Fluid Mech.,* 47 (3), 437, 1971.

Barndorff-Nielsen, O., Exponentially decreasing distributions of the logarithm of particle size, *Proc. Res. Soc. Lond. A,* 353, 401, 1977.

Bauckhage, K., The phase-Doppler-difference-method, a new laser-Doppler technique for simultaneous size and velocity measuremets. *Part. Part. Syst. Charact.* 5, 16-22, 1988.

Beck, M.S. and Plaskowski, A., *Cross Correlation Flowmeters: Their Design and Application.* Adam Hilger, Bristol, 1987.

Bellan, J. and Harstad, K., The details of convective evaporation of dense and dilute clusters of droplets, *Int. J. Heat Mass Transfer*, 30, 1083, 1987.

Berlemont, A., Desjonqueres, P. and Gouesbet, G., Particle lagrangian simulation in turbulent flows, *Int. J. Multiphase Flows*, 16, 19, 1990.

Bird, G.A., *Molecular Gas Dynamics*, Clarendon, Oxford, 1976.

Bird, R.B., Stewart, W.E. and Lightfoot, E.N., *Transport Phenomena*, Wiley & Sons, New York, 1960.

Bohnet, M., Staubgehaltsbestimmung in strömenden Gasen. *Chemie-Ing.-Techn.*, 45, 18, 1973.

Bolio, E.J. and Sinclair, J.L., Gas turbulence modulation in the pneumatic conveying of massive particles in vertical tubes, *Intl. J. Multiphase Flow*, 21, 985, 1995.

Börner, Th., Durst, F. and Manero, E., LDV measurements of gas-particle confined jet flow and digital data processing, *Proc. 3rd International Symposium on Applications of Laser Anemometry to Fluid Mechanics,* Paper 4.5, 1986.

Bouillard, J.X., Lyczkowski, R.W. and Gidaspow, D., Porosity distributions in a fluidized bed with an immersed obstacle, *AIChE J.*, 35, 908, 1989.

Brady, J., Stokesian dynamics simulation of particulate flows, *Particulate Two-Phase Flow,* Roco, M. (Ed.), 912, Butterworth-Heinemann, Boston MA, 1993.

Brock, J.R., On the theory of thermal forces acting on aerosol particles, *J. of Colloid Science*, 17, 768, 1962.

Broßmann, R., Die Lichtstreuung an kleinen Teilchen als Grundlage einer Teilhengrößenbestimmung. Doctoral Thesis, University of Karlsruhe, Faculty of Mechanical and Process Engineering, 1966.

Chavez, A. and Mayinger, F., Evaluation of pulsed laser holograms of spray droplets using digital image processing. *Proc. 2nd Int. Congress on Optical Particle Sizing*, Arizona, 462-471, 1990.

Chein, R. and Chung, J.N., Simulation of particle dispersion in a two-dimensional mixing layer, *AIChE J.*, 34, 946, 1988.

Chen, X-Q. and Pereira, J.C.F., Numerical study of a nonevaporating polydispersed turbulent hollow-cone spray, *Num. Meth. for Multiphase Flows*, ASME FED Vol. 236, 41, 1996.

Cheremisinoff, N.P. and Cheremisinoff, P.N., *Hydrodynamics of Gas-Solid Fluidization,* Gulf Publishing Co., Houston TX, 1984.

Chigier, N. A., Ungut, A. and Yule, A. J., Particle size and velocity measurements in planes by laser anemometer, *Proc. 17th Symp. (Intl.) on Combustion*, 315, 1979.

Chigier, N. A., Optical imaging of sprays. *Prog. Energy Combust. Sci.*, 17, 211, 1991.

Chigier, N.A., Spray Combustion, *Proc. ASME Fluids Engr. Div.*, FED Vol. 223, 3, 1995.

Christiansen, J.P., Numerical simulation of hydrodynanics by the method of point vortices, *J. Compt. Physics*, 13, 363, 1973.

Chu, C.M. and Churchill, S.W., Numerical solution of problems in multiple scattering of electromagnetic radiation, *J. Phys. Chem.*, 59, 955, 1955.

Chung, M.K., Sung, H.J. and Lee, K.B., Computational study of turbulent gas-particle flow in a venturi, *J. Fluids Engr.*, 108, 248, 1986.

Chung, J.N. and Troutt, T.R., Simulation of particle dispersion in an axisymmetric jet, *J. Fluid Mech.*, 186, 199, 1988.

Ciccone, A. D., Kawall, J. G. and Keffer, J. F., Flow visualization/digital image analysis of saltating particle motions, *Experiments in Fluids*, 9, 65, 1990.

Clamen, A. and Gauvin, W.H., Effects of turbulence on the drag coefficients of spheres in a supercritical flow regime, *AIChE J.*, 15, 184, 1969.

Clift, R. and Gauvin, W.H., The motion of particles in turbulent gas streams, *Proc. Chemeca '70*, 1, 14, 1970.

Clift, R., Grace, J.R. and Weber, M.E., *Bubbles, Drops and Particles*, Academic Press, San Diego, CA, 1978.

Crowe, C.T., Drag coefficients of inert, burning or evaporating particles accelerating in gas streams, PhD Dissertation, U. of Michigan, 1961.

Crowe, C.T., REVIEW-Numerical models for dilute gas-particle flows, *J. Fluids Engr.*, 104, 297, 1982.

Crowe, C.T., Babcock, W.R., Willoughby, P.G. and Carlson, R.L., Measurement of particle drag coefficients in flow regimes encountered by particles in a rocket nozzle, United Technology Report 2296-FR, 1969.

Crowe, C.T., Babcock, W. and Willoughby, P.G., Drag coefficient for particles in rarefied, low Mach number flows, *Prog. Heat and Mass Trans.*, 6, 419, 1973.

Crowe, C.T., Sharma, M.P. and Stock, D.E., The Particle-Source-in-Cell method for gas droplet flow, *J. Fluid Engr.*, 99, 325, 1977.

Crowe, C.T., Droplet-gas interaction in counter-current spray dryers, *Drying Technology*, Vol. 1, 35, 1983-84.

Crowe, C.T., Chow, L.C. and Chung, J.N., An assessment of steam operated spray dryers, *Drying '85* (Ed. Mujumdar, A.), Hemisphere Pub. Corp., Washington DC, 261, 1985.

Crowe, C.T., Gore, R.A. and Troutt, T.R., Particle dispersion by coherent structures in free shear flows, *Part. Sci. Tech.* 3, 149, 1985

Crowe, C.T., Chung, J.N. and Troutt, T.R., Particle mixing in free shear flows, *Prog. Engr. Combust. Sci.*, 14, 171, 1988.

Crowe, C.T., The state-of-the-art in the development of numerical models for dispersed phase flows, *Proc. First Intl. Conf. on Multiphase Flows -'91-Tsukuba*, Vol. 3, 49, 1991.

Crowe, C.T., Modeling turbulence in multiphase flows, *Proc. 2nd Intl. Symp. on Engr. Turbulence Modeling and Measurements,* Elsevier, Amsterdam, 899, 1993

Crowe, C.T., Troutt, T.R., Chung, J.N., Davis, R.W. and Moore, E.F., A turbulent flow with particle mixing, *Aerosol Sci. & Tech.*, 22, 135, 1995.

Crowe, C.T., Chung, J.N. and Troutt, T.R., Numerical Models for Two-Phase Turbulent Flows, *Ann. Rev. Fluid Mech., 28, 11,* 1996.

Csandy, G.T., Turbulent diffusion of heavy particles in the atmosphere, *J. Atmos. Sci.*, 20, 201, 1963.

Cundall, P.A. and Strack, O.D., A discrete numerical model for granular assemblies, *Geotechnique*, 29, 47, 1979.

Czarnecki, J. and Dabros, T., Attenuation of the van der Waals attraction energy in the particle/semi-infinite medium systems due to the roughness of the particle surface, *J. Colloid Interface*, 78, 25, 1980.

Dandy, D.S. and Dwyer, H.A., A sphere in shear flow at finite Reynolds number: effect of particle lift, drag and heat transfer, *J. Fluid Mech.*, 216, 381, 1990.

DANTEC/Invent, STREU, A computational code for the light scattering properties of spherical particles. Instruction Manual, 1994.

Dasgupta, A., Shih, T. I-P., Kundu, K. and Deur, J.M., Numerical simulation of flow past an array of moving spherical particles, *AIAA paper 94-3283,* 1994.

Davis, J.M., The aerodynamics of golf balls, *J. Appl. Physics*, 20, 821, 1949.

Denis, S.C.R., Singh, S.N. and Ingham, D.B., The steady flow due to a rotating sphere at low and moderate Reynolds numbers, *J. Fluid Mech.*, 101(2), 257, 1980.

Di Felice, R., The voidage function for fluid-particle interaction systems, *Intl. J. Multiphase Flow*, 20, 153, 1994.

Dilber, I. and Mereu, S., Effect of turbulent fluctuations in the numerical simulation of spray dryers, *Num. Meth. for Multiphase Flows*, ASME FED Vol. 236, 291, 1996.

Ding, D. and Gidaspow, D., A bubbling fluidization model using kinetic theory of granular flow, *AIChE J.*, 36, 523, 1990.

Ding, J., Lyczkowski, R.W. Sha, W.T., Altobelli, S.A. and Fukushima, E., Numerical analysis of liquid-solids suspension velocities and concentrations obtained by NMR imaging, *Powder Tech.*, 77, 301, 1993.

Dukowicz, J.K., A particle-fluid numerical model for liquid sprays, *J. Comput. Phys.*, 33, 229, 1980.

Durst, F. and Zaré, M., Laser-Doppler measurements in two-phase flows. *Proc. of the LDA-Symposium*, University of Denmark, 1975.

Durst, F., Melling, A. and Whitelaw, J.H., *Principles and Practice of Laser-Doppler Anemometry.* 2nd Edition, Academic Press, London, 1981.

Durst, F., Review-combined measurements of particle velocities, size distribution and concentration, *J. of Fluids Eng.*, 104, 284, 1982.

Durst, F., Melling, A. and Whitelaw, J.H. *Theorie und Praxis der Laser-Doppler Anemometrie.* G. Braun, Karlsruhe, 1987.

Eisenklam, P., Arunachalam, S.A. and Weston, J.A., Evaporation rates and drag resistance of burning drops, *Eleventh (Intl.) Symp. on Combustion,* 715, 1967.

Elghobashi, S. and Truesdell, G.C., Direct simulation of particle dispersion in a decaying isotropic turbulence, *J. Fluid Mech.,* 242, 655, 1992.

Elghobashi, S. and Truesdell, G.C., On the two-way interaction between homogeneous turbulence and dispersed solid particles I: turbulence modulation. *Phys. Fluids A,* Vol. 5, 1790, 1993.

Epstein, P.S., Zur Theorie des Radiometers, *Zeit. für Physik,* 54, 537, 1929.

Ergun, S., Fluid flow through packed columns, *Chem. Engr. Prog.,* 48, 89, 1952.

Farmer, W. M., Measurement of particle size, number density and velocity using a laser interferometer. *Applied Optics,* 11, 2603, 1972.

Farmer, W. M., Observation of large particles with a laser interferometer. *Applied Optics,* 13, 610, 1974.

Feng, J., Hu, H.H. and Joseph, D.D., Direct simulation of initial value problems for the motion of solid bodies in a Newtonian fluid. Part 2: Couette and Poiseuille flows, *J. Fluid Mech.,* 277, 271, 1994.

Finnie, I., Some observations on the erosion of ductile materials, *Wear,* Vol. 19, 81, 1972.

Fischer, K, Leithner, R., Müller, H. and Schiller, A., On the simulation of gas-solid flow in pulverized coal-fired furnaces, *Num. Meth. for Multiphase Flows,* ASME FED Vol. 236, 239, 1996.

Fiveland, W.A., Three-dimensional radiative heat-transfer solutions by discrete-ordinates method, *J. Thermophysics,* 2, 309, 1988.

Founti, M. and Klipfel, A.S., Numerical simulation of particle-particle collisions in nearly dense two-phase flows, *Num. Meth. for Multiphase Flows,* ASME FED Vol. 236, 89, 1996.

Founti M. and Klipfel, A.S., Numerical simulation of pneumatic transport and erosion wear in the distribution ducts in large lignite power plants, *Erosion Processes,* ASME FED Vol. 236, 717, 1996.

Frank, Th., Schade, K.P. and Petrak, D., Numerical simulation and experimental investigation of gas-solid two-phase flow in a horizontal channel, *Intl. J. Multiphase Flow,* 19, 187, 1993.

Frank, Th. and Wassen, E., Parallel solution algorithms for lagrangian simulation of disperse multiphase flows, *Num. Meth. for Multiphase Flows,* ASME FED Vol. 236, 11, 1996.

Frössling, N., Über die Verdüngtung fallendenTropfen, *Gerlands Beitr. Zur Geophysik,* 52, 170, 1938.

Gidaspow, D., *Multiphase Flow and Fluidization,* Academic Press, San Diego, CA, 1994.

Giridharan, M.G., Lee, J.G., Krishnan, A., Przekwas, A.J. and Gross, K., A numerical model for coupling between atomization and spray dynamics in liquid rocket thrust chambers, AIAA Paper 92-3768, 1992.

Gore, R.A. and Crowe, C.T., The effect of particle size on modulating turbulent intensity, *Intl. J. Multiphase Flow,* 15, 279, 1989.

Gosman, A.D., Pun, W.M., Runchal, A.K., Spalding, D.B. and Wolfshtein, M., *Heat and mass transfer in recirculating flows,* Academic Press, London, 1969.

Gosman, A.D. and Ioannides, E., Aspects of computer simulation of liquid-fueled combustors, AIAA Paper 81-0323, 1981.

Govan, A.H., Hewitt, G.F. and Ngan, G.F., Particle motion in a turbulent pipe flow, *Intl. Jnl. Multiphase Flow,* 15, 471, 1989.

Graham, D.I and James, P.W., Turbulent dispersion of particles using eddy interaction models, *Int. J. Multiphase Flow,* 22, 157, 1996.

Grehan, G. and Gouesbet, G., Simultaneous measurements of velocities and size of particles in flows using a combined system incorporating a top-hat beam technique. *App. Opt.* 25, 3527, 1986.

Grehan, G., Gouesbet, G., Nagwi, A. and Durst, F., On elimination of the trajectory effects in phase-Doppler systems. *Proc. 5th European Symp. Particle Characterization (PARTEC 92),* 309, 1992.

Griffiths, W.D. and Boyson, F., Computational fluid dynamics (CFD) and empirical modelling of the performance of a number of cyclone separators, *J. Aerosol Sci.,* 27, 281,1996.

Hamaker, H.C., The London-Van der Waals' attraction between spheroid particles, *Physica,* 4, 1058, 1937.

Hamed, A. and Tabakoff, W., Erosion in turbomachine, *Erosion Processes,* ASME FED Vol. 236, 743, 1996.

Happel, J. and Brenner, H., *Low Reynolds Number Hydrodynamics,* Noordhoff Intl. Pub., Leiden, 1973.

Hardalupas, Y. and Taylor, A.M.K.P., On the measurement of particle concentration near a stagnation point. *Exp. in Fluids,* 8, 113, 1989.

Hardalupas, Y., Hishida, K., Maeda, M., Morikita, H., Taylor, A.M.K.P. and Whitelaw, J.H., Shadow Doppler technique for sizing particles of arbitrary shape. *Appl. Optics,* 33, 8417, 1994.

Harlow, F.H. and Amsden, A.A., Numerical calculation of multiphase fluid flow, *J. Comput. Physics,* 17, 19, 1975.

Harris, S.E. and Crighton, D.G., Solitons, solitary waves, and voidage distrubances in gas-fluidized beds, *J. Fluid Mech.,* 266, 243, 1994.

Harlow, F.H. and Fromm, J.E., Computer experiments in fluid mechanics, *Sci. Am.,* 212, 104, 1965.

Hassan, Y. A., Blanchat, T.K., Seeley, C.H. and Canaan, R.E., Simultaneous velocity measurements of both comonents of a two-phase flow using particle image velocimetry. *Int. J. Multiphase Flow,* 18, 371, 1992.

He, J. and Simonin, O., Non-equilibrium prediction of the particle-phase stress tensor in vertical pneumatic conveying, *Gas-Solid Flows,* ASME FED Vol. 166, 253, 1993.

Hedley, A.B., Nuruzzaman, A.S.M. and Martin, G.F., Progress review No. 62 - Combustion of single droplets and simplified spray systems, *J.*

Inst. Fuel, 44, 38, 1971.

Henderson, C.B., Drag coefficients of spheres in continuum and rarefied flows, *AIAA J.* 14, 707, 1976.

Hermsen, R.W., Review of particle drag models, JANAF Performance Standardization Subcommittee 12th Meeting Minutes, CPIA, 113, 1979.

Hess, C. F., Nonintrusive optical single-particle counter for measuring the size and velocity of droplets in a spray, *Applied Optics,* 23, 4375, 1984.

Hess, C. F. and Espinosa, V. E., Spray characterization with a nonintrusive technique using absolute scattered light. *Optical Eng.,* 23, 604, 1984.

Hetsroni, G. and Sokolov, M., Distribution of mass, velocity and intensity of turbulence in a two-phase turbulent jet, *J. Appl. Mech.,* 38,315,1971.

Hetsroni, G., Particles-turbulence interaction, *Intl. J. Multiphase Flow,* 15, 735, 1989.

Hewitt, G.F. and Roberts, D.N., Investigation of Interfacial Phenomena in Annual Two-Phase Flow by Flash Photography, AERE-M2159, UKAEA, Harwell, 1969.

Hibbeler, R.C., *Dynamics,* Prentice-Hall, Englewood Cliffs, NJ, 1995.

Hill, S.C. and Smoot, L.D., Comprehensive three dimensional model for the simulation of combustion systems: PCGC-3, *Energy and Fuels,* 7, 874, 1993.

Hirleman, D.E., Oechsle, V. and Chigier, N.A., Response characteristics of laser diffraction particle size analyzers: Optical sample volume extent and lens effects. *Optical Engineering,* 23, 610, 1984.

Hishida, K. and Maeda, M., Application of laser/phase Doppler anemometry to dispersed two-phase flow. *Part. Part. Syst. Charact.,* 7, 152, 1990.

Hjelmfelt, A.T. and Mockros, L.F., Motion of discrete particles in a turbulent fluid, *App. Sci. Res.,* 16, 149, 1966.

Horner, S.F., *Fluid-Dynamic Drag,* published by author, Midland Park, NJ, 1965.

Howell, J.R., Application of Monte Carlo to heat trasnfer problems, *Advances in Heat Transfer,* 5, 1, Academic Press, Orlando FL,1968.

Hu, H.H., Numerical simulation of channel Poiseuille flow of solid-liquid mixtures, ASME FED Vol. 236, 97, 1996.

Huber, N. and Sommerfeld, M., Characterization of the cross-sectional particle concentration distribution in pneumatic conveying systems, *Powder Tech.,* 79, 191, 1994.

Hussaini, M.Y. and Zang, T.A., Spectral method in fluid mechanics, *Annu. Rev. Fluid Mech.,* 19, 339, 1987.

Iller, R. and Neunzert, H., On simulation methods for the Boltzmann equation, *Trans. Theory and Stat. Phys.,* Vol. 16, 141, 1987.

Ingebo, R.D., Drag coefficients for droplets and solid spheres in clouds accelerating in airstreams, NACA TN 3762, 1956.

Ishii, M., *Thermo-fluid Dynamic Theory of Two-Phase Flow,* Eyrolles, Paris, 1975.

Issa, R.I., and Oliveira, P.J., Numerical prediction of phase separation in two-phase flow through T-junctions, *Computers Fluids,* 23, 347, 1994.

Johnson, A.A. and Tezduyar, T.E., Fluid particle simulations reaching 100 particles, *ASME FEDSM'97-3184*, 1997.

Jorgensen, S.E. and Johnsen, I., *Principles of Environmental Science and Technology*, Elsevier Scientific Publishing Co. 1981.

Jurewicz, J.T. and Stock, D.E., Numerical model for turbulent diffusion in gas-particle flows, ASME 76-WA/FE-33, 1976.

Kaufman, K.C. and Fiveland, W.A., Combustion modeling of coal-fired, aerodynamically-staged burners, *Gas-Particle Flows*, ASME FED Vol. 228, 317, 1995.

Kavanau, L.L., Heat transfer from spheres to a rarefied gas in subsonic flow, *ASME Trans.*, 77, 617, 1955.

Kawaguchi, T., Yamamoto, Y., Tanaka, T. and Tsuji, Y., Numerical simulation of a single rising bubble in a two-dimensional fluidized bed, *Proc. 2nd Intl. Conf. on Multiphase Flows*, FB2-17, 1995.

Kenning, V.M. and Crowe, C.T., Effect of particles on the carrier phase turbulence in gas-particle flows, *Intl. J. Multiphase Flows,* 23, 403, 1997.

Kim, S.J. and Crowe, C.T., A model of cyclone separators with applications to hot-gas cleanup, *Gas-Solid Flows*, ASME FED Vol. 10, 57, 1984.

Kim, J.C. and Lee, K.W., Experimental study of particle collection by small cyclones, *Aerosol Sci. Technol.*, 12, 1003, 1990.

Kipphan, H., Bestimmung von Transportkenngrößen bei Mehrphasenströmungen mit Hilfe der Korrelationstechnik. *Chem.-Ing.-Techn.*, 49, 695, 1977.

Kliafas, Y., Taylor, A.M.K.P. and Whitelaw, J.H., Errors due to turbidity in particle sizing using laser-Doppler anemometry, *J. Fluid Engineering*, 112, 142, 1990.

Klinzing, G. E., *Gas-Solid Transport*, McGraw-Hill, New York, 1981.

Kreith, F., *Principles of Heat Transfer*, Intext Educational Pub., New York, 1973.

Launder, B.E. and Spalding, D.B., *Mathematical Models of Turbulence*, Academic Press, London, 1972.

Lee, M.M., Hanratty, T.J. and Adrian, R.J., An axial viewing photographic technique to study turbulence characteristics of particles, *Intl. J. Multiphase Flow*, 15, 787, 1989.

Leith, D. and Licht, W., Collection efficiency of cyclone type particle collectors - a new theoretical approach, *Am. Inst. Chem. Eng. Symp. Ser.* (Air-1971), Vol. 68, 196, 1972.

Leonard, A., Computing three dimensional incompressible flows with vortex elements, *Annu. Rev. Fluid Mech.*, 17, 523, 1985.

Liljegren, L.M., On modeling particles as point forces in a gas-particle flow, *Numerical Methods for Multiphase Flows*, ASME FED Vol. 236, 173, 1996.

Liljegren, L.M., Ensemble-average equations of a particulate mixture, *J. Fluids Engr.*, 119, 428, 1997.

Ling, W., Chung, J.N., Troutt, T.R. and Crowe, C.T., Numerical simulation of particle dispersion in a three-dimensional temporal mixing layer, *ASME FEDSM97-3182*, 1997.

Lischer, J. and Louge, M.Y., Optical fiber measurements of particle concentration in dense suspensions: Calibration and Simulation, *Appl. Optics*, 30, 8, 1991.

Litchford, R.J. and Jeng, S-M., Efficient statistical transport model for particle dispersion in sprays, *AIAA J.*, 29, 1443, 1991.

Longmire, E. K. and Eaton, J. K., Structure of a particle-laden round jet, *Jnl. Fluid Mech.*, 236, 217, 1992.

Longmire, E.K. and Anderson, S.L., Effects of vortices on particle motion in a stagnation zone. Gas-Particle Flows, ASME FED, Vol. 228, 89, 1995.

Lockwood, F.C., Salooja, A.P. and Syed, S.A., A prediction method for coal-fired furnaces, *Combust. Flame*, 38, 1, 1980.

Lu, Q.Q., Fontaine, J.R. and Aubertin, G., Numerical study of the solid particle motion in grid generated turbulence, *Int. J. Heat Mass Trans.*, 36, 79, 1993.

McLaury, B.S., Shirazi, S.A., Shadley, J.R. and Rybicki, E.F., Modeling erosion in chokes, *Erosion Processes*, ASME FED Vol. 236, 773, 1996.

Maccoll, J.W., Aerodynamics of a spinning sphere, *J. Roy. Aero. Soc.*, 32, 777, 1928.

Magnée, A., Generalized law of erosion: application to various alloys and intermetallics, *Wear*, Vol. 181, 500, 1995.

Marcus, R.D., Leung, L.S., Klinzing, G.E. and Rizk, F., *Pneumatic conveying of solids*, Chapman and Hall, New York, 1990.

Martin, J. and Meiburg, E., The accumulation of dispersion of heavy particles in forced two-dimensional mixing layers. Part I: The fundamental and harmonic cases. *Phys. of Fluids A*, 26(4), 883, 1994.

Masters, K., *Spray Drying: an introduction to principles, operational practice and applications*, Leonard Hill Books, London, 1972.

Matsumoto, S. and Saito, S., On the mechanism of suspensions in horizontal pneumatic conveying: Monte Carlo simulation based on the irregular bouncing model, *J. Chem. Engr. Japan*, 3, 83, 1970a.

Matsumoto, S. and Saito, S., Monte Carlo simulation of horizontal pneumatic conveying based on the rough wall model, *J. Chem. Engr. Japan*, 3, 223, 1970b.

Maxey, M.R. and Riley, J.J., Equation of motion for a small rigid sphere in a nonuniform flow, *Phys. Fluids*, 26(4), 883, 1983.

MacInnes, J.M. and Bracco, F.V., Stochastic particle dispersion and the tracer particle limit, *Phys. Fluids A*, 4, 2809, 1992.

McLaughlin, J.B., Inertial migration of small sphere in linear shear flows, *J. Fluid Mech.*, 224, 261, 1991.

Mehta, A., (Ed.), *Granular Matter*, Springer-Verlag, New York, 1994.

Mei, R., Lawrence, C.J. and Adrian, R., Unsteady drag on a sphere at finite Reynolds number with small fluctuations in the free-stream velocity, *J. Fluid Mech.*, 223, 613, 1991.

Mei, R., An approximate expression for the shear lift on a spherical particle at finite Reynolds number, *Intl. J. Multiphase Flow*, 18, 145, 1992.

Michaelides, E.E. and Feng, Z-G., Analogies between the transient momentum and energy equations of particles, *Prog. Energy and Comb. Sci.*, 22, 147, 1996.

Mie, G., Beiträge zur Optik trüber Medien, speziell kolloidaler Metallösungen. *Ann. der Physik*, 25, 377, 1908.

Millikan, R.A., The general law of fall of a small spherical body through a gas, and its bearing upon the nature of molecular reflection from surfaces, *Phys. Rev.*, 22, 1, 1923.

Mindlin, R.D., Compliance of elastic bodies in contact, *J. Appl. Mech. (Trans. ASME)*, 16, 259, 1949.

Mindlin, R.D. and Deresiewicz, H., Elastic spheres in contact under varying oblique forces, *J. Appl. Mech. (Trans. ASME)*, 20, 327, 1953.

Modarress, D. and Tan, H., LDA signal discrimination in two-phase flows, *Experiments in Fluids,* 1, 129, 1983.

Morikawa, Y., Tanaka, T., Nakatsukasa, N. and Nakatani, M., Numerical simulation of gas-solid two-phase flow in a two-dimensional horizontal channel, *Intl. J. Multiphase Flow,* 13, 671, 1987.

Mugele, R.A. and Evans, H.D., Droplet size distribution in sprays, *Ind. Engr. Chem.* 43, 1317, 1951.

Muoio, N.G., Crowe, C.T., Fritsching, U. and Bergmann, D., Effect of thermal coupling on numerical simulations of the spray forming process, *Num. Meth. for Multiphase Flows*, ASME FED Vol. 236, 233, 1996.

Murai, Y. and Matsumoto, Y., Numerical simulation of turbulent bubble plumes using Eulerian-Lagrangian bubbly flow model equations, *Num. Meth. for Multiphase Flows*, ASME FED Vol. 236, 67, 1996.

Nam, S., Computational prediction of the delivered water flux through fire plumes and comparison with measurements, *ASME HTD* Vol. 321, 97, 1995.

Naqwi, A.A. and Durst, F., Light scattering applied to LDA and PDA measurements. Part 1: Theory and numerical treatments. *Part. Part. Syst. Charact.*, 8, 245, 1991.

Negus, C. R. and Drain, L. E., Mie calculations of scattered light from a spherical particle traversing a fringe pattern produced by two intersecting laser beams, *J. Phys. D: Applied Physics*, 15, 375, 1982.

Neve, R.S. and Jaafar, F.B., The effects of turbulence and surface roughness on the drag of spheres in thin jets, *The Aero. J.*, 86, 331, 1982.

Neve, R.S. and Shansonga, T., The effects of turbulence characteristics on sphere drag, *Int. J. Heat and Fluid Flow*, 10, 318, 1989.

Oakley, D.E. and Bahu, R.E., Spray/gas mixing behavior within spray dryers, *Drying '91*, Mujumdar, A.S. and Filkova, I. (Eds.), Elsevier, Amsterdam, 1991.

Odar, F. and Hamilton, W.S., Forces on a sphere accelerating in a viscous fluid. *J. Fluid Mech.*, 18, 302, 1964.

Odar, F., Verification of the proposed equation for calculation of the forces on a sphere accelerating in a viscous flow, *J. Fluid Mech.*, 25, 591, 1966.

Ory, E. and Perkins, R.J., Numerical study of particle motion in a turbulent mixing layer using the discrete vortex method, ASME FEDSM'97-3614,

1997.

Patankar, S.V., *Numerical Heat Transfer and Fluid Flow*, McGraw-Hill, New York, 1980.

Patnaik, P.C., Vittal, N. and Pande, P.K., Drag coefficient of a stationary sphere in gradient flow, *J. Hydraulic Research*, 30, 389, 1992.

Perkins, R.J. and Hunt, J.C.R., Particle tracking in turbulent flows. In *Advances in Turbulence*, 2, Springer-Verlag, Berlin, 286, 1989.

Petrak, D. and Hoffmann, A., The properties of a new fiberoptic measuring technique and its application to fluid-solid flows. *Advances in Mechanics*, 8, 59, 1985.

Pfender, E. and Lee, Y.C., Particle dynamics and particle heat and mass transfer in thermal plasmas. Part I. The motion of a single particle without thermal effects. *Plasma Chem. and Plasma Proc.*, 5, 211, 1985.

Philip, O.G., Schmidt, W.D. and Hassan, Y.A., Development of a high speed particle image velocimetry technique using fluorescent tracers to study steam bubble collapse, *Nuclear Eng. and Design* 149, 375, 1994.

Picard, A., Berlemont, A. and Gouesbet, G., Modeling and predicting turbulence fields and the dispersion of discrete particles transported in turbulent flows, *Int. J. Multiphase Flows*, 12, 237, 1986

Putnam, A., Integrable form of droplet drag coefficient, *ARS Jnl.*, 31, 1467, 1961.

Qiu, H.-H. and Sommerfeld, M., A reliable method for determining the measurement volume size and particle mass fluxes using phase-Doppler anemometry. *Experiments in Fluids*, 13, 393, 1992.

Raithby, G.D. and Eckert, E.R.G., The effect of turbulence parameters and support position on the heat transfer from spheres, *Int. J. Heat and Mass Transfer*, 11, 1233, 1968.

Ranz, W.E. and Marshall, W.R., Evaporation from drops - I and II, *Chem. Engr. Prog.* 48, 141 and 173, 1952.

Reeks, M.W., On the dispersion of small particles suspended in isotropic turbulence fields, *J. Fluid Mech.* 83, 529, 1977.

Reeks, M.W. and McKee, S., The dispersive effects of Basset history forces on particle motion in a turbulent flow, *Phys. Fluids*, 27(7), 1573, 1984.

Reeks, M.W., On the constitutive relations for dispersed particles in nonuniform flows. 1: Dispersion in simple shear flow, *Phys. Fluids* A, 5, 750, 1993.

Renksizbulut, M. and Yuen, M.C., Experimental study of droplet evaporation in a high-temperature air stream, *J. of Heat Transfer*, 105, 384, 1983.

Rensner, D. and Werther, J., Estimation of the effective measuring volume of single fiber reflection probes for solids concentration measurements. *Preprints of the 5th European Symposium Particle Characterization (PARTEC 92)*, 107, 1992.

Reynolds, W.C., The potential and limitations of direct and large eddy simulation. In *Turbulence at the Crossroads*, (ed. Lumley, J.L.), 313, Springer Verlag, New York, 1990.

Rhie, C.M. and Chow, W.L., Numerical study of the turbulent flow past an airfoil with trailing edge separation, *AIAA J.*, 21, 1525, 1983.

Richardson, J.F. and Zaki, W.N., Sedimentation and Fluidization - Part 1, *Trans. Inst. Chem. Engr.*, 32, 35-53, 1954.

Rivard, W.C. and Torrey, M.D., K-FIX,: A computer code for transient two-dimensional flow, *LA-NUREG-6623*, Los Alamos, 1977.

Rizk, M.A. and Elghobashi, S.E., A two-equation turbulence model for disperse, dilute, confined two-phase flows, *Int. J. Multiphase Flows*, 15, 119, 1989.

Roberts, D.W., Particle sizing using laser interferometry. *Applied Optics*, 16, 1861, 1977.

Roco, M. and Shook, C.A., Turbulent flow of incompressible mixtures, *J. Fluids Engr.*, 107, 224, 1985.

Rodi, W., Ferziger, J.H., Breuer, M. and Pourquié, M., Status of large eddy simulation: Results of a workshop, *J. Fluids Engr.*, 119, 248, 1997.

Rogallo, R.S. and Moin, P., Numerical simulation of turbulent flow, *Annu. Rev. Fluid Mech.*, 16, 99, 1984.

Rowe, P.N., Claxton, K.T. and Lewis, J.B., Heat and mass transfer from a single sphere in an extensive flowing fluid, *Trans. Instn. Chem. Engrs.* 43, T14, 1965.

Rubinow, S.I. and Keller, J.B., The transverse force on spinning sphere moving in a viscous fluid, *J. Fluid Mech.*, 11, 447, 1961.

Rudinger, G., Fundamentals of Gas-Particle Flow, *Handbook of Powder Technology*, Vol. 2, Elsevier Scientific Publishing Co., Amsterdam, 1980.

Rudoff, R.R. and Bachalo, W.D., Measurements of droplet drag coefficients in polydispersed turbulent flow field, *AIAA Paper 88-0235*, 1988.

Saffman, M., Automatic calibration of LDA measurement volume size., *Appl. Optics*, 26, 2592, 1987.

Saffman, M., Optical particle sizing using the phase of LDA signals. Dantec Information, No. 05, 8, 1987.

Saffman, P.G., The lift on a small sphere in a slow shear flow, *J. Fluid Mech.*, 22, 385, 1965.

Saffman, P.G., Corrigendum to "The lift on a small sphere in a slow shear flow", *J. Fluid Mech.*, 31, 624, 1968.

Sankar, S.V. and Bachalo, W.D., Response characteristics of the phase-Doppler particle analyzer for sizing spherical particles larger than the wavelength. *Appl. Optics*, 30, 1487, 1991.

Sarpkaya, T., Freeman scholar lecture: Computational methods with vortices, *J. Fluids Engr.*, 111, 5, 1985.

Sato, Y. and Hishida, K., Transport process of turbulent energy in particle-laden turbulent flow, *Int. J. Heat and Fluid Flow*, 17, 202, 1996.

Savage, S.B., Granular flows at high shear rates, in *Theory of Dispersed Multiphase Flow*, R.E. Meyer (Ed.), Academic Press, 339, 1983.

Schaaf, S.A. and Chambré, P.L., Fundamentals of gas dynamics, *High Speed Aerodynamics and Jet Propulsion*, Emmons H.W., (Ed.), Vol. 3, pp. 687-739, Princeton Univ. Press, 1958.

Shanov, V., Tabakoff, W. and Gunaraj, J.A., Study of CVD coated and uncoated INCO 718 exposed to particulate flow, *Transport Phenomena in*

403

Materials Processing and Manufacturing, ASME HTD-Vol. 336, FED-Vol. 240, 227, 1996.

Sharma, M.P. and Crowe, C.T., A novel physico-computational model for quasi one-dimensional gas-particle flow, *J. Fluids Engr.*, 100, 343, 1978.

Shen, N., Tsuji, T. and Morikawa, Y., Numerical simulation of gas-solid two-phase flow in a horizontal pipe, *JSME*, 55, 2294, 1989.

Shiller L. and Naumann, A., Über die grundlegenden Berechungen bei der Schwerkraftaufbereitung, *Ver. Deut. Ing.*, 77, 318, 1933.

Shimizu, A., Numerical prediction of erosion for suspension flow duct, *Gas-Solids Flows - 1993*, ASME FED, Vol. 166, 237, 1993.

Shook, C.A. and Roco, M.C., *Slurry Flow: Principles and Practice*, Butterworth-Heinemann, Boston MA, 1991.

Schöneborn, P.-R., The interaction between a single particle and an oscillating fluid, *Int. J. Multiphase Flow*, 2, 307-317, 1975.

Siegel, R. and Howell, J.R., *Thermal Radiation Heat Transfer*, McGraw Hill, 1981.

Simonin, O., Prediction of the dispersed phase turbulence in particle-laden flows, *ASME FED Vol. 121*, 197, 1991.

Simonin, O., Deutsch, E. and Minier, J.P., Eulerian prediction of the fluid/particle correlated motion in turbulent two-phase flows, *App. Sci. Res.*, 51, 275, 1993.

Sinclair, J.L. and Jackson, R., Gas-particle flow in a vertical pipe with particle-particle interactions, *AIChE J.*, 35, 1473, 1989.

Sirignano, W.A., Fluid dynamics of sprays - 1992 Freeman Scholar Lecture, *J. Fluids Engr.*, 115, 345, 1993.

Slattery, J.C., *Momentum, Energy, and Mass Transfer in Continua*, McGraw-Hill, New York, 1972.

Smith, P.J., Fletcher, T.J. and Smoot, L.D., Model for pulverized coal-fired reactors, *18th Intl. Symp. on Combustion*, 1285, 1981.

Smoot, L.D. and Pratt, D.T., *Pulverized-Coal Combustion and Gasification*, Plenum Press, New York, 1979.

Smoot, L.D. and Smith, P.J., *Coal combustion and gasification*, Plenum Press, New York, 1985.

Sommerfeld, M. and Zivkovic, G., Recent Advances in the simulation of pneumatic conveying through pipe systems, *Computational Methods in Applied Sciences*, Hirsch C., (Ed.) 201, 1992.

Sommerfeld, M., Huber, N. and Wächter, P., Particle-wall collisions: Experimental studies and numerical models. *Gas-Solid Flows 1993*, ASME FED 166, 183, 1993.

Sommerfeld, M., Modellierung und numerische Berechnung von Partikel-beladenen turbulenten Stromung mit Hilfe des Euler/Lagrange Verfahrens, Habilitationsschrift, University of Erlangen-Nuernburg, Shaker-Verlag, Aachen, 1994.

Sommerfeld, M. and Qiu, H.H., Particle concentration measurements by phase-Doppler anemometry in complex dispersed two-phase flows. *Exper. in Fluids*, 18, 187, 1995.

Soo, S. L., Stukel, J. J. and Hughes, J. M., Measurement of mass flow and density of aerosols in transport. *Environ. Sci. and Tech.*, 3, 386, 1969.

Squires, K.D. and Eaton, J.K., Particle response and turbulence modification in isotropic turbulence, *Phys. Fluids*, A2(7), 1191, 1990.

Squires, K.D. and Eaton, J.K., Preferential concentration of particles by turbulence, *Phys. Fluids* A, 3, 130, 1991.

Squires, K.D. and Wang, Q., On the large eddy simulation of particle-laden turbulence, *Proc. Eighth Workshop on Two-Phase Predictions*, Erlangen, 1996.

Stock, D.E., Particle dispersion in flowing gases - 1994 Freeman Scholar Lecture, *J. Fluids Engr.*, 118, 4, 1996.

Strehlow, R.A., *Combustion fundamentals*, McGraw-Hill, New York, 1984.

Synder, W.H. and Lumley, J.L., Some measurements of particle velocity and auto-correlation functions in a turbulent flow, *J. Fluid Mech.* 48, 41, 1971.

Swithenbank, J., Beer, J. U., Taylor, D. S., Abbot, D. and Mc-Creath, G. C., A laser diagnostic technique for the measurement of droplet and particle size distribution. *Prog. in Astro. and Aero., AIAA*, 421, 1977.

Tabakoff, W., Performance deterioration on turbomachinery with the presence of particles, *Particulate Laden Flows in Turbomachinery*, AIAA/ASME Joint Fluids, Plasma, Thermodynamics and Heat Transfer Conference, 3, 1982.

Talbot, L., Thermophoresis - A review, *Rarefied Gas Dynamics, Part 1,* Fisher, S. (Ed.), 74, 467, 1981.

Tanaka, T., Yamagata, K and Tsuji, Y., Experiment on fluid forces on a rotating sphere and spheroid, *Proc. Second KSME-JSME Fluids Engr. Conf.*, 1, 366, 1990.

Tanaka, T and Tsuji, Y., Numerical simulation of gas-solid two-phase flow in a vertical pipe: on the effect of interparticle collision, ASME FED Vol. 121, 123, 1991.

Tanaka, T., Yonemura, S. and Tsuji, Y., Effects of particle properties on the structure of clusters, *ASME FED* Vol. 228, 297, 1995.

Tanaka, T., Yonemura, S., Kiribayashi, K. and Tsuji, Y., Cluster formation and particle-induced instability in gas-solid flows predicted by the DSMC method, *JSME Intl. J.*, 39(B), 239,1996.

Tang, L., Crowe, C.T., Chung, J.N. and Troutt, T.R., Effect of momentum coupling on the development of shear layers in gas-particle mixtures, *Proc. Intl. Conf. on Mech. of Two-Phase Flows*, National Taiwan Univ., 387, 1989.

Tang, L., Wen, F., Yang, Y., Crowe, C.T., Chung, J.N. and Troutt, T.R., Self-organizing particle dispersion mechanism in free shear flows, *Phys. Fluids* A, 4, 2244, 1992.

Tayali, N. E. and Bates, C. J., Particle sizing techniques in multiphase flows: A review. *Flow Meas. Instrum.*, 1, 77, 1990.

Taylor, G.I., Diffusion by continuous movements, *Proc. London Math. Soc.* Ser. 2, 196, 1921.

Tchen, C.M., Mean values and correlation problems connected with the motion of small particles suspended in a turbulent fluid, Doctoral Dissertation,

Delft, Holland, 1949.

Thomas, M.E. and Dionne, P.J., Turbine engine inlet protection system concept design using CFD technology, AIAA-94-2813, 1994.

Thorpe, R.D., Dash, S.M. and Pergamet, H.S., Inclusion of gas/particle interactions in a shock-capturing model for nozzle and exhaust plumes, *AIAA Paper 79-1288*, 1979.

Tokuhiro, A., Maekawa, M., Iizuka, K., Hishida, K. and Maeda, M., The effect of a single bubble on turbulence structure in grid turbulence flow by combined shadow-image and PIV technique. *Proc. of the 8th Int. Symp. on Applications of Laser Techniques to Fluid Mechanics*, Lisbon, 1996.

Tong, X-L. and Wang, L-P., Direct simulations of particle transport in two and three dimensional mixing layers, *ASME FEDSM'97-3635*, 1997.

Torobin, L.B. and Gauvin, W.H. The drag coefficients of single spheres moving in steady and accelerated motion in a turbulent fluid, *AIChE J.*, 7, 615, 1961.

Tropea, C. Laser Doppler anemometry: Recent developments and future challenges. *Meas. Sci. Techn.* 6, 605, 1995.

Tropea, C., Xu, T.-H., Onofri, F., Grehan, G., Haugen, P. and Stieglmeier, M., Dual mode phase Doppler anemometer. *PARTEC 95*, Preprints 4th Int. Congress Optical Particle Sizing, 287, 1995.

Tsuji, Y., Shen, N-Y. and Morikawa, Y., Numerical simulation of gas-solid flows. I. (Particle-to-wall collisions), Tech. Rpt. of Osaka University, 39, 233, 1975.

Tsuji, Y., Morikawa, Y. and Shiomi, H., LDV measurements of an air-solid two-phase flow in a vertical pipe, *J. Fluid Mech.*, 139, 417, 1984.

Tsuji, Y., Shen, N-Y. and Morikawa, Y., Lagrangian simulation of dilute gas-solids flows in a horizontal pipe, *Adv. Powder Tech.*, 2, 63, 1991.

Tsuji, Y., Tanaka, T. and Ishida, T., Lagrangian numerical simulation of plug flow of collisionless particles in a horizontal pipe, *Powder Tech.*, 71, 239, 1992.

Tsuji, Y., Kawaguchi, T. and Tanaka, T., Discrete particle simulation of a two-dimensional fluidized bed, *Powder Tech.* 77, 79, 1993.

Tsuji, Y., 1994, Flow patterns in circulating fluidized beds, IFPRI Annual Research report ARR 28-1.

Tsuo, Y.P. and Gidaspow, D., Computation of flow patterns in circulating fluidized beds, *AIChE J.*, Vol. 36, 885, 1990.

Tu, J.Y., Lee, B.E. and Fletcher, C.A.J., Eulerian modeling of particle-wall collisions in confined gas-particle flows via a Lagrangian approach, *Erosion Processes*, ASME FED Vol. 236, 751, 1996.

Uhlherr, P.H.T. and Sinclair, C.G., The effect of free stream turbulence on the drag coefficient of spheres, *Proc. Chemca '70*, 1, 1, 1970.

Umhauer, H., Particle size distribution analysis by scattered light measurements using an optically defined measuring volume. *J. Aerosol Sci.*, 14, 765, 1983.

Umhauer, H., Löffler-Mang, M., Neumann, P. und Leukel, W., Pulse holography and phase-Doppler technique. A comparison when applied to swirl pressure-jet atomizers. *Particle and Particle Systems Characterization*, 7, 226, 1990.

Ushimaru, K. and Butler, G.W., Numerical simulation of gas-solid flow in an electrostatic precipitator, *Gas-Solid Flows*, ASME FED Vol. 10, 87, 1984.

Valentine, J.R. and Decker, R.A., Application of Lagragnian particle tracking scheme in a body-fitted coordinate system, *Num. Meth. for Multiphase Flows*, ASME FED Vol. 185, 39, 1994.

Van Beek, J. P. A. J. and Riethmuller, M. L., Rainbow phenomena applied to the measurement of droplet size and velocity and to the detection of non sphericity, *Applied Optics*, 35, 2259, 1996.

Van de Hulst, H.C., *Light Scattering by Small Particles*. Dover Publ., Inc., New York, 1981.

Van Dyke, M., *An Album of Fluid Motion*. The Parabolic Press, Stanford CA, 1982.

Versteeg, H.K. and Malalasekera, W., *An Introduction to Computational Fluid Dynamics*, Longman Scientific and Technical, Essex, England, 1995.

Voir, D.J. and Michaelides, E.E., The effect of the history term on the motion of rigid spheres in a viscous fluid, *Int. J. Multiphase Flow*, 20, 547, 1994.

Walsh, M.J., Drag coefficient equations for small particles in high speed flows, *AIAA Jnl.*, 13, 1526, 1975.

Walton, O.R. and Braun, R.L., Viscosity, granular temperature and stress calculations for shearing assemblies of inelastic, friction disks, *J. Rheology*, 30, 949, 1986.

Walton, O.R., Numerical simulation of inelastic , frictional, particle-particle interaction, *Particulate two-phase flow*, Roco, M., (Ed.), Butterworth-Heinemann, Boston MA, 884, 1993.

Wang, L-P. and Stock, D.E., Stochastic trajectory models for turbulent diffusion: Monte Carlo process versus Markov chains, *Atmospher. Envir.*, 26A, 1599, 1992.

Wang, H., Strunck, V., Müller, H. and Dopheide, D., New technique for multi-component velocity measurements using a single HF-pulsed diode laser and a single photodetector. *Experiments in Fluids*, 18, 36, 1994.

Wang, Q. and Squires, K.D., Large eddy simulation of particle deposition in a vertical turbulent channel, *Int. J. Multiphase Flow*, 22, 667, 1996a.

Wang, Q. and Squires, K.D., Large eddy simulation of particle dispersion in a three-dimensional turbulent mixing layer, *ASME FED* Vol. 236, 33, 1996b.

Wang, L-P., Wexler, A.S. and Zhou, Y., On the collision rate of small particles in isotropic turbulence: Party 1. Zero-inertia case, *ASME FEDSM'97-3636*, 1997

Warnica, W.D., Renksizbulut, M. and Strong, A.B., Drag coefficient of spherical liquid droplets. Part 2: Turbulent gaseous fields, *Exp. Fluids*, 18, 265, 1994.

Weiner B.B., Particle and droplet sizing using Fraunhofer diffraction. In *Modern Methods of Particle Size Analysis*, Barth, H. G. (Ed.), J. Wiley, New York, 135, 1984.

Wells, M.R. and Stock, D.E., The effects of crossing trajectories on the dispersion of particles in turbulent flow, *J. Fluid Mech.* 136, 31,1983.

Wen, F., Kamalu, N., Chung, J. N., Crowe, C. T. and Troutt, T. R., Particle dispersion by vortex structures in plane mixing layers, *J. Fluids Engr.*, 114, 657, 1992.

Wen, C.Y. and Yu, Y.H., Mechanics of Fluidization, *Chem. Engr. Prog. Sump. Series,* 62, 100, 1966.

White, H.J., *Industrial Electrostatic Precipitation,* Addison-Wesley, Reading PA, 1963.

Wilcox, D.C., *Turbulence Modeling for CFD,* DEC Ind., La Cañada, CA, 1993.

Williams, A., *Combustion of liquid fuel sprays,* Butteworths, 1990.

Williams, R.A., Xie, C.G., Dickin, F.J., Simons, S.J.R. and Beck, M.S., Review: Multi-phase flow measurement in powder processing. *Powder Tech.*, 66, 203, 1991.

Williams, R. and Mustoe, G.W. (Ed.), Proceedings of the Second International Conference on Discrete Element Methods, *First IESL Publication Ed.*, 1993.

Williams, R.A. and Beck, M.S. (Eds.) *Process Tomography,* Butterworth-Heinemann Ltd., Oxford, 1995.

Wilson, K.C., Addie, G.R. and Clift R., *Slurry transport using centrifugal pumps,* Elsevier, Amsterdam, 1992.

Xu, T.-H., Durst, F. and Tropea, C., The three-parameter log-hyperbolic distribution and its application to particle sizing, *Atom. and Sprays,* 3, 109, 1993.

Yang, Y., Crowe, C.T. and Chung, J.N., Experiments on particle dispersion in a plane wake, WSU MMO Rep. 95-7, 1995.

Ye, J. and Richards, C.D., Droplet and vapor transport in a turbulent jet, *Proc. of 26th Intl. Symp. on Combustion,* 1679, 1997.

Yearling, P.R. and Gould, R.D., Convective heat and mass transfer from a single evaporating water, methanol and ethanol droplet, *ASME FED* Vol. 233, 33, 1995.

Yeoman, M. L., Azzopardi, B. J., White, H. J., Bates, C. J. and Roberts, P. J., Optical development and application of a two-colour LDA system for the simultaneous measurement of particle size and particle velocity. *Engineering Applications of Laser Velocimetry,* 127, ASME Winter Annual Meeting, 1982.

Yuen, M.C. and Chen, L.W., On drag of evaporating liquid droplets, *Comb. Sci. and Tech.,* 14, 147, 1976.

Yuu, S., Yasukouchi, N., Hirosawa, Y. and Jotaki, T., Particle diffusion in a dust laden round jet, *AIChE J.,* 24, 509, 1978.

Zarin, N.A. and Nicholls, J.A., Sphere drag in solid rockets - non continuum and turbulence effects, *Comb. Sci. and Tech.,* 3, 273, 1971.

Zhou, Q. and Leschziner, M.A., A time-correlated stochastic model for particle dispersion in isotropic turbulence, *Eighth Symp. on Turb. Shear Flows,* Tech. Univ. Munich, 1, 1031, 1991.

Zhou, Q. and Yao, S.C., Group modeling of impacting spray dynamics, *Int. J. Heat Mass Trans.*, 35, 121, 1992.

Zhou, Y., Wexler, A.S. and Wang, L-P., On the collision rate of small particles in isotropic turbulence: Part 2. Finite-inertia case. *ASME FEDSM'97-3685*, 1997.

Appendix A

Single Particle Equations

The objective of this appendix is to present a derivation for the governing equations for a single particle or droplet. These equations are fundamental to the development of numerical models for dispersed-phase flows. These derivations result from the application of the Reynolds Transport Theorem which is presented below.

A.1   Reynolds transport theorem

Newton's second law and the first law of thermodynamics are well known for a fluid element of constant mass moving through a flow field. Applying these laws to identifiable fluid elements is the Lagrangian approach to fluid mechanics. However, the most useful and popular approach to describe the dynamic and thermal state of a fluid is the Eulerian approach. In this case the state of the flow is described by the properties of fluid elements at the time they pass a given point (window) in space. The Reynolds transport theorem provides the link between the Lagrangian and Eulerian approaches.

Fundamental to the Reynolds transport theorem is the mathematical description of the mass flux of a fluid. Consider an element of area ΔS in Figure A.1 through which fluid passes with velocity w_i with respect to the surface. The mass flux is given by

$$\Delta \dot{M} = \rho w_i n_i \Delta S \tag{A.1}$$

where n_i is a unit vector normal to the area element.

For a closed surface, as shown in Figure A.1, the unit vector has a positive sense if pointing outward from the enclosed volume. Thus the dot product in Equation A.1 represents mass flow *from the control volume* or a mass *efflux.* Integrating the mass flux over the entire surface of the volume gives the net mass efflux from the volume

$$\dot{M} = \int_A \rho w_i n_i dS \tag{A.2}$$

409

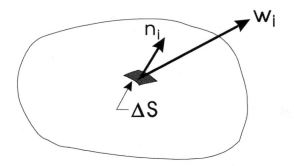

Figure A.1: Mass flow through an element of surface area.

The same approach can be used to determine the net efflux of other quantities such as energy or momentum. If B is the extensive property of a fluid and β is the corresponding intensive property, then the net efflux of property B from the volume is

$$\dot{B} = \int_A \rho \beta w_i n_i dS \tag{A.3}$$

Consider an arbitrary volume in space through which fluid can pass as shown in Figure A.2. This volume is the control volume. It can move through space, rotate or change size. The surface around the control volume is the control surface. Now consider a mass element which moves through the control volume as shown in Figure A.3. This fluid element will be referred to as the *system*. At time t the boundaries of the system are the control surface. At this time the mass of the system is the mass in the control volume. At time $t + \delta t$ the system moves out of the control volume. The mass of the system which has left the volume is δM_{out} while the mass which has entered the volume is δM_{in}. The amount of property B associated with the mass (system) which has left is δB_{out}, while the corresponding amount of B that has entered is δB_{in}.

By definition the rate of change of property B of the system is given by

$$\frac{dB_{sys}}{dt} = \lim_{\delta t \to 0} \frac{B_{sys}(t + \delta t) - B_{sys}(t)}{\delta t} \tag{A.4}$$

The amount of B of the system at time t corresponds to the amount of B in the control volume at time t because at this instant the system is entirely surrounded by the control surface. Thus

$$B_{sys}(t) = B_{cv}(t) \tag{A.5}$$

At time $t + \delta t$ some mass has left while some mass has entered the control volume. The mass of the system now consists of the mass which has left plus the mass in the control volume minus the mass which has entered. Thus the value of the property B of the system at time $t + \delta t$ is

$$B_{sys}(t + \delta t) = \delta B_{out} + B_{cv}(t + \delta t) - \delta B_{in} \tag{A.6}$$

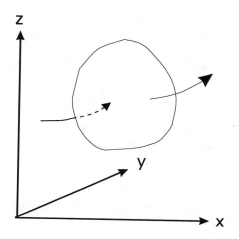

Figure A.2: Mass passing through a control volume.

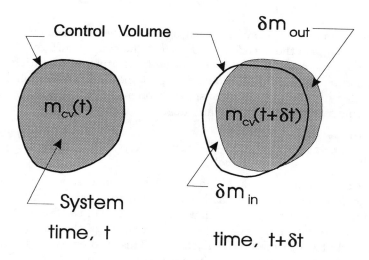

Figure A.3: System moving through a control volume at two successive times.

Substituting the expressions for B_{sys} at t and $t + \delta t$ into Equation (A.4) and rearranging yields

$$\frac{dB_{sys}}{dt} = \lim_{\delta t \to 0} \left[\frac{B_{cv}(t + \delta t) - B_{cv}(t)}{\delta t} + \frac{\delta B_{out}}{\delta t} - \frac{\delta B_{in}}{\delta t}\right] \qquad (A.7)$$

The first two terms are simply the rate of change of the property B in the control volume; that is,

$$\lim_{\delta t \to 0} \left[\frac{B_{cv}(t + \delta t) - B_{cv}(t)}{\delta t}\right] = \frac{dB_{cv}}{dt} \qquad (A.8)$$

The amount of B in the control volume is obtained by integrating the product of the corresponding intensive property β and fluid density over the control volume. Thus

$$\frac{dB_{cv}}{dt} = \frac{d}{dt} \int_{cv} \rho\beta dV \qquad (A.9)$$

The other two terms represent the net mass flux of B out of the control volume or, in other words, the net efflux of property B across the control surface. Thus

$$\lim_{\delta t \to 0} \left[\frac{\delta B_{out}}{\delta t} - \frac{\delta B_{in}}{\delta t}\right] = \int_{c} \rho\beta w_i n_i dS \qquad (A.10)$$

It must be emphasized that the velocity w_i is measured *with respect to the control surface* because the integral is the property flux across the control surface.

Finally the Reynolds transport theorem becomes

$$\frac{dB_{sys}}{dt} = \frac{d}{dt} \int_{cv} \rho\beta dV + \int_{cs} \rho\beta w_i n_i dS \qquad (A.11)$$

Thus the rate of change of B of a system is equal to the rate of change of B in the control volume plus the net efflux of B across the control surface. There is no restriction on the control volume. It can be translating, accelerating, rotating or changing in size and the Reynolds transport theorem is still applicable.

A corollary to the Reynolds transport theorem can be obtained by applying Leibnitz rule for differentiation of an integral; namely,

$$\frac{d}{dt} \int_{X_1(t)}^{X_2(t)} f(x, t)dx = \int_{X_1(t)}^{X_2(t)} \frac{\partial f(x, t)}{\partial t}dx + f(X_1, t)\frac{dX_1}{dt} - f(X_2, t)\frac{dX_2}{dt} \quad (A.12)$$

where X_1 and X_2 are the spatial limits of integration as shown in Figure A.4a. One notes that the integrand has been changed to a partial derivative of the function with respect to time and that the derivatives of the limits, X_1 and X_2, are taken with respect to time. The time derivative of the limits corresponds to the speed at which the boundaries move. This theorem can be extended to a function of three space variables in which the integral is taken over a volume enclosed by a surface as shown in Figure A.4b. For consistency, the volume and surface will be the control volume and control surface. Leibnitz rule is

$$\frac{d}{dt} \int_{cv} f(x_i, t)dV = \int_{cv} \frac{\partial f(x_i, t)}{\partial t}dV + \int_{cs} f(x_i, t)U_{i,s} n_i dS \qquad (A.13)$$

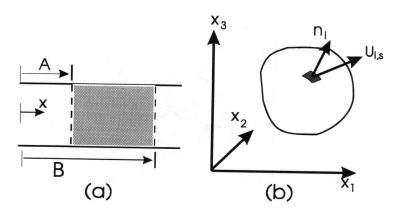

Figure A.4: Translating boundaries in relation to Leibnitz rule: a) one-dimensional system, b) control surface.

where $U_{i,s}$ is the velocity of the control surface with respect to the coordinate reference frame.

Applying Leibnitz rule to the Reynolds transport theorem in Equation (A.11) one has

$$\frac{dB_{sys}}{dt} = \int_{cv} \frac{\partial(\rho\beta)}{\partial t}dV + \int_{cs} \rho\beta(w_i + U_{i,s})n_i dS \qquad (A.14)$$

However the vector sum of the velocity at the control surface and the velocity of the fluid with respect to the control surface yields the velocity of the fluid with respect to the coordinate reference frame.

$$w_i + U_{i,s} = u_i \qquad (A.15)$$

Thus the alternate form of the Reynolds transport theorem is

$$\frac{dB_{sys}}{dt} = \int_{cv} \frac{\partial(\rho\beta)}{\partial t}dV + \int_{cs} \rho\beta u_i n_i dS \qquad (A.16)$$

In this form the velocity of the fluid at the control surface is the velocity *with respect to the coordinate reference frame*, not the control surface. This form is most useful for deriving the fundamental equations for fluid flow.

The original form of the Reynolds transport theorem will now be applied to the property conservation for particles and droplets conveyed by a fluid.

The fundamental equations for dispersed phase flows are formulated from the equations describing the conservation of mass, momentum and energy for an individual droplet or particle. The resulting equations might be regarded as "Lagrangian" because they apply to moving mass elements and do not describe the change of conditions at a point. The following discussion applies to "particles" or "droplets" although the word "particle" will be used throughout.

Consider the particle shown in Figure A.5. The particle is moving through the fluid with a velocity v_i with respect to an inertial reference frame. Assume

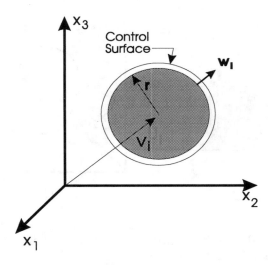

Figure A.5: Moving control surface enclosing particle.

for the present that the particle is not rotating. Assume also that the particle is spherical in order to avoid the complexities associated with aspherical particles yet still retain the essence of the equations.

The control surface is located adjacent to, and just outside of, the particle surface. It thus lies within the boundary layer of the particle. The transition from particle surface properties to local fluid properties takes place across the boundary layer. As the particle burns, evaporates or condenses, the control surface moves to remain adjacent to and just outside of the particle surface. The velocity, w_i, is the velocity through the control surface with respect to the control surface. Thus as the mass of the particle changes, due to evaporation for example, the mass passes through the control surface with a velocity w_i.

A.2   Mass conservation

The equation for mass conservation of a system is

$$\frac{dM}{dt} = 0 \qquad (A.17)$$

because, by definition, the mass of a system, M, is constant. The intensive property corresponding to mass is the mass per unit mass or, simply, unity. Thus the Reynolds transport theorem applied to mass conservation is

$$\frac{d}{dt} \int_{cv} \rho_d dV + \int_{cs} \rho_s w_i n_i dS = 0 \qquad (A.18)$$

where ρ_d is the material density of the particle[1] and ρ_s is the density of the fluid at the control surface.

The integral of the density over the volume of the particle is simply the instantaneous mass of the particle, m, so the continuity equation can be written as

$$\frac{dm}{dt} = -\int_{cs} \rho_s w_i n_i dS \qquad (A.19)$$

If the magnitude of the efflux velocity w_i and fluid density are uniform over the control surface the continuity equation simplifies to

$$\frac{dm}{dt} = -\rho_s w S_d \qquad (A.20)$$

where w is the magnitude of the efflux velocity vector. Obviously if the particle is evaporating, the efflux velocity is positive and the particle mass decreases with time. On the other hand, condensation gives rise to a negative efflux velocity and the particle mass increases with time. This equation forms the basis for the nondimensional numbers relating to mass transfer discussed in Chapter 4.

A.3 Momentum conservation

A.3.1 Linear momentum

Newton's second law states that the net force acting on a system is equal to the rate of change of momentum of the system, or

$$F_i = \frac{d(MU_i)}{dt} \qquad (A.21)$$

In this case the extensive variable is the momentum, MU_i, so the corresponding intensive variable is the velocity, U_i, measured with respect to an inertial reference frame. The application of Reynolds transport theorem for rate of change of momentum yields

$$F_i = \frac{d}{dt} \int_{cv} \rho_d U_i dV + \int_{cs} \rho_s U_{i,s} n_j w_j dS \qquad (A.22)$$

where $U_{i,s}$ is the velocity of the fluid at the control surface with respect to the inertial reference frame. This is a vector equation for the direction i.

In general, the particle may be rotating and there may be an internal circulation within the particle (or droplet). However, it will be assumed here that there is no internal motion and the particle is rotating about an axis through its center of mass. The velocity inside the particle is then given by

$$U_i = v_i + \epsilon_{ijk}\omega_j \xi_k \qquad (A.23)$$

---

[1] The subscript "d" represents the dispersed phase.

where v_i is the velocity of the particle's center of mass with respect to an inertial reference frame, ω_i is the particle's rotation vector and ξ_i is the distance from the center of mass. Thus the integral for the momentum in the control volume can be written as

$$\frac{d}{dt}\int_{cv}\rho_d U_i dV = \frac{d}{dt}\int_{cv}\rho_d(v_i + \epsilon_{ijk}\omega_j\xi_k)dV \qquad (A.24)$$

Because v_i and ω_i are constant over the interior of the particle, the above equation becomes

$$\frac{d}{dt}\int_{cv}\rho_d U_i dV = \frac{d}{dt}(v_i m) + \frac{d}{dt}\epsilon_{ijk}\omega_j\int_{cv}\xi_k\rho_d dV \qquad (A.25)$$

By definition of the center of mass, the integral on the right side is zero so the equation can be reduced to

$$\frac{d}{dt}\int_{cv}\rho_d U_i dV = \frac{d}{dt}(v_i m) = v_i\frac{dm}{dt} + m\frac{dv_i}{dt} \qquad (A.26)$$

If the particle is evaporating or condensing, then the velocity of the surface with respect to the center of the particle is $\dot{r}n_i + \epsilon_{ijk}\omega_j r_k$ where $\dot{r}$ is the regression rate of the surface. Thus the velocity of the fluid crossing the control surface with respect to an inertial reference frame is

$$U_{i,s} = v_i + \epsilon_{ijk}\omega_j r_k + (\dot{r} + w)n_i = v_i + \epsilon_{ijk}\omega_j r_k + w'n_i \qquad (A.27)$$

where, for convenience, the sum of the regression rate and the efflux velocity, $(\dot{r} + w)n_i$ has been replaced by $w'n_i$.

The net efflux of momentum from the control surface is given by

$$\int_{cs}\rho_s U_{i,s}w_j n_j dS = \int_{cs}\rho_s(v_i + \epsilon_{ijk}\omega_j r_k + w'n_i)w_l n_l dS \qquad (A.28)$$

In that v_i is not a function of the surface position, one can write

$$\int_{cs}\rho_s v_i w_j n_j dS = v_i\int_{cs}\rho_s w_j n_j dS \qquad (A.29)$$

By using the continuity equation, Equation A.19, the integral is reduced to

$$v_i\int_{cs}\rho_s w_j n_j dS = -v_i\frac{dm}{dt} \qquad (A.30)$$

Finally the addition of Equations A.26 and A.30 yields the momentum equation for the particle

$$F_i = m\frac{dv_i}{dt} + \int_{cs}\rho_s w'n_i w_j n_j dS + \epsilon_{ijk}\omega_j\int_{cs}\rho_s r_k w_l n_l dS \qquad (A.31)$$

For an evaporating or burning particle, the efflux velocity and regression velocity will be normal to the surface at the surface so the first integral in Equation A.31 becomes

$$\int_{cs} \rho_s w' n_i w_j n_j dS = \int_{cs} \rho_s w' w n_i dS \qquad (A.32)$$

This term represents the thrust on the particle due to mass efflux from the surface. For example, if a particle were a flat disk as shown in Figure A.6, and the mass was rejected only from the leeward face, the particle would experience a thrust as evaluated with the above integral. In some instances like a burning coal particle, the mass efflux is not uniform over the surface and spurious lateral motions are observed. However, for most applications involving droplet burning or evaporation, the nonuniformity of momentum flux is small compared to the other forces acting on a droplet. If the momentum efflux is uniform over the surface, then

$$\int_{cs} \rho_s w' w n_i dS = \rho_s w w' \int_{cs} n_i dS = 0 \qquad (A.33)$$

Also, if the mass flux is uniform over the surface and the particle is spherical, the integral associated with the particle rotation is also zero.

$$\epsilon_{ijk}\omega_j \int_{cs} \rho_s r_k w_l n_l dA = \epsilon_{ijk}\omega_j \int_{cs} \rho_s w r n_k dA = 0$$

The momentum equation finally simplifies to

$$m\frac{dv_i}{dt} = F_i \qquad (A.34)$$

or, in words, the forces acting on the particle are equal to the product of the instantaneous mass of the particle and the acceleration at the particle's center of mass.

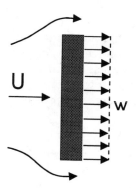

Figure A.6: "Flat disk" particle with mass efflux from one side.

The forces acting on a particle can be subdivided into two sources: surface forces and body forces.

$$F_i = F_{i,s} + F_{i,b} \qquad (A.35)$$

Body forces

Body forces are forces that act on the mass of the particle. The most common is the gravitational force which is the product of the particle mass and local acceleration due to gravity.

$$F_{i,g} = mg_i \qquad (A.36)$$

where g_i is the component of the gravitational acceleration in direction i.

Another body force could be the Coulomb force or the force due to charges in an electric field. This force is the most important in the dynamics of particles in an electrostatic precipitator. The Coulomb force is given by

$$F_{i,c} = -qE_i \qquad (A.37)$$

where q is the charge on the particle and E_i is the potential gradient of the electric field. Typically the charge on a particle accumulates on the particle surface and there is a saturation charge beyond which no additional charges can be accommodated on the surface.

Another body force may be due to a magnetic field in which case the force depends on the gradient in the magnetic field and the permeability of the particle mass.

Surface forces

The surface forces are due to pressure and shear stress acting on the surface. The net pressure force is obtained from

$$F_{i,p} = -\int_{cs} p_s n_i dS \qquad (A.38)$$

where n_i is the unit outward normal vector from the surface and p_s is the pressure at the surface. The minus sign is needed to provide the pressure force *on the particle*. This pressure force may have components which are in the direction of particle motion (drag) and normal to the particle motion (lift). The pressure force component in the drag direction is called *form drag*.

The net force due to shear stress is represented by

$$F_{i,\tau} = \int_{cs} \tau_{j,i} n_j dS \qquad (A.39)$$

where τ_{ij} is the shear stress tensor at the particle surface. The component of this force in the drag direction is the *viscous* or *shear drag*.

The sum of the pressure and shear forces acting on the particle constitutes the surface forces

$$F_{i,s} = \int_{cs} (-p_s n_i + \tau_{ji} n_j) dS \qquad (A.40)$$

The representation of the surface forces by drag and lift coefficients is discussed in Chapter 4.

The final form for the particle momentum equation is

$$m\frac{dv_i}{dt} + \int_{cs} \rho_s w' w n_i dA = F_{i,b} + F_{i,s} \tag{A.41}$$

which reduces to

$$m\frac{dv_i}{dt} = F_{i,b} + F_{i,s} \tag{A.42}$$

if the momentum efflux from the surface is uniform.

A.3.2   Moment of momentum

The equation for the rotation is obtained by equating the time rate of change of the moment of momentum to the applied torque.

$$T_i = \frac{d}{dt}(MH_i) \tag{A.43}$$

where T_i is the applied torque and H_i is the moment of momentum per unit mass or

$$H_i = \epsilon_{ijk} r_j U_k \tag{A.44}$$

Applying Reynolds transport theorem to the rotating particle, the rotation equation becomes

$$T_i = \frac{d}{dt}\int_{cv} \rho H_i dV + \int_{cs} \rho_s H_i w_j n_j dS \tag{A.45}$$

The momentum of momentum within the control volume is given by

$$\int_{cv} \rho H_i dV = \int_{cv} \rho \epsilon_{ijk} \xi_j U_k dV \tag{A.46}$$

where ξ_i is the vector from the center of mass of the particle. Using the equation for velocity, Equation A.23, one can express the particle's moment of momentum as

$$\int_{cv} \rho H_i dV = \int_{cv} \rho \epsilon_{ijk} \xi_j (v_k + \epsilon_{klm} \omega_l \xi_m) dV \tag{A.47}$$

Because U_i^c is constant over the volume,

$$\int_{cv} \rho \epsilon_{ijk} \xi_j v_k dV = \epsilon_{ijk} v_k \int_{cv} \rho \xi_j dV = 0 \tag{A.48}$$

by definition of the center of mass. The integral over the second term becomes[2]

---

[2] This manipulation follows from the identity from tensor analysis.

$$\epsilon_{ijk} \epsilon_{ilm} = \delta_{jl} \delta_{km} - \delta_{jm} \delta_{kl}$$

$$\int_{cv} \rho \epsilon_{ijk} \xi_j \epsilon_{klm} \omega_l \xi_m) dV = \omega_i \int_{cv} \rho \xi_j \xi_j dV - \omega_j \int_{cv} \rho \xi_j \xi_i dV \tag{A.49}$$

Taking the subscript i as the 1-direction, the equation becomes

$$\omega_i \int_{cv} \rho \xi_j \xi_j dV - \omega_j \int_{cv} \rho \xi_j \xi_i dV =$$
$$\omega_1 \int_{cv} \rho(\xi_2^2 + \xi_3^2) dV - \omega_2 \int_{cv} \rho \xi_1 \xi_2 dV - \omega_3 \int_{cv} \rho \xi_1 \xi_3 dV \tag{A.50}$$

The first integral on the right is the moment of inertia, I_1, about the axis in the 1-direction. If the particle is symmetric about the center or mass (a sphere), the other integrals on the right side vanish and the momentum of inertia is the same about any axis so the first term of Equation A.45 becomes

$$\frac{d}{dt} \int_{cv} \rho H_i dV = \frac{d}{dt} (\omega_i I) \tag{A.51}$$

The net efflux of the moment of momentum through the surface is given by

$$\int_{cs} \rho_s H_i w_j n_j dS = \int_{cs} \rho_s \epsilon_{ijk} r_j (v_k + \epsilon_{klm} \omega_l r_m + n_k w') w dS \tag{A.52}$$

The first integral on the right side is equal to zero,

$$\int_{cs} \rho_s \epsilon_{ijk} r_j v_k w dS = v_k \rho_s w \epsilon_{ijk} \int_{cs} r_j dS = 0 \tag{A.53}$$

provided the particle is spherical and the mass efflux is constant over the surface. The last integral is

$$\int_{cs} \rho_s \epsilon_{ijk} r_j n_k w' w dS = \rho_s w w' r \int_{cs} \epsilon_{ijk} n_j n_k dS = 0 \tag{A.54}$$

if the particle is spherical and the mass flux is uniform. The second integral can be evaluated as

$$\int_{cs} \rho_s \epsilon_{ijk} \epsilon_{klm} r_j \omega_l r_m w dS = -\omega_i \frac{dI}{dt} \tag{A.55}$$

for a spherical particle because $\rho_s w dS$ represents the mass reduction rate at the surface due to mass transfer. Finally, the net efflux of moment of momentum from a spherical, uniformly evaporating droplet is

$$\int_{cs} \rho_s H_i w_j n_j dS = -\omega_i \frac{dI}{dt} \tag{A.56}$$

Incorporating Equations A.51 and A.56 into Equation A.45 yields

$$T_i = I \frac{d\omega_i}{dt} \tag{A.57}$$

as the equation for rotation of a spherical, uniformly evaporating droplet.

The body forces due to gravity do not contribute to the applied torque. Also the forces due to pressure on the surface of a spherical droplet do not contribute to a torque since the force passes through the center of mass (no moment arm). The torque due to the surface shear stress is given by

$$T_i = \int_{cs} \epsilon_{ijk} r_j \tau_{lk} n_l dS \tag{A.58}$$

The evaluation of the torque on a particle is addressed in Chapter 4.

A.4 Energy conservation

The first law of thermodynamics for a system states that the rate of change of energy is equal to the rate at which heat is transferred to the system and the rate at which work is done by the system,

$$\frac{dE}{dt} = \dot{Q} - \dot{W} \tag{A.59}$$

where, E the energy, includes both internal and external (mechanical) energies as well as the energy associated with surface tension.

The internal energy is written as

$$I = Mi \tag{A.60}$$

where i is the internal energy per unit mass (specific internal energy). The external energy is the kinetic energy,

$$KE = M \frac{U_i U_i}{2} \tag{A.61}$$

where U_i is the velocity of the matter with respect to an inertial reference frame. Thus the specific external energy[3] is $U_i U_i / 2$. The energy associated with surface tension is the product of the surface area of the particle and the surface tension

$$E_\sigma = S\sigma \tag{A.62}$$

Thus the total energy can be written as

$$E = M(i + \frac{U_i U_i}{2}) + S\sigma \tag{A.63}$$

which, for convenience, will be abbreviated to

$$E = Me + S\sigma \tag{A.64}$$

---

[3] The energy due to rotation is not included here.

where e is the sum of the specific internal and kinetic energy.

Application of the Reynolds transport theorem to relate the change of internal and external energies of the system to the change in the control volume and the net efflux through the control surface gives

$$\frac{d(Me)}{dt} = \frac{d}{dt} \int_{cv} \rho_d edV + \int_{cs} \rho_s ew_i n_i dS \tag{A.65}$$

Thus the first law of thermodynamics applied to the control volume is

$$\dot{Q} - \dot{W} = \frac{d}{dt} \int_{cv} \rho_d edV + \int_{cs} \rho_s e_s w_i n_i dS + \frac{d}{dt}(S\sigma) \tag{A.66}$$

where $\dot{Q}$ and $\dot{W}$ are the heat transfer rate to the system and the rate at which work is done by the system the instant it occupies the control volume.

As with the momentum equation, it will be assumed that the particle is neither rotating nor has internal motion. In this case, the velocity of the matter inside the droplet is v_i. The rate of change of energy within the control volume becomes

$$\begin{aligned}
\frac{d}{dt} \int_{cv} \rho_d edV + \frac{d}{dt}(S\sigma) &= \frac{d}{dt} \int_{cv} \rho_d \left(i + \frac{v_i v_i}{2} \right) dV + \sigma \frac{dS}{dt} \\
&= \frac{d}{dt} \left(m \frac{v_i v_i}{2} \right) + \frac{d}{dt}(m\bar{i}) + \frac{2\sigma}{r} \dot{r} S
\end{aligned} \tag{A.67}$$

where $\bar{i}$ is the average specific internal energy of the particle and the surface tension is assumed invariant with time.

Differentiation of the kinetic energy term yields two terms

$$\frac{d}{dt}\left(m \frac{v_i v_i}{2} \right) = mv_i \frac{dv_i}{dt} + v^2 \frac{dm}{dt} \tag{A.68}$$

where v is simply the magnitude of the velocity vector v_i. This expression will be utilized later in the final assembly of the terms in the Reynolds transport theorem.

Evaluation of the energy flux term in the Reynolds transport equation requires the energy of the matter crossing the control surface; namely,

$$e_s = i_s + \frac{U_{i,s} U_{i,s}}{2} \tag{A.69}$$

where i_s is the specific internal energy and $U_{i,s}$ is the velocity with respect to an inertial reference frame. This velocity can be expressed as

$$U_{i,s} = v_i + w_i + \dot{r} n_i \tag{A.70}$$

It is convenient to use the sum of the efflux velocity at the surface and the surface regression velocity as a single variable $w_i' = (w + \dot{r})n_i$ as was done in the momentum equation. Thus the specific kinetic energy at the surface is

$$\frac{U_{i,s} U_{i,s}}{2} = \frac{(v_i + w'n_i)(v_i + w'n_i)}{2} = \frac{v^2}{2} + \frac{w'^2}{2} + w'n_i v_i \tag{A.71}$$

Substituting these expressions into the efflux term of the Reynolds transport theorem yields

$$\int_{cs} \rho_s e_s w_j n_j dS = \left(\frac{v^2}{2} + \frac{w'^2}{2}\right) \int_{cs} \rho_s w_j n_j dS +$$
$$v_i \int_{cs} \rho_s w' w n_i dS + \int_{cs} i_s \rho_s w_j n_j dS \tag{A.72}$$

From the continuity equation, Equation A.18, the integral of the mass flux over the surface is equal to $-(dm/dt)$ so the above integral reduces to

$$\int_{cs} \rho_s e_s w_j n_j dS = -\left(\frac{v^2}{2} + \frac{w'^2}{2}\right)\frac{dm}{dt} +$$
$$v_i \int_{cs} \rho_s w' w n_i dS + \int_{cs} \rho_s i_s w_j n_j dS \tag{A.73}$$

Adding Equations A.67 and A.73 gives the rate of change of energy of the system at the instant it is in the control volume

$$\frac{dE}{dt} = m v_i \frac{dv_i}{dt} - \frac{(w')^2}{2}\left(\frac{dm}{dt}\right) + \frac{d}{dt}(m\bar{i})$$
$$+ v_i \int_{cs} \rho_s w' w n_i dS + \int_{cs} \rho_s i_s w_j n_j dS + \frac{2\sigma}{r}\dot{r}S \tag{A.74}$$

This expression must now be equated to the rate of heat transfer to the particle and the rate at which work is done by the particle on the surroundings.

A.4.1 Heat transfer to particle

Heat transfer takes place through two mechanisms: radiation and conduction.

The radiative heat transfer is the net energy due to absorption and emission. If the droplet or the particle is considered a "grey" body, the net radiative heat transfer to the particle is

$$\dot{Q}_r = 4\pi r^2 (\alpha J - \epsilon \sigma T_d^4) \tag{A.75}$$

where α is the absorptivity, ϵ is the emissivity, J is the radiosity and σ is the Stephan-Boltzmann constant. In general, a solution for the radiation field has to be obtained before the local radiosity can be determined.

The net conductive heat transfer can be evaluated by integrating the heat flux vector, $q_{i,c}$, over the surface as shown in Figure A.7. The net heat transfer due to conduction is

$$\dot{Q}_c = -\int_{cs} q_{i,c} n_i dS \tag{A.76}$$

where the minus sign is needed to give the net heat transfer rate to the droplet.

A.4.2 Work rate of particles on surroundings

Work arises from displacement of both body and surface forces. The work rate due to body forces is simply the negative product of the velocity of the particle and the body force,

$$\dot{W}_b = -F_{i,b} v_i \tag{A.77}$$

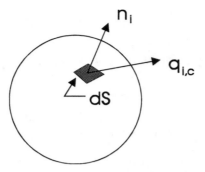

Figure A.7: Heat transfer on particle surface.

The work rate due to pressure forces is the dot product of the pressure force acting on an element of area and the velocity of matter passing through the element

$$d\dot{W}_p = (v_i + w_i')n_i p_s dS \qquad (A.78)$$

where p_s is the pressure at the control surface. By definition the pressure force is acting on the element so the motion of the matter against the pressure force represents work done by the system. The work per unit time done by the system over the entire control surface is

$$\dot{W}_p = \int_{cs} (v_i + w_i')n_i p_s dS \qquad (A.79)$$

The shear stress force acting on the area element in the i-direction is $\tau_{ji}n_j dS$ where τ_{ij} is the shear stress tensor. The work rate associated with this force and the velocity at the surface is

$$d\dot{W}_s = -(v_i + w_i')\tau_{ji}n_j dS \qquad (A.80)$$

In this case the dot product of the force and the velocity represents work done by the surroundings on the system, hence the negative sign. The total work due to shear stress forces acting on the system is

$$\dot{W}_s = -\int_{cs} (v_i + w_i')\tau_{ji}n_j dS \qquad (A.81)$$

Adding the pressure and shear work rate integrals and realizing the velocity v_i is constant with respect to the integration results in

$$\int_{cs}(v_i + w_i')(n_i p_s - \tau_{ji}n_j)dS = v_i \int_{cs}(n_i p_s - \tau_{ji}n_j)dS +$$

$$\int_{cs} p_s w_i' n_i dS - \int_{cs} w_i' \tau_{ji}n_j dS \qquad (A.82)$$

The first integral on the right represents the surface forces acting on the droplet. Note that this integral is the negative of the integral in Equation A.40 so

$$v_i \int_{cs} (n_i p_s - \tau_{ji}n_j)dS = -v_i F_{i,s} \qquad (A.83)$$

The second integral on the right of Equation A.82 can be expressed as

$$\int_{cs} p_s w'_i n_i dS = \int_{cs} p_s \left(w_i + \dot{r} n_i \right) n_i dS$$
$$= \int_{cs} \rho_s \frac{p_s}{\rho_s} w_i n_i dS + \int_{cs} \dot{r} p_s dS \tag{A.84}$$

The third integral is zero

$$\int_{cs} w'_i \tau_{ji} n_i dS = \int_{cs} w' n_i \tau_{ji} n_j dS = 0 \tag{A.85}$$

because the vector $\tau_{ji} n_j$ lies in the local tangent plane of the surface and the vector n_i is normal to the surface. Hence the dot product of these two vectors is zero.

Combining all the terms (Equations A.74, A.77, A.83 and A.84) together to form the energy equation

$$\frac{dE}{dt} = \dot{Q} - \dot{W}$$

results in

$$\frac{d}{dt}(mi) + mv_i \frac{dv_i}{dt} - \frac{(w')^2}{2} \frac{dm}{dt} + \frac{2\sigma}{r} \dot{r} S$$
$$+ \int_{cs} i_s \rho_s w_i n_i dS + v_i \int_{cs} \rho_s w' w n_i dS$$
$$= \dot{Q}_r + \dot{Q}_c + v_i (F_{i,b} + F_{i,s}) \tag{A.86}$$
$$- \int_{cs} \rho_s \frac{p_s}{\rho_s} w_i n_i dS - \int_{cs} \dot{r} p_s dS$$

This equation can be rewritten as

$$\frac{d}{dt}(mi) + v_i \left(m \frac{dv_i}{dt} + \int_{cs} \rho_s w' w n_i dA - F_{i,b} - F_{i,s} \right)$$
$$- \frac{(w')^2}{2} \frac{dm}{dt} + \frac{2\sigma}{r} rS + \int_{cs} \left(i_s + \frac{p_s}{\rho_s} \right) \rho_s w_i n_i dS \tag{A.87}$$
$$= \dot{Q}_r + \dot{Q}_c - \int_{cs} \dot{r} p_s dS$$

With reference to the momentum equation for the droplet, Equation A.31, the factor multiplying the v_i in the second term is zero so the energy equation further reduces to

$$\frac{dm\bar{i}}{dt} - \frac{w'^2}{2} \frac{dm}{dt} + \int_{cs} h_s \rho_s w_i n_i dS = \dot{Q}_r + \dot{Q}_c - \dot{r} S (\bar{p}_s + \frac{2\sigma}{r}) \tag{A.88}$$

where h_s is the enthalpy of the fluid crossing the control surface and $\bar{p}_s$ is the average pressure on the particle surface. The regression rate has been assumed uniform over the surface. The combination $\bar{p}_s + 2\sigma/r$ is the pressure inside the droplet, p_d.[4] Thus, the energy equation further simplifies to

$$m\frac{d\bar{i}}{dt} + \bar{i}\frac{dm}{dt} - \frac{w'^2}{2}\frac{dm}{dt} + \int_{cs} h_s\rho_s w_i n_i dS = \dot{Q}_r + \dot{Q}_c - \dot{r}\rho_d\frac{p_d}{\rho_d}S \qquad (A.89)$$

The factor $\rho_d \dot{r} A$ is the rate of change of droplet mass dm/dt so the second term on the left side can be combined with the last term on the right side which results in

$$m\frac{d\bar{i}}{dt} = \dot{Q}_r + \dot{Q}_c + \frac{dm}{dt}(\bar{h}_s - \bar{h}_d + \frac{w'^2}{2}) \qquad (A.90)$$

where $\bar{h}_d$ is the enthalpy of the particle phase. Usually the kinetic energy associated with the mass efflux velocity is very small compared with the enthalpy difference between phases

$$\bar{h}_s - \bar{h}_d >> \frac{w'^2}{2}$$

so the conventional form of the particle energy equation is

$$m\frac{d\bar{i}}{dt} = \dot{Q}_r + \dot{Q}_c + \frac{dm}{dt}(\bar{h}_s - \bar{h}_d) \qquad (A.91)$$

This equation states that the energy of the particle is controlled by the heat transfer to the particle and the energy associated with the change of phase. For example, if there is no heat transfer to an evaporating droplet, the internal energy (temperature) of the droplet will decrease as mass is evaporated ($dm/dt < 0$) from the surface. The application of this equation is addressed in Chapter 4 and subsequent chapters.

Other equations can be developed for the single particles such as the entropy equation. The development of these equations follows the same approach as used above.

---

[4] Actually there will be a pressure jump across the surface due to the momentum flux so p_d is not the actual pressure inside the droplet. Under most conditions, this pressure jump is small.

Appendix B

Volume Averaged Conservation Equations

In dealing with a mixture of droplets or particles it is impractical to solve for the fluid properties at every point in the mixture. Thus one is driven to consider the average properties in a volume containing many particles and to ascribe average values to a point in the flow enclosed by the volume. For example, the concept of bulk density as the density of the phase per unit volume of mixture is an average property. A formal approach is now sought to express the conservation laws for the continuous phase in terms of average properties.

B.1   Volume averaging

Consider a mixture of fluid and particles enclosed in the spherical volume shown in Figure B.1. The average properties within the volume are ascribed to a point in the volume which, in this case, is the sphere center. The volume has to be large enough such that a small increase in the volume will not affect the value of the average; that is, the same value for bulk density would be obtained if the volume were slightly changed. Still, the volume has to be small compared to the system dimensions or it will not be possible to write differential equations for the conservation laws.

The formalism used for volume averaging can be found in Slattery (1972). The averaging volume (V) is composed of the volume of the continuous phase (V_c) and the volume occupied by the dispersed phase (V_d). Based on the volume over which averages are taken, there are two types of averages to be considered. Let B be some property of the continuous phase per unit volume. The *phase average* of B is the average over the volume of the continuous phase and is defined as

$$\langle B \rangle = \frac{1}{V_c} \int_{V_c} B \, dV \qquad (B.1)$$

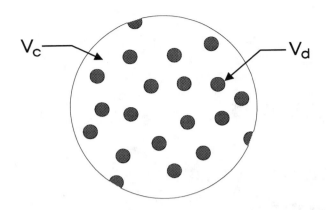

Figure B.1: Particles in an averaging volume.

If B were the continuous phase density, ρ_c, then $\langle \rho_c \rangle$ is the average material density of the continuous phase in the volume. Obviously, if the continuous phase density is constant then $\langle \rho_c \rangle = \rho_c$.

The *local volume average* of the property B is given by

$$\overline{B} = \frac{1}{V} \int_{V_c} B dV \tag{B.2}$$

Once again if B is the continuous phase density, the local volume average density is

$$\overline{\rho_c} = \frac{1}{V} \int_{Vc} \rho_c dV \tag{B.3}$$

The equation for local average density can be rewritten as[1]

$$\overline{\rho_c} = \frac{V_c}{V} \frac{1}{V_c} \int_{V_c} \rho_c dV \tag{B.4}$$

The ratio V_c/V is the volume fraction, α_c, of the fluid so the relationship between phase average and local average continuous phase density is given by

$$\overline{\rho_c} = \alpha_c \langle \rho_c \rangle \tag{B.5}$$

A more formal statement regarding the minimum size of the averaging volume is now possible. If the averaging volume is displaced over a distance corresponding to the linear dimension of the volume ($\sim V^{1/3}$), then the phase average properties should not change. This implies that

$$\langle \overline{B} \rangle = \overline{B} \tag{B.6}$$

---

[1] The local volume average of the density is the same as the bulk density introduced in Chapter 2.

B.1.1 Volume average of the gradient operation

All the conservation laws for a flowing fluid involve a gradient term. To form the average of the conservation equations over the averaging volume requires a relationship for the local volume average of the gradient. The local volume average of the gradient over the continuous phase is defined as

$$\overline{\frac{\partial B}{\partial x_i}} = \frac{1}{V} \int_{V_c} \frac{\partial B}{\partial x_i} dV \tag{B.7}$$

We now seek a formal method to interchange the volume and gradient operation.

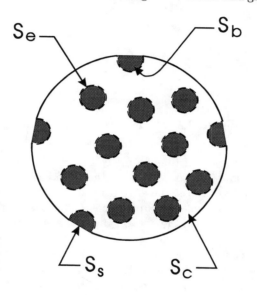

Figure B.2: Definition of surfaces.

Consider the averaging volume containing particles as shown in Figure B.2. Some of the particles are completely enclosed by the volume while others (boundary particles) have been severed by the surface enclosing the volume. The portion of the surface through the continuous phase is designated as S_c while the surfaces which sever the particles is S_s. The complete surface of the control volume, S, is equal to

$$S = S_c + S_s$$

The surface surrounding the particles completely inside the volume is S_e while the inside surface of the boundary particles is S_b. The total surface surrounding the particles inside the averaging volume, S_d, is

$$S_d = S_b + S_e$$

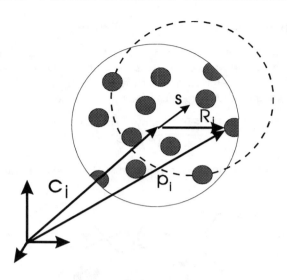

Figure B.3: Moving averaging volume.

Assume that the center of the volume can be displaced along the path s as shown in Figure B.3 and that the volume neither rotates nor deforms. The vector p_i represents a point on the surface with respect to the coordinate reference frame. The vector R_i is a vector from the center of the averaging to the surface. As the control volume moves, the dispersed phase elements are fixed. Consider the displaced averaging volume and an R_i vector terminating on the surface of a dispersed phase element shown in Figure B.4. As the volume is moved (and the dispersed phase element is fixed), the vector R_i moves along the surface as shown so the vector p_i is incremented by an amount δp_i which lies tangent to the dispersed phase surface. Thus dp_i/ds is a vector tangent to the surface on all the S_b surfaces.

The change in the volume average of B as the averaging volume translates along the path s can be determined by application of Leibnitz rule for the derivative of an integral. The change with s of the integral of B over V_c can be expressed as

$$\frac{d}{ds}\int_{V_c} B dV = \int_{V_c} \frac{\partial B}{\partial s} dV + \int_{S_b+S_c} B\frac{dp_i}{ds} n_i dS \tag{B.8}$$

where n_i is the unit outward normal vector from the averaging volume. We now require that B be explicit functions of time and position only so that B is not a function of where the center of the averaging volume is located. Thus

$$\frac{\partial B}{\partial s} = 0 \tag{B.9}$$

Also, on the surface S_b, the dot product of the vector dp_i/ds and n_i is zero since δp_i lies along the tangent to the surface S_b. Thus the change with s of the

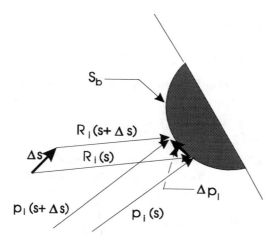

Figure B.4: Details of vectors in neighborhood of a particle.

integral of B over the continuous phase volume reduces to

$$\frac{d}{ds}\int_{V_c} B\,dV = \int_{S_c} B\frac{dp_i}{ds}n_i\,dS \qquad (B.10)$$

The path of s is given by the vector function $c_i(s)$. The coordinate of a general point on the surface of the averaging volume, $p_i(s)$, is the vector sum of the center coordinate and the volume radius, R_i,

$$p_i(s) = c_i(s) + R_i \qquad (B.11)$$

The change of $p_i(s)$ with s on the surfaces S_c is obtained from

$$\frac{dp_i}{ds} = \frac{dc_i}{ds} \qquad (B.12)$$

since the vector R_i is constant; that is, the averaging volume neither deforms nor rotates. Because dc_i/ds is not a function of the position on the surface S_c, the integral in Equation B.10 becomes

$$\int_{S_c} B\frac{dp_i}{ds}n_i\,dS = \frac{dc_i}{ds}\int_{S_c} Bn_i\,dS \qquad (B.13)$$

Also, the derivative with respect to s can be written using the chain rule as

$$\frac{d}{ds} = \frac{dc_i}{ds}\frac{\partial}{\partial x_i} \qquad (B.14)$$

Finally Equation B.8 can be expressed as

$$\frac{dc_i}{ds}\left(\frac{\partial}{\partial x_i}\int_{V_c} B\,dV - \int_{S_c} Bn_i\,dS\right) = 0 \qquad (B.15)$$

Since dc_i/ds is not zero,

$$\frac{\partial}{\partial x_i} \int_{V_c} B dV = \int_{S_c} B n_i dS \tag{B.16}$$

Applying Green's transformation to the definition of the local volume average of the gradient allows us to write

$$\int_{V_c} \frac{\partial B}{\partial x_i} dV = \int_{S_d + S_c} B n_i dS \tag{B.17}$$

where S_d is the surface of all the dispersed phase elements inside the averaging volume including the interior surface of those elements intersected by the surface of the averaging volume. Using Equation B.15 for the surface integral over S_c results in

$$\frac{1}{V} \int_{V_c} \frac{\partial B}{\partial x_i} dV = \frac{1}{V} \frac{\partial}{\partial x_i} \int_{V_c} B dV + \frac{1}{V} \int_{S_d} B n_i dS \tag{B.18}$$

or

$$\overline{\frac{\partial B}{\partial x_i}} = \frac{\partial \overline{B}}{\partial x_i} + \frac{1}{V} \int_{S_d} B n_i dS \tag{B.19}$$

which is the theorem for the volume average of the gradient. The unit normal vector n_i is defined as "outward" from the continuous phase and "inward" to the dispersed phase. It is more convenient to define n_i as the unit normal vector directed outward from the dispersed phase which changes the volume averaging theorem to

$$\overline{\frac{\partial B}{\partial x_i}} = \frac{\partial \overline{B}}{\partial x_i} - \frac{1}{V} \int_{S_d} B n_i dS \tag{B.20}$$

One notes that the volume average of the gradient is the gradient of the volume average plus an additional term which involves the integral of B over the interface between the dispersed and continuous phases. The property B can be a scalar, vector or second-order tensor field.

B.1.2 Volume averaging of the time derivative

One must also consider the volume average of the time derivative when applying the volume averaging equations to a dispersed phase flow. If the volume of the dispersed phase is not changing with time, then one can write

$$\overline{\frac{\partial B}{\partial t}} = \frac{\partial \overline{B}}{\partial t} \tag{B.21}$$

However if the volume does change with time, then a corollary to the volume averaging of the spatial derivative is needed for the temporal derivative.

Consider as before the averaging volume enclosing the dispersed phase elements. The continuous phase is "enclosed" by the surface through the continuous phase, S_c, and the surfaces adjacent to the dispersed phase, S_d. Applying the corollary of the Leibnitz theorem for partial derivatives one has

$$\frac{\partial}{\partial t} \int_{V_c} B dV = \int_{V_c} \frac{\partial B}{\partial t} dV + \int_{S_d + S_c} B \frac{\partial p_i}{\partial t} n_i dS \tag{B.22}$$

where p_i is the vector representing the surface enclosing the continuous phase. There is no change of the surface S_c with time so

$$\int_{S_e} B \frac{\partial p_i}{\partial t} n_i dS = 0 \tag{B.23}$$

The surface of every dispersed phase element can be described as

$$p_i = l_i + r_i(t) \tag{B.24}$$

where l_i is the location of the center of the dispersed phase element and r_i is the vector from the center to the surface of the element. The partial derivative of the vector on the surface with respect to time is

$$\frac{\partial p_i}{\partial t} = \dot{r} n_i \tag{B.25}$$

since the location of the center is a function only of spatial coordinates. Thus the derivative becomes

$$\frac{\partial}{\partial t} \int_{V_c} B dV = \int_{V_c} \frac{\partial B}{\partial t} dV + \int_{S_d} B \dot{r} dS \tag{B.26}$$

and the volume average of the time derivative is given by

$$\overline{\frac{\partial B}{\partial t}} = \frac{\partial \overline{B}}{\partial t} + \frac{1}{V} \int_{S_d} B \dot{r} dS \tag{B.27}$$

This completes the formulation of the volume averaging equations for application to the continuous phase flow equations. The application of this theorem to derive the volume averaged forms of the continuity, momentum and energy equations for fluid-particle mixtures is provided below.

B.2 Continuity equation

The continuity equation for the continuous phase is

$$\frac{\partial \rho_c}{\partial t} + \frac{\partial}{\partial x_i} (\rho_c u_i) = 0 \tag{B.28}$$

Taking the local volume average of each term

$$\overline{\frac{\partial \rho_c}{\partial t}} + \overline{\frac{\partial}{\partial x_i} (\rho_c u_i)} = 0 \tag{B.29}$$

and applying the volume averaging equations gives

$$\frac{\partial \overline{\rho_c}}{\partial t} + \frac{1}{V} \int_{S_d} \rho_c \dot{r} dS + \frac{\partial}{\partial x_i} (\overline{\rho_c u_i}) - \frac{1}{V} \int_{S_d} \rho_c u_i n_i dS \tag{B.30}$$

At the droplet surface, the velocity of the gas crossing the surface is

$$u_i = v_i + (\dot{r} + w)n_i \qquad (B.31)$$

where v_i is the velocity of the particle center and wn_i is the velocity of the gases with respect to the droplet surface.[2] Substituting this equation into Equation B.30 results in

$$\frac{\partial \overline{\rho_c}}{\partial t} + \frac{\partial}{\partial x_i}(\overline{\rho_c u_i}) = \frac{1}{V}\int_{S_d} \rho_c (v_i + wn_i)\, n_i dS \qquad (B.32)$$

The average density $\overline{\rho_c}$ can be written as $\alpha_c \langle \rho_c \rangle$. The average mass flux is

$$\overline{\rho_c u_i} = \alpha_c \frac{1}{V_c}\int_{V_c} \rho_c u_i dV \qquad (B.33)$$

The average velocity $\tilde{u}_i$ is defined such that

$$\tilde{u}_i = \frac{1}{\langle \rho_c \rangle}\frac{1}{V_c}\int_{V_c} \rho_c u_i dV \qquad (B.34)$$

which is the *mass-averaged* velocity. For a constant density fluid the mass averaged velocity is equal to the local volume average, $\tilde{u}_i = \langle u_i \rangle$. The velocity u_i can be expressed as the sum of the mass-averaged velocity and the deviation therefrom.

$$u_i = \tilde{u}_i + \delta u_i \qquad (B.35)$$

Substituting this equation into Equation B.34 shows

$$\int_{V_c} \rho_c \delta u_i dV = 0 \qquad (B.36)$$

The integral over the droplet surfaces

$$\frac{1}{V}\int_{S_d} \rho_c (v_i + wn_i)\, n_i dS = \frac{1}{V}\int_{S_b + S_e} \rho_c (v_i + wn_i)\, n_i dS \qquad (B.37)$$

For all droplets internal to the volume, the integral

$$\int_{S_e} \rho_c w dS = \rho_{cs} w S_d \qquad (B.38)$$

represents the mass addition to the volume due to mass flux from the droplets where ρ_{cs} is the average continuous phase density at the droplet surface. The integral

$$\int_{S_e} \rho_c v_i n_i dS = \rho_{cs} v_i \int_{S_e} n_i dS = 0 \qquad (B.39)$$

makes no contribution to mass coupling.

---

[2] Droplet rotation could also be included but the resulting continuity equation will be unchanged.

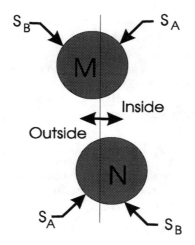

Figure B.5: Complementary boundary droplets.

The integral over the boundary droplets can be assessed in the following way. Consider two droplets which are intersected by the boundary as shown in Figure B.5. Both droplets are the same size and have the same distribution of properties (density, velocity, etc.) over the surface. The portion of the surface of droplet M inside the volume is S_A and the area outside is S_B. Droplet N is a *complementary* droplet with an area S_B inside the volume and S_A outside the volume. The integral of a function φ over the surface S_A of droplet M is related to the integral over the complete droplet surface by

$$\int_{S_{A,M},} \varphi dS = \int_{S_D} \varphi dS - \int_{S_{B,M}} \varphi dS \tag{B.40}$$

where S_D is the total surface of the droplet. The integral of the function φ over the surface S_A of droplet N is

$$\int_{S_{A,N}} \varphi dS = \int_{S_D} \varphi dS - \int_{S_{B,N}} \varphi dS \tag{B.41}$$

The complementary droplet is defined as having such a position with respect to the boundary that

$$\int_{S_{B,M}} \varphi dS + \int_{S_{B,N}} \varphi dS = \int_{S_D} \varphi dS \tag{B.42}$$

Thus adding Equations B.40 and B.41 gives

$$\int_{S_{A,M}} \varphi dS + \int_{S_{A,N}} \varphi dS = \int_{S_D} \varphi dS \tag{B.43}$$

so the sum of a *complementary pair* of droplets is equivalent to one droplet completely inside the control volume. Thus one can say that complementary pairs of *boundary droplets* can be replaced with complete droplets inside the control volume. The integral $\int_{S_b} \rho_c w dS$ can be regarded as the mass flux from droplets completely inside the volume which replace pairs of complementary boundary droplets. Also, the integral $\int_{S_b} \rho_c v_i n_i dS$ is zero.

Thus the continuity equation for the carrier phase becomes

$$\frac{\partial}{\partial t} \left(\alpha_c \langle \rho_c \rangle \right) + \frac{\partial}{\partial x_i} \left(\alpha_c \langle \rho_c \rangle \tilde{u}_i \right) = s_{mass} \tag{B.44}$$

where s_{mass} is the mass source term which is the mass addition per unit volume due to all the droplets inside the control volume or

$$s_{mass} = -\frac{1}{V} \sum_k \dot{m}_k \tag{B.45}$$

where $\dot{m}_k$ is the rate of change of the mass of droplet k and the summation is carried out over all droplets inside the control volume including the boundary droplets. If all droplets evaporate at the same rate, the source term can be expressed as

$$s_{mass} = -n \dot{m} \tag{B.46}$$

where n is the local droplet number density and $\dot{m}$ is the evaporation rate.

B.3   Momentum equation

The momentum equation for the carrier phase is

$$\frac{\partial}{\partial t} \left(\rho_c u_i \right) + \frac{\partial}{\partial x_j} \left(\rho_c u_i u_j \right) = -\frac{\partial p}{\partial x_i} + \frac{\partial \tau_{ij}}{\partial x_j} + \rho_c g_i \tag{B.47}$$

Applying the averaging equation to the time derivative gives

$$\overline{\frac{\partial}{\partial t} \left(\rho_c u_i \right)} = \frac{\partial}{\partial t} \left(\overline{\rho_c u_i} \right) + \frac{1}{V} \int_{S_d} \rho_c u_i \dot{r} dS \tag{B.48}$$

and averaging the convection term results in

$$\overline{\frac{\partial}{\partial x_j} \left(\rho_c u_i u_j \right)} = \frac{\partial}{\partial x_j} \left(\overline{\rho_c u_i u_j} \right) - \frac{1}{V} \int_{S_d} \rho_c u_i u_j n_j dS \tag{B.49}$$

Substituting the velocity at the surface, Equation B.31, into the above equations and adding yields the following expression

$$\overline{\frac{\partial}{\partial t} \left(\rho_c u_i \right)} + \overline{\frac{\partial}{\partial x_j} \left(\rho_c u_i u_j \right)} =$$

$$\frac{\partial}{\partial t} \left(\overline{\rho_c u_i} \right) + \frac{\partial}{\partial x_j} \left(\overline{\rho_c u_i u_j} \right) - \frac{1}{V} \int_{S_d} \rho_c \left(v_i + w' n_i \right) \left(v_j + w n_j \right) n_j dS \tag{B.50}$$

where, now, $w' = w + \dot{r}$ which is the velocity of the fluid at the surface with respect to the droplet center.

As with the continuity equation, the volume average of the mass flux can be expressed as

$$\overline{\rho_c u_i} = \alpha_c \langle \rho_c \rangle \tilde{u}_i \tag{B.51}$$

The volume average of the convection term becomes

$$\overline{\rho_c u_i u_j} = \alpha_c \langle \rho_c u_i u_j \rangle \tag{B.52}$$

Using Equations B.35 and B.36 the convection term becomes

$$\overline{\rho_c u_i u_j} = \alpha_c \langle \rho_c \rangle \tilde{u}_i \tilde{u}_j + \alpha_c \langle \rho_c \delta u_i \delta u_j \rangle \tag{B.53}$$

The last term, which is the stress due to sub-volume fluctuations, is equivalent to the *Reynolds stress* in a single-phase flow. The flow does not have to be turbulent to create this "stress" since fluctuations can be caused by the flow around the particles. For convenience here this stress will be incorporated with the laminar stress and the combination will simply be τ_{ij}.

Expanding the surface integral in Equation B.50 results in

$$\frac{1}{V} \int_{S_d} \rho_c \left(v_i + w' n_i \right) \left(v_j + w n_j \right) n_j dS = \overbrace{\frac{1}{V} \int_{S_d} \rho_c v_i v_j n_j dS}^{(a)}$$

$$\overbrace{\frac{1}{V} \int_{S_d} \rho_c v_i w dS}^{(b)} + \overbrace{\frac{1}{V} \int_{S_d} \rho_c v_j w' n_i n_j dS}^{(c)} + \overbrace{\frac{1}{V} \int_{S_d} \rho_c w' w n_i dS}^{(d)} \tag{B.54}$$

Assuming the density is uniform over the droplet surface term (a) is zero since v_i is constant with respect to the integral over the surface. If the mass flux from the surface is uniform, the integrals (c) and (d) are zero. Term (b) can be evaluated by summing over all the droplets in the volume including the boundary droplets,

$$\frac{1}{V} \int_{S_d} v_i \rho_c w dS = -\frac{1}{V} \sum_k v_{i,k} \dot{m}_k \tag{B.55}$$

where $v_{i,k}$ is the velocity of droplet k. If the evaporation rates and velocities of all the droplets in the volume are the same, this source term simplifies to

$$\frac{1}{V} \int_{S_d} v_i \rho_c w dS = -n v_i \dot{m} \tag{B.56}$$

The average of the pressure gradient term in the momentum equation is

$$\overline{\frac{\partial p}{\partial x_i}} = \frac{\partial \bar{p}}{\partial x_i} - \frac{1}{V} \int_{S_d} p n_i dS = \frac{\partial}{\partial x_i} \left(\alpha_c \langle p \rangle \right) - \frac{1}{V} \int_{S_d} p n_i dS \tag{B.57}$$

It is convenient to decompose the pressure on the droplet surface into the sum of the local average pressure and the deviation therefrom.

$$p = \langle p \rangle + \delta p \tag{B.58}$$

The integral of the pressure over the droplets inside the averaging volume is

$$\frac{1}{V} \int_{S_e} p n_i dS = \frac{1}{V} \int_{S_e} \delta p n_i dS \tag{B.59}$$

which is the "form force" (a part of the drag and lift force) on the droplets. The integral over the boundary droplets is

$$\frac{1}{V} \int_{S_b} p n_i dS = \frac{1}{V} \int_{S_b} (\langle p \rangle + \delta p) n_i dS$$
$$= -\frac{1}{V} \int_{S_s} \langle p \rangle \, n_i dS + \frac{1}{V} \int_{S_b} \delta p n_i dS \tag{B.60}$$

where S_s is the surface severed by the averaging volume. The sign on the integral over the severed surface is changed because n_i over the surface S_b is directed inward and n_i over the surface S_s is directed outward from the control volume. The integral over the severed surfaces can be replaced by[3]

$$\frac{1}{V} \int_{S_s} \langle p \rangle \, n_i dS = \frac{1}{V} \int_{S} \alpha_d \langle p \rangle \, n_i dS = \frac{\partial}{\partial x_i} (\alpha_d \langle p \rangle) \tag{B.61}$$

and the integral over the inside surfaces, S_b, is the sum of the "form forces" on the boundary droplets.

The same procedure can be carried out with the shear stress term resulting in

$$\overline{\frac{\partial \tau_{ij}}{\partial x_j}} = \frac{\partial}{\partial x_j} (\alpha_c \langle \tau_{ij} \rangle) - \frac{1}{V} \int_{S_d} \delta \tau_{ij} n_i dS \tag{B.62}$$

where the last term is the force due to shear on the droplets inside (including boundary droplets) the averaging volume. The integral of the pressure force and the shear stress on all the interior surfaces S_d is equal to the hydrodynamic forces (lift and drag) on the droplets in the volume. Thus

$$\frac{1}{V} \int_{S_d} (-\delta p n_i + \delta \tau_{ij} n_j) \, dS = \frac{1}{V} \sum_k F_{i,k} \tag{B.63}$$

where $F_{i,k}$ is the force *on* the droplet k. If all the droplets in the control volume have the same force, the term becomes

$$\frac{1}{V} \int_{S_d} (-p n_i + \tau_{ij} n_j) \, dS = n F_i \tag{B.64}$$

This is a momentum source term due to drag-lift interaction between the droplets and the carrier phase.

---

[3] The local area of severed surfaces ΔS_s is related to the local area ΔS by $\Delta S_s = \alpha_d \Delta S$.

Collecting all the terms, the momentum equation for the carrier phase assumes the form

$$\frac{\partial}{\partial t}\left(\langle\rho_c\rangle\,\tilde{u}_i\right) + \frac{\partial}{\partial x_j}\left(\langle\rho_c\rangle\,\tilde{u}_i\tilde{u}_j\right) = -\frac{1}{V}\sum_k v_{i,k}\dot{m}_k$$
$$-\frac{\partial}{\partial x_i}\langle p\rangle + \frac{\partial}{\partial x_j}\langle\tau_{ij}\rangle - \frac{1}{V}\sum_k F_{i,k} + \alpha_c\langle\rho_c\rangle\,g_i \tag{B.65}$$

The hydrodynamic interaction term can also be rewritten to include the pressure gradient and shear stress gradient. From Chapter 5 the drag force on a particle can be expressed as

$$F_i = -V_d\frac{\partial\langle p\rangle}{\partial x_i} + V_d\frac{\partial\langle\tau_{ij}\rangle}{\partial x_j} + L_i \tag{B.66}$$

where V_d is the droplet volume and L_i is the rest of the surface forces acting on the particle; namely, the lift force, steady-state drag, virtual mass and Basset force. Substituting the above terms in the momentum equation results in the final form of the momentum equation

$$\frac{\partial}{\partial t}\left(\alpha_c\langle\rho_c\rangle\,\tilde{u}_i\right) + \frac{\partial}{\partial x_j}\left(\alpha_c\langle\rho_c\rangle\,\tilde{u}_i\tilde{u}_j\right) = -\frac{1}{V}\sum_k v_{i,k}\dot{m}_k$$
$$-\alpha_c\frac{\partial}{\partial x_i}\langle p\rangle + \alpha_c\frac{\partial}{\partial x_j}\langle\tau_{ij}\rangle - \frac{1}{V}\sum_k L_{i,k} + \alpha_c\langle\rho_c\rangle\,g_i \tag{B.67}$$

If the unsteady forces on the particle are excluded, $L_{i,k}$ becomes[4]

$$L_{i,k} = 3\pi\mu_c D_k f_k(\tilde{u}_i - v_{i,k}) \tag{B.68}$$

and the momentum equation can be written as

$$\frac{\partial}{\partial t}\left(\alpha_c\langle\rho_c\rangle\,\tilde{u}_i\right) + \frac{\partial}{\partial x_j}\left(\alpha_c\langle\rho_c\rangle\,\tilde{u}_i\tilde{u}_j\right) = -\frac{1}{V}\sum_k v_{i,k}\dot{m}_k$$
$$-\alpha_c\frac{\partial}{\partial x_i}\langle p\rangle + \alpha_c\frac{\partial}{\partial x_j}\langle\tau_{ij}\rangle - \frac{3\pi\mu_c}{V}\sum_k D_k f_k(\tilde{u}_i - v_{i,k}) + \alpha_c\langle\rho_c\rangle\,g_i \tag{B.69}$$

If the conditions are such that the dispersed phase volume fraction is small and its effect negligible, the momentum equation reduces to

$$\frac{\partial}{\partial t}\left(\langle\rho_c\rangle\,\tilde{u}_i\right) + \frac{\partial}{\partial x_j}\left(\langle\rho_c\rangle\,\tilde{u}_i\tilde{u}_j\right) = -\frac{1}{V}\sum_k v_{i,k}\dot{m}_k$$
$$-\frac{\partial}{\partial x_i}\langle p\rangle + \frac{\partial}{\partial x_j}\langle\tau_{ij}\rangle - \frac{3\pi\mu_c}{V}\sum_k D_k f_k(\tilde{u}_i - v_{i,k}) + \langle\rho_c\rangle\,g_i \tag{B.70}$$

The momentum flux due to mass transfer from the droplet surface, $-\frac{1}{V}\sum_k v_{i,k}\dot{m}_k$, and the hydrodynamic interaction forces, $-\frac{3\pi\mu_c}{V}\sum_k D_k f_k(\tilde{u}_i - v_{i,k})$, are the sources of momentum coupling between phases. If all the droplets

---

[4]The steady-state drag force is proportional to the difference between the local carrier phase and particle velocity. The choice of the local carrier phase velocity is somewhat arbitrary (volume or mass average). In this formulation the mass average velocity is used.

have the same mass (and size), evaporate at the same rate and move at the same local velocity, the momentum equation becomes

$$\frac{\partial}{\partial t}\left(\alpha_c \langle \rho_c \rangle \tilde{u}_i\right) + \frac{\partial}{\partial x_j}\left(\alpha_c \langle \rho_c \rangle \tilde{u}_i \tilde{u}_j\right) = -n\dot{m}v_i$$

$$-\alpha_c \frac{\partial}{\partial x_i}\langle p \rangle + \alpha_c \frac{\partial}{\partial x_j}\langle \tau_{ij} \rangle - \beta_V(\tilde{u}_i - v_i) + \alpha_c \langle \rho_c \rangle g_i \tag{B.71}$$

where v_i is the droplet velocity and β_V is the proportionality constant introduced in Chapter 6.

B.4 Energy equation

The total energy equation includes both the internal and external energies (kinetic energy) of the carrier phase. The thermal energy equation is obtained by subtracting the kinetic energy from the total energy equation.

B.4.1 Total energy equation

The total energy equation for the carrier phase is

$$\frac{\partial}{\partial t}\left[\rho_c\left(i_c + \frac{U^2}{2}\right)\right] + \frac{\partial}{\partial x_i}\left[\rho_c u_i\left(i_c + \frac{U^2}{2}\right)\right] =$$

$$-\frac{\partial}{\partial x_i}\left(u_i p\right) + \frac{\partial}{\partial x_j}\left(u_i \tau_{ji}\right) - \frac{\partial q_i}{\partial x_i} + \rho_c u_i g_i \tag{B.72}$$

where i is the specific internal energy and q_i is the heat transfer. The kinetic energy per unit mass, $U^2/2$, corresponds to the dot product $u_i u_i/2$. Combining the flow work with the internal energy in the convection term introduces the enthalpy.

$$\overbrace{\frac{\partial}{\partial t}\left[\rho_c\left(i_c + \frac{U^2}{2}\right)\right]}^{(1)} + \overbrace{\frac{\partial}{\partial x_i}\left[\rho_c u_i\left(h_c + \frac{U^2}{2}\right)\right]}^{(2)} =$$

$$+ \overbrace{\frac{\partial}{\partial x_j}\left(u_i \tau_{ji}\right)}^{(3)} - \overbrace{\frac{\partial q_i}{\partial x_i}}^{(4)} + \overbrace{\rho_c u_i g_i}^{(5)} \tag{B.73}$$

Local energy change and energy flux

Applying volume averaging to term (1) results in

$$\overline{\frac{\partial}{\partial t}\left[\rho_c\left(i_c + \frac{U^2}{2}\right)\right]} = \frac{\partial}{\partial t}\left[\overline{\rho_c\left(i_c + \frac{U^2}{2}\right)}\right] + \frac{1}{V}\int_{S_d} \rho_c \dot{r}\left(i_c + \frac{U^2}{2}\right) dS \tag{B.74}$$

and to term (2) yields

$$\overline{\frac{\partial}{\partial x_i} \left[\rho_c u_i \left(h_c + \frac{U^2}{2} \right) \right]} = \frac{\partial}{\partial x_i} \left[\overline{\rho_c u_i \left(h_c + \frac{U^2}{2} \right)} \right]$$

$$- \frac{1}{V} \int_{S_d} \rho_c u_i \left(h_c + \frac{U^2}{2} \right) n_i dS \tag{B.75}$$

The volume average of the energy density in Equation B.74 becomes

$$\overline{\rho_c \left(i_c + \frac{U^2}{2} \right)} = \alpha_c (\langle \rho_c i_c \rangle + \frac{\langle \rho_c U^2 \rangle}{2}) \tag{B.76}$$

The volume average of the product of density and internal energy is written in terms of average values as

$$\langle \rho_c i \rangle = \langle \rho_c \rangle \tilde{i}_c \tag{B.77}$$

which defines the internal energy as the *mass-averaged* internal energy in the volume. The average of the product of the density and kinetic energy be expressed as

$$\frac{\langle \rho_c U^2 \rangle}{2} = \frac{1}{2} \langle \rho_c u_i u_i \rangle \tag{B.78}$$

Once again using Equations B.35 and B.36 the kinetic energy becomes

$$\frac{1}{2} \langle \rho_c u_i u_i \rangle = \frac{1}{2} \langle \rho_c \rangle \tilde{u}_i \tilde{u}_i + \frac{1}{2} \langle \rho_c \delta u_i \delta u_i \rangle \tag{B.79}$$

In most applications, the second term is insignificant, especially in low Mach number flows where the kinetic energy is small compared to the thermal energies. Thus the following approximation will be used.

$$\frac{\langle \rho_c U^2 \rangle}{2} \simeq \frac{1}{2} \langle \rho_c \rangle \tilde{u}_i \tilde{u}_i \tag{B.80}$$

The same arguments apply to the energy convection so the following approximate convection term will utilized,

$$\overline{\rho_c u_i \left(h_c + \frac{U^2}{2} \right)} \simeq \alpha_c \langle \rho_c \rangle \tilde{u}_i \left(\tilde{h}_c + \frac{\tilde{u}_j \tilde{u}_j}{2} \right) \tag{B.81}$$

where $\tilde{h}$ is the mass-averaged enthalpy. Of course, if the density of the fluid is constant, the mass averaged properties are simply the volume-averaged values.

Using Equation B.15 for the gas velocity at the droplet surface and combining Equations B.74 and B.75 shows

$$\overline{\frac{\partial}{\partial t} \left[\rho_c \left(i_c + \frac{U^2}{2} \right) \right]} + \overline{\frac{\partial}{\partial x_i} \left[\rho_c u_i \left(h_c + \frac{U^2}{2} \right) \right]} =$$

$$\frac{\partial}{\partial t} \left[\alpha_c \langle \rho_c \rangle \left(\tilde{i}_c + \frac{1}{2} \tilde{u}_i \tilde{u}_i \right) \right] + \frac{\partial}{\partial x_i} \left[\alpha_c \langle \rho_c \rangle \tilde{u}_i \left(\tilde{h}_c + \frac{1}{2} \tilde{u}_j \tilde{u}_j \right) \right] \tag{B.82}$$

$$- \frac{1}{V} \int_{S_d} \left[\dot{r} p n_i + \rho_c \left(v_i + w n_i \right) \left(i_c + \frac{U^2}{2} \right) \right] n_i dS$$

Consider now each of the terms in the integrals identified as follows

$$\int_{S_d} \dot{r} p dS + \int_{S_d} \rho_c (v_i + wn_i)(i_c + \frac{p}{\rho_c} + \frac{u_j u_j}{2}) n_i dS =$$

$$\underbrace{\int_{S_d} v_i p n_i dS}_{(a)} + \underbrace{\int_{S_d} p w dS}_{(b)} + \underbrace{\int_{S_d} \dot{r} p dS}_{(c)} + \underbrace{\int_{S_d} v_i \rho_c i n_i dS}_{(d)} \quad \text{(B.83)}$$

$$+ \underbrace{\int_{S_d} \rho_c w i dS}_{(e)} + \underbrace{\int_{S_d} \rho_c v_i \frac{u_j u_j}{2} n_i dS}_{(f)} + \underbrace{\int_{S_d} \rho_c w \frac{u_j u_j}{2} dS}_{(g)}$$

The integral in term (a) is carried out over the surfaces of the internal droplets, S_e, and the boundary droplets, S_b. The integral over the internal droplets becomes the dot product of the droplet velocity and "form force" on the fluid due to the droplets..

$$\frac{1}{V} \int_{S_e} v_i p n_i dS = \frac{1}{V} v_i \int_{S_e} p n_i dS \quad \text{(B.84)}$$

This term will ultimately be combined with the shear stress over the surface to yield the work rate due to the fluid dynamic force on the droplet. The pressure on the boundary droplets also contributes to the energy equation through a flow work component. The integral over the boundary droplets associated with term (a) gives

$$\frac{1}{V} \int_{S_b} v_i p n_i dS = \frac{1}{V} \int_{S_b} v_i \langle p \rangle n_i dS + \frac{1}{V} \int_{S_b} v_i \delta p n_i dS \quad \text{(B.85)}$$

which is equal to

$$\frac{1}{V} \int_{S_b} v_i p n_i dS = -\frac{\partial}{\partial x_i} (\alpha_d \langle v_i \rangle \langle p \rangle) + \frac{1}{V} \int_{S_b} v_i \delta p n_i dS \quad \text{(B.86)}$$

where $\langle v_i \rangle$ is the droplet volume (or mass) averaged velocity[5]. The first integral represents that force due to pressure on the portion of the boundary droplet which is inside the averaging volume.[6] This can be regarded as a flow work on the carrier phase associated with the motion of the particles. The second integral is dot product of the droplet velocity and form force acting on the fluid due to the boundary droplets.

Terms (b) and (e) combine to give the enthalpy flux from the droplet surface.

---

[5] In terms of the particle size distribution, the droplet mass average velocity is

$$\langle v_i \rangle = \int_0^\infty f_m(D) v_i(D) dD$$

[6] The sign on this integral is changed for the same reason presented for Equation B.60.

$$\int_{S_d} pwdS + \int_{S_d} \rho_c wi_c dS = \int_{S_d} \rho_c h_c wdS \tag{B.87}$$

Term (d) is zero when the product $\rho_c i$ is uniform over the droplet surface.

Term (c) represents the work rate associated with contraction or dilation of the particle volume and be replaced by

$$\int_{S_d} \dot{r}pdS = \sum_k \dot{V}_{d,k} p_{s,k} \tag{B.88}$$

where $\dot{V}_{d,k}$ is the dilation rate of droplet k and $p_{s,k}$ is the pressure on the surface of droplet k.

Terms (f) and (g) require further development. The kinetic energy associated with the fluid velocity at the surface is

$$\frac{u_i u_i}{2} = \frac{(v_i + w'n_i)(v_i + w'n_i)}{2} = \frac{v_i v_i}{2} + v_i w'n_i + \frac{w'^2}{2} \tag{B.89}$$

For a uniform mass flux over the surface term (f) can now be written as

$$\int_{S_d} \rho_c v_i \frac{u_j u_j}{2} n_i dS = \rho_c \left(\frac{v_j v_j}{2} + \frac{w'^2}{2}\right) v_i \int_{S_d} n_i dS + v_i w' v_j \int_{S_d} n_i n_j dS \tag{B.90}$$

and both integrals on the right side are zero. Term (g) becomes

$$\int_{S_d} \rho_c w_i \frac{u_j u_j}{2} n_i dS = \int_{S_d} \rho_c w \left(\frac{v_i v_i}{2} + v_i w' n_i + \frac{w'^2}{2}\right) dS$$

$$= -\sum_k \dot{m}_k \left(\frac{v_{i,k} v_{i,k}}{2} + \frac{w_k'^2}{2}\right) \tag{B.91}$$

which is the flux of kinetic energy into the carrier phase from the droplets.

Work rate due to shear stress

Applying the averaging formalism to term (3) in Equation B.73 gives

$$\overline{\frac{\partial}{\partial x_j}(u_i \tau_{ji})} = \frac{\partial}{\partial x_j}\left(\alpha_c \langle u_i \tau_{ji} \rangle\right) - \frac{1}{V}\int_{S_d} u_i \tau_{ij} n_j dS \tag{B.92}$$

The integral over surface S_e results in

$$\frac{1}{V}\int_{S_e} u_i \tau_{ij} n_i dS = \frac{1}{V}\int_{S_e} (v_i + w'n_i)\tau_{ij} n_j dS = \frac{1}{V}\int_{S_d} v_i \tau_{ij} n_j dS \tag{B.93}$$

The second term disappears because, by definition, the shear stress is perpendicular to the outward normal vector, $n_i \tau_{ij} n_j = 0$. It is also necessary to integrate over the surface of the boundary droplets so the integral becomes

$$\frac{1}{V} \int_{S_b} v_i \tau_{ij} n_j dS = -\frac{\partial}{\partial x_j} \left(\alpha_d \langle v_i \tau_{ij} \rangle \right) + \frac{1}{V} \int_{S_b} v_i \delta \tau_{ij} n_j dS \tag{B.94}$$

where the last term is the work rate associated with shear stress force on the boundary droplets. The average of the product of velocity a shear stress will be approximated as

$$\langle u_i \tau_{ji} \rangle \simeq \tilde{u}_i \langle \tau_{ji} \rangle$$

$$\langle v_i \tau_{ij} \rangle \simeq \langle v_i \rangle \langle \tau_{ij} \rangle \tag{B.95}$$

Combining terms in Equations B.84, B.86, B.93 and B.94 yields

$$\frac{1}{V} v_i \int_{S_e} (-pn_i + \tau_{ij} n_j) dS$$

$$\frac{1}{V} v_i \int_{S_b} (-\delta p n_i + \delta \tau_{ij} n_j) dS = \frac{1}{V} \sum_k v_{i,k} F_{i,k} \tag{B.96}$$

where the summation is performed over all droplets inside the volume together with the boundary droplets.

Heat transfer

The volume average of the heat transfer term (4) results in

$$\overline{\frac{\partial q_i}{\partial x_i}} = \frac{\partial \bar{q}_i}{\partial x_i} - \frac{1}{V} \int_{S_d} q_i n_i dS \tag{B.97}$$

The heat transfer flux around a droplet can be expressed as the sum of an average value over the surface and a deviation therefrom,

$$q_i = \hat{q} n_i + \delta q_i, \tag{B.98}$$

so the integral over the interior droplets becomes

$$\frac{1}{V} \int_{S_e} q_i n_i dS = \frac{1}{V} \int_{S_e} \hat{q} dS \tag{B.99}$$

which is the net heat transfer to the fluid from the droplets. The integral over the boundary droplets is

$$\frac{1}{V} \int_{S_b} q_i n_i dS = \frac{1}{V} \int_{S_b} \hat{q} dS + \frac{1}{V} \int_{S_e} \delta q_i n_i dS \tag{B.100}$$

where the first term yields the net heat transfer to the fluid by the boundary droplets and the second term is the transfer *through* the droplets. Thus the heat transfer term can be reduced to the following form

$$\overline{\frac{\partial q_i}{\partial x_i}} = \frac{\partial}{\partial x_i} \left(\alpha_c \langle q_i \rangle + \alpha_d \langle q_{i,d} \rangle \right) - \frac{1}{V} \sum_k \dot{Q}_k \tag{B.101}$$

where $\dot{Q}_k$ is the net heat transfer rate from droplet k to the fluid inside the averaging volume and the summation includes the boundary droplets. For convective heat transfer between the carrier phase and the particles $\dot{Q}_k$ becomes

$$\dot{Q}_k = 2\pi k_c (Nu/2)_k D_k (T_{d,k} - \langle T_c \rangle) \tag{B.102}$$

The composite heat transfer through the carrier phase and the droplets will be expressed as

$$q_i^{eff} = \alpha_c \langle q_i \rangle + \alpha_d \langle q_{i,d} \rangle . \tag{B.103}$$

or, in terms of a temperature gradient,

$$q_i^{eff} = -k_{eff} \frac{\partial \langle T_c \rangle}{\partial x_i} \tag{B.104}$$

where k_{eff} is the effective thermal conductivity for the mixture.

Work due to body force

Finally, the volume average of term (5) in the energy equation is

$$\overline{\rho_c u_i g_i} = \alpha_c \langle \rho_c \rangle \tilde{u}_i g_i \tag{B.105}$$

Assembling all the terms, the total energy equation becomes

$$\frac{\partial}{\partial t}\left[\alpha_c \langle \rho_c \rangle \left(\tilde{i}_c + \frac{\tilde{u}_i \tilde{u}_i}{2}\right)\right] + \frac{\partial}{\partial x_j}\left[\alpha_c \langle \rho_c \rangle \tilde{u}_j \left(\tilde{h}_c + \frac{\tilde{u}_i \tilde{u}_i}{2}\right)\right] =$$

$$-\frac{1}{V}\sum_k \dot{m}_k \left(h_{s,k} + \frac{v_{i,k} v_{i,k}}{2} + \frac{w_k'^2}{2}\right) - \frac{1}{V}\sum_k v_{i,k} F_{i,k}$$

$$-\frac{\partial}{\partial x_i}\left(\alpha_d \langle v_i \rangle \langle p \rangle\right) + \frac{\partial}{\partial x_j}\left(\alpha_d \langle v_i \rangle \langle \tau_{ij} \rangle\right) + \frac{\partial}{\partial x_j}\left(\alpha_c \tilde{u}_i \langle \tau_{ij} \rangle\right) \tag{B.106}$$

$$+\frac{1}{V}\sum_k \dot{V}_{d,k} p_{s,k} - \frac{\partial q_i^{eff}}{\partial x_i} + \frac{1}{V}\sum_k \dot{Q}_k + \alpha_c \langle \rho_c \rangle \tilde{u}_i g_i$$

where $h_{s,k}$ is the enthalpy of the gas at the surface of droplet k. Taking the forces due to the pressure gradient and shear stress gradient out of the interaction force $F_{i,k}$ yields

$$\frac{\partial}{\partial t}\left[\alpha_c \langle \rho_c \rangle \left(\tilde{i}_c + \frac{\tilde{u}_i \tilde{u}_i}{2}\right)\right] + \frac{\partial}{\partial x_j}\left[\alpha_c \langle \rho_c \rangle \tilde{u}_j \left(\tilde{h}_c + \frac{\tilde{u}_i \tilde{u}_i}{2}\right)\right] =$$

$$-\frac{1}{V}\sum_k \dot{m}_k \left(h_{s,k} + \frac{v_{i,k} v_{i,k}}{2} + \frac{w_k'^2}{2}\right) - \frac{3\pi \mu_c}{V}\sum_k D_k f_k (\tilde{u}_i - v_{i,k}) v_{i,k}$$

$$- \langle p \rangle \frac{\partial}{\partial x_i}\left(\alpha_d \langle v_i \rangle\right) + \frac{1}{V}\sum_k \dot{V}_{d,k} p_{s,k} + \langle \tau_{ij} \rangle \frac{\partial}{\partial x_j}\left(\alpha_d \langle v_i \rangle\right) + \frac{\partial}{\partial x_j}\left(\alpha_c \tilde{u}_i \langle \tau_{ij} \rangle\right)$$

$$+\frac{\partial}{\partial x_i}\left(k_{eff} \frac{\partial \langle T_c \rangle}{\partial x_i}\right) + \frac{2\pi k_c}{V}\sum_k \left(\frac{Nu}{2}\right)_k D_k (T_{d,k} - \langle T_c \rangle) + \alpha_c \langle \rho_c \rangle \tilde{u}_i g_i \tag{B.107}$$

If all the droplets have the same mass (and size), move at the same local velocity and evaporate at the same rate, the total energy equation becomes

$$\frac{\partial}{\partial t}\left[\alpha_c \langle \rho_c \rangle \left(\tilde{i}_c + \frac{\tilde{u}_i \tilde{u}_i}{2}\right)\right] + \frac{\partial}{\partial x_j}\left[\alpha_c \langle \rho_c \rangle \tilde{u}_j \left(\tilde{h}_c + \frac{\tilde{u}_i \tilde{u}_i}{2}\right)\right] =$$

$$-n\dot{m}\left(h_s + \frac{v_{i,}v_{i,}}{2} + \frac{(w')^2}{2}\right) - v_i \beta_V (\tilde{u}_i - v_i)$$

$$\hspace{8cm}\text{(B.108)}$$

$$- \langle p \rangle \frac{\partial}{\partial x_i}(\alpha_d v_i) + n\dot{V}_d p_s + \langle \tau_{ij} \rangle \frac{\partial}{\partial x_j}(\alpha_d v_i) + \frac{\partial}{\partial x_j}(\alpha_c \tilde{u}_i \langle \tau_{ij} \rangle)$$

$$+ \frac{\partial}{\partial x_i}(k_{eff} \frac{\partial \langle T_c \rangle}{\partial x_i}) + \beta_T (T_d - T_c) + \alpha_c \langle \rho_c \rangle \tilde{u}_i g_i$$

where the coefficients β_V and β_T have been defined in Chapter 6. In low Mach number flows, the total energy equation can be simplified by neglecting the kinetic energy terms compared to the enthalpy and internal energy. Also other terms, such as the work associated with droplet dilation and the heat transfer through the fluid and droplets, are generally small. Obviously the simplification of the energy equation depends on the specific application.

For the situation where the volume fraction of the dispersed phase is small (and $\alpha_c \simeq 1$), the total energy equation simplifies to

$$\frac{\partial}{\partial t}\left[\langle \rho_c \rangle \left(\tilde{i}_c + \frac{\tilde{u}_i \tilde{u}_i}{2}\right)\right] + \frac{\partial}{\partial x_i}\left[\langle \rho_c \rangle \tilde{u}_i \left(\tilde{h}_c + \frac{\tilde{u}_i \tilde{u}_i}{2}\right)\right] =$$

$$-n\dot{m}\left(h_s + \frac{v_{i,}v_{i,}}{2} + \frac{(w')^2}{2}\right) - v_i \beta_V (\tilde{u}_i - v_i) \hspace{2cm}\text{(B.109)}$$

$$+ \frac{\partial}{\partial x_j}(\tilde{u}_i \langle \tau_{ij} \rangle) + \frac{\partial}{\partial x_i}(k_c \frac{\partial \langle T_c \rangle}{\partial x_i}) + \beta_T (T_d - T_c) + \langle \rho_c \rangle \tilde{u}_i g_i$$

where k_c is the thermal conductivity of the continuous phase.

B.4.2 Thermal energy equation

The equation for kinetic energy is obtained by multiplying the momentum equation, Equation B.67 by $\tilde{u}_i$

$$\tilde{u}_i \frac{\partial}{\partial t}(\alpha_c \langle \rho_c \rangle \tilde{u}_i) + \tilde{u}_i \frac{\partial}{\partial x_j}(\alpha_c \langle \rho_c \rangle \tilde{u}_i \tilde{u}_j) = -\frac{\tilde{u}_i}{V}\sum_k v_{i,k} \dot{m}_k$$

$$-\tilde{u}_i \alpha_c \frac{\partial}{\partial x_i} \langle p \rangle + \tilde{u}_i \alpha_c \frac{\partial}{\partial x_j} \langle \tau_{ij} \rangle - \frac{1}{V}\sum_k \tilde{u}_i L_{i,k} + \alpha_c \langle \rho_c \rangle g_i \tilde{u}_i. \hspace{1cm}\text{(B.110)}$$

The first two terms can be manipulated first taking the derivative by parts and then using the continuity equation, Equation B.44,

$$\tilde{u}_i \frac{\partial}{\partial t}(\alpha_c \langle \rho_c \rangle \tilde{u}_i) + \tilde{u}_i \frac{\partial}{\partial x_j}(\alpha_c \langle \rho_c \rangle \tilde{u}_i \tilde{u}_j) = -\frac{\tilde{u}_i \tilde{u}_i}{V}\sum_n \dot{m}_n$$

$$+ \alpha_c \langle \rho_c \rangle \frac{\partial}{\partial t}\frac{\tilde{u}_i \tilde{u}_i}{2} + \alpha_c \langle \rho_c \rangle \tilde{u}_i \frac{\partial}{\partial x_j}\frac{\tilde{u}_i \tilde{u}_i}{2} \hspace{1cm}\text{(B.111)}$$

Also, by further manipulation one has

$$\frac{\partial}{\partial t}\left(\alpha_c \left\langle \rho_c \right\rangle \frac{\tilde{u}_i \tilde{u}_i}{2}\right) + \frac{\partial}{\partial x_j}\left(\alpha_c \left\langle \rho_c \right\rangle \tilde{u}_i \frac{\tilde{u}_i \tilde{u}_i}{2}\right) = -\frac{1}{V}\frac{\tilde{u}_i \tilde{u}_i}{2}\sum_k \dot{m}_k$$

$$\alpha_c \left\langle \rho_c \right\rangle \frac{\partial}{\partial t}\frac{\tilde{u}_i \tilde{u}_i}{2} + \alpha_c \left\langle \rho_c \right\rangle \tilde{u}_i \frac{\partial}{\partial x_j}\frac{\tilde{u}_i \tilde{u}_i}{2}$$

(B.112)

Combining Equations B.111 and B.112 for the local and convective derivatives of kinetic energy and substituting into Equation B.110 gives

$$\frac{\partial}{\partial t}\alpha_c \left\langle \rho_c \right\rangle \frac{\tilde{u}_i \tilde{u}_i}{2} + \frac{\partial}{\partial x_j}\alpha_c \left\langle \rho_c \right\rangle \tilde{u}_i \frac{\tilde{u}_i \tilde{u}_i}{2} =$$

$$\frac{1}{V}\frac{\tilde{u}_i \tilde{u}_i}{2}\sum_k \dot{m}_k - \frac{\tilde{u}_i}{V}\sum_k v_{i,k}\dot{m}_k - \tilde{u}_i \alpha_c \frac{\partial}{\partial x_i}\left\langle p \right\rangle$$

$$+\tilde{u}_i \alpha_c \frac{\partial}{\partial x_j}\left\langle \tau_{ij} \right\rangle - \frac{1}{V}\sum_k \tilde{u}_i L_{i,k} + \alpha_c \left\langle \rho_c \right\rangle g_i \tilde{u}_i.$$

(B.113)

Subtracting this equation from the total energy equation, Equation B.107, yields the following form of the thermal energy equation.

$$\frac{\partial}{\partial t}\left[\alpha_c \left\langle \rho_c \right\rangle \tilde{i}_c\right] + \frac{\partial}{\partial x_i}\left[\alpha_c \left\langle \rho_c \right\rangle \tilde{u}_i \tilde{h}_c\right] = \tilde{u}_i \alpha_c \frac{\partial}{\partial x_i}\left\langle p \right\rangle$$

$$-\sum_n \dot{m}_k \left(h_{s,k} + \frac{|v_{i,k}-\tilde{u}_i|^2}{2} + \frac{w_k'^2}{2}\right) + \frac{1}{V}\sum_k \left(\tilde{u}_i - v_{i,k}\right) L_{i,k}$$

$$-\left\langle p \right\rangle \frac{\partial}{\partial x_i}\left(\alpha_d \left\langle v_i \right\rangle\right) + \frac{1}{V}\sum_k \dot{V}_{d,k}p_{s,k} + \left\langle \tau_{ij} \right\rangle \left[\frac{\partial}{\partial x_j}\left(\alpha_d \left\langle v_i \right\rangle + \alpha_c \tilde{u}_i\right)\right]$$

$$-\frac{\partial q_i^{eff}}{\partial x_i} + \frac{1}{V}\sum_k \dot{Q}_k$$

(B.114)

If the unsteady drag and heat transfer terms are excluded, the thermal energy equation becomes

$$\frac{\partial}{\partial t}\left[\alpha_c \langle \rho_c \rangle \tilde{i}_c\right] + \frac{\partial}{\partial x_i}\left[\alpha_c \langle \rho_c \rangle \tilde{u}_i \tilde{h}_c\right] = \overbrace{\tilde{u}_i \alpha_c \frac{\partial}{\partial x_i} \langle p \rangle}^{a}$$

$$\overbrace{- \sum_n \dot{m}_k \left(h_{s,k} + \frac{|v_{i,k} - \tilde{u}_i|^2}{2} + \frac{w_k'^2}{2} \right)}^{b} + \overbrace{\frac{3\pi \mu_c}{V} \sum_k D_k f_k |\tilde{u}_i - v_{i,k}|^2}^{c}$$

$$\overbrace{- \langle p \rangle \frac{\partial}{\partial x_i} \left(\alpha_d \langle v_i \rangle \right)}^{d} + \overbrace{\frac{1}{V} \sum_k \dot{V}_{d,k} p_{s,k}}^{e} + \overbrace{\langle \tau_{ij} \rangle \left[\frac{\partial}{\partial x_j} \left(\alpha_d \langle v_i \rangle + \alpha_c \tilde{u}_i \right) \right]}^{f}$$

$$\overbrace{+ \frac{\partial}{\partial x_i} \left(k_{eff} \frac{\partial \langle T_c \rangle}{\partial x_i} \right)}^{g} + \overbrace{\frac{2\pi k_c}{V} \sum_k \left(\frac{Nu}{2}\right)_k D_k (T_{d,k} - \langle T_c \rangle)}^{h}$$

(B.115)

The terms in this equation can be identified as follows:

- (a) reversible work,
- (b) energy influx from the dispersed phase,
- (c) energy dissipation (always positive) from the particle-fluid interaction,
- (d) flow work due to motion of dispersed phase,
- (e) work rate associated with droplet dilation,
- (f) energy dissipation due to shear stress in the continuous phase,
- (g) heat transfer through the mixture, and
- (h) heat transfer from the dispersed phase.

If all the droplets move at the same local velocity and evaporate at the same rate, the thermal energy equation simplifies to

$$\frac{\partial}{\partial t}\left[\alpha_c \langle \rho_c \rangle \tilde{i}_c\right] + \frac{\partial}{\partial x_i}\left[\alpha_c \langle \rho_c \rangle \tilde{u}_i \tilde{i}_c\right] =$$

$$-n\dot{m}\left(h_s + \frac{|v_i - \tilde{u}_i|^2}{2} + \frac{w_k'^2}{2} \right) + \beta_V |\tilde{u}_i - v_i|^2$$

$$- \langle p \rangle \frac{\partial}{\partial x_i} \left(\alpha_d v_i + \alpha_c \tilde{u}_i \right) + n\dot{V}_d p_s$$

$$+ \langle \tau_{ij} \rangle \left[\frac{\partial}{\partial x_j} \left(\alpha_d v_i + \alpha_c \tilde{u}_i \right) \right] + \frac{\partial}{\partial x_i}\left(k_{eff} \frac{\partial \langle T_c \rangle}{\partial x_i} \right) + \beta_T (T_d - \langle T_c \rangle)$$

(B.116)

where the unsteady drag and heat transfer terms have been excluded.

If the volume fraction of the dispersed phase is negligibly small, the thermal energy equation reduces to

$$\frac{\partial}{\partial t}\left[\langle\rho_c\rangle\,\tilde{i}_c\right] + \frac{\partial}{\partial x_i}\left[\langle\rho_c\rangle\,\tilde{u}_i\tilde{i}_c\right] =$$

$$-n\dot{m}\left(h_s + \frac{|v_i - \tilde{u}_i|^2}{2} + \frac{w'^2}{2}\right) + \beta_V\,|\tilde{u}_i - v_i|^2 - \langle p\rangle\,\frac{\partial\langle\tilde{u}_i\rangle}{\partial x_i} \qquad (B.117)$$

$$+ \langle\tau_{ij}\rangle\left[\frac{\partial}{\partial x_j}\tilde{u}_i\right] + \frac{\partial}{\partial x_i}(k_c\frac{\partial\langle T_c\rangle}{\partial x_i}) + \beta_T(T_d - \langle T_c\rangle)$$

If there are no particles or droplets in the carrier flow, Equation B.117 reduces to that for a single-phase flow.

The above energy equations do not include a radiation term. If absorption of energy by radiation is important, it is included by adding an additional term to the right side of the energy equation. Also, the energy equation is for a homogeneous continuous phase. Modifications would be necessary to account for a mixture of chemical species.

Appendix C

Brownian Motion

If the size of a particle suspended in a fluid is very small (less than a micron), the motion of the particle is affected by the discrete nature of molecular motion, exhibiting a random motion due to collisions of molecules with the particle as shown in Figure C.1. This is called Brownian motion which occurs in both gases and liquids, the amplitude of the fluctuating motion being smaller in a liquid. Since the particle mass is much larger than that of the impacting molecule, the velocity of the particle motion is small compared to the molecular motion. If the particle spatial concentration is not uniform, the particles migrate toward the region of smaller concentration due to *Brownian motion*. The variation of spatial concentration due to one-dimensional Brownian diffusion is shown in Figure C.2. One-dimensional diffusion means that the concentration is a function of time and one spatial coordinate (x).

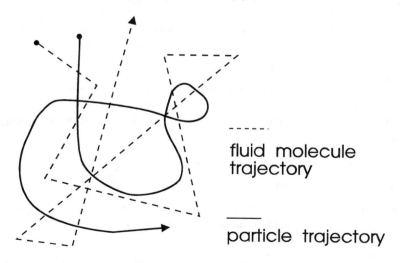

fluid molecule
trajectory

particle trajectory

Figure C.1: Particle movement through Brownian motion.

motion. Particles undergoing Brownian motion are moved by random molecular impact force and resisted by fluid drag. For simplicity, consider one-dimensional motion as shown in Figure C.2. If fluid is assumed to be a gas at rest, the equation of motion of a small particle is given by

$$m\frac{dv}{dt} = F(t) - \frac{1}{B}v \qquad (C.1)$$

where m and v are particle mass and velocity, respectively. This is referred to as the *Langevin equation*. The first term on the right side, $F(t)$, is the random impact force due to the molecules and the second term is the Stokes drag force modified for rarefaction effects. The factor B, defined as the *mobility*, is expressed as

$$B = \frac{C_c}{6\pi\mu_c a} \qquad (C.2)$$

where C_c is the Cunningham correction factor, μ_c the fluid viscosity and a the particle radius. The Cunningham correction factor, which corrects for rarefied flow effects, is a function of the Knudsen number and is defined in Chapter 4.

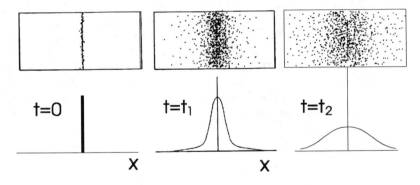

Figure C.2: One-dimensional diffusion from a line source.

Substituting the relation $v = \frac{dx}{dt}$ into Equation C.1 and multiplying both sides by x yields,

$$\frac{1}{2}m\frac{d^2x^2}{dt^2} - m\left(\frac{dx}{dt}\right)^2 = xF(t) - \frac{1}{2B}\frac{dx^2}{dt} \qquad (C.3)$$

Taking the average of the above equation over many particles and realizing that the average of $xF(t)$ is zero because the random force has no correlation with particle position,[1] one obtains

---

[1] In Brownian motion, the response time of the particles is much smaller than the time between molecular impacts. Thus the particles have no "memory" of the previous impact so the particle position and force are uncorrelated.

$$\frac{1}{2}m\frac{d^2\langle x^2\rangle}{dt^2} + \frac{1}{2B}\frac{d\langle x^2\rangle}{dt} = m\langle v^2\rangle \tag{C.4}$$

where $\langle\ \rangle$ signifies average value. The right side of the above equation is the kinetic energy of particles. With no loss in translation energy with molecular impact, the kinetic energy of the particle motion should equal that of the gas which is related to the absolute temperature T of gas through the Boltzman constant k by

$$m\langle v^2\rangle = kT \tag{C.5}$$

Substituting the above relation into Equation C.4 and integrating results in

$$\frac{d\langle x^2\rangle}{dt} = 2kBT + C\exp\left(-\frac{t}{mB}\right) \tag{C.6}$$

where C is the constant of integration. The second term of the right side of Equation C.6 becomes negligible after a certain time and the equation simplifies to[2]

$$\langle x^2\rangle = 2kBT \cdot t \tag{C.7}$$

Another approach is based on the diffusion or *Fokker-Plank* equation. Again using the one-dimensional formulation for simplicity, the diffusion process is expressed by the equation

$$\frac{\partial P}{\partial t} = D_f\frac{\partial^2 P}{\partial x^2} \tag{C.8}$$

where D_f is the diffusion coefficient and $P(x,t)$ is the probability that a particle which was initially $(t=0)$ located at $x=0$ reaches position x at time t. The solution to Equation C.8 with the initial condition[3] $P(0,0) = \delta(0)$ and boundary condition $P(\infty,t) = 0$ is

$$P(x,t) = \frac{1}{2\sqrt{\pi D_f t}}\exp\left(-\frac{x^2}{4D_f t}\right) \tag{C.9}$$

The mean value of x^2, namely $\langle x^2\rangle$, can be evaluated from

$$\langle x^2\rangle = \int_{-\infty}^{\infty} x^2 P dx = 2D_f t \tag{C.10}$$

and the mean value of x is

$$\langle x\rangle = \int_{-\infty}^{\infty} x P dx = \sqrt{\frac{4D_f t}{\pi}} \tag{C.11}$$

---

[2] This equation was derived by Einstein in 1905.

[3] $\delta(0)$ is the Dirac delta function at $x = 0$. Thus all the particles are initially at position $x = 0$ and diffuse away from this point with time.

By comparing Equation C.10 with the Einstein relation in Equation C.7, the diffusion coefficient D_f is found to be

$$D_f = kBT \qquad (C.12)$$

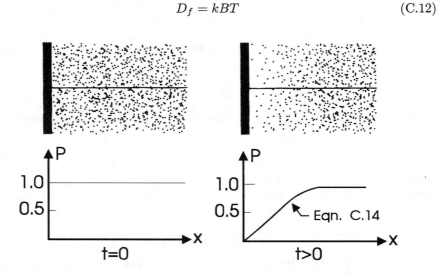

Figure C.3: Particle diffusion in the presence of a wall.

Solutions of the diffusion equation have various forms depending on boundary conditions. The above solution corresponds to particle diffusion into an infinite space with no boundary. If, for example, the space were bounded by a solid wall as shown in Figure C.3, the boundary and initial conditions would be

$$\begin{array}{ll} P(x,0) = 1 & \text{for } x > 0 \\ P(0,t) = 0 & \text{for } t \geq 0 \end{array} \qquad (C.13)$$

and the solution to the diffusion equation is

$$P(x,t) = \text{erf}\left\{\frac{x}{2\sqrt{D_f t}}\right\} \qquad (C.14)$$

where $\text{erf}(x)$ is the error function defined by

$$\text{erf}(z) = \frac{2}{\sqrt{\pi}} \int_0^z e^{-\zeta^2} d\zeta$$

The distribution of particle concentration predicted by Equation C.14 is shown in Figure C.3. The boundary condition at the wall signifies that particles impacting the wall are all deposited on the wall (with no re-entrainment) and, thus, the concentration at the wall is zero. The rate of deposition, or deposition flux, is defined as the number of particles per time per unit area which deposit on the wall. The deposition flux is given by

$$J = -n_0 D_f \left(\frac{\partial P}{\partial x} \right)_{x=0} \tag{C.15}$$

where n_0 is particle concentration far from the wall. The particles diffuse from regions of high concentration to low concentration so $\frac{\partial P}{\partial x}$ in the above application is negative.

Appendix D

Program Listing

This appendix provides the computer program for a gas-particle flow in a quasi-one-dimensional duct using conservative variables.

The program is based on the use of conservative variables introduced in Chapter 8. The approach is a marching solution in which the inlet conditions are specified and the program calculates the flow properties in the duct. The solution proceeds cell by cell. Position 1 is always the upstream side of the cell and position 2, the downstream side. When a solution is obtained for a given cell, the calculation proceeds to the next cell and the values at position 2 for the old cell become those for position 1 of the new cell.

D.1 Program structure

The structure of the program is shown in Figure D.1 and is described as follows.

Block 1: The physical inlet parameters are specified as well as the computational parameters.

Block 2: The variables needed in the computation are calculated. The subroutine DUCTGEO is called to provide the cross-sectional area as a function of distance.

Block 3: The initial values of the conservative variables are calculated (encoding) and the values for printout at the first station are loaded into an array.

Block 4: The velocity and pressure are first estimated at position 2 assuming incompressible flow. Also the average density, velocity, temperature as well as the friction factor are calculated. Subroutine PARTICLE is called to obtain the particle velocity and temperature at position 2 as well as the momentum and energy coupling terms.

Block 5: New conservative variables are obtained at position 2 by incorporating source terms. The factor "sonic" is evaluated to assure that the flow is still subsonic.

Block 6: The new conservative variables at position 2 are decoded for velocity and pressure. If the velocity iteration is not converged, the new velocities

457

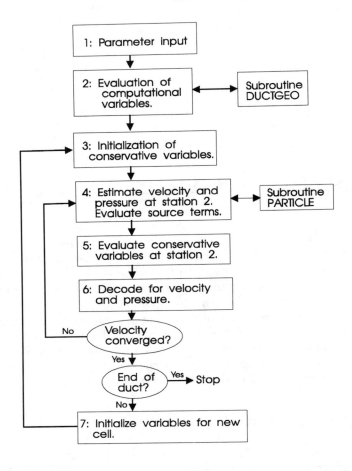

Figure D.1: Flow diagram for computer program based on conservative variables.

and pressures are used to repeat the iteration until the velocity at station 2 is converged.

Block 7: The calculation moves to the adjacent downstream cell, station 2 of the old cell becomes station 1 of the new cell and the calculation is repeated. The variables are also stored in an array for printout. The procedure returns to Block 4 until the end of the duct is reached.

Subroutine PARTICLE: This subroutine integrates the particle momentum and energy equations and evaluates the source terms.

Subroutine DUCTGEO: This subroutine provides the geometry of the duct; that is, the cross-sectional area as a function of distance.

D.2 Particle velocity and temperature equations

The particle velocity equation is

$$m\frac{dV}{dt} = 3\pi\mu D f(U - V) + mg - 2m f_s \frac{V^2}{D_d}$$

which can be written as

$$\frac{dV}{dt} = \frac{f}{\tau_V}(U - V) + g - 2f_s \frac{V^2}{D_d}$$

It is more convenient to have the velocity a function of distance instead of time so the equation is written as

$$\frac{dV}{dt} = V\frac{dV}{ds} = \frac{1}{2}\frac{dV^2}{ds}$$

$$\frac{V_{i+1}^2 - V_i^2}{\Delta s} = \frac{2f}{\tau_V}(U - V) + 2g - 4f_s\frac{V^2}{D_d}$$

Using an implicit formulation, the difference equation becomes

$$V_{i+1}^2\left(1 + 4f_s\frac{\Delta s}{D_d}\right) + \frac{2f\Delta s}{\tau_V}V_{i+1} - \frac{2f\Delta s}{\tau_V}U_{i+1} - V_i^2 - 2g\Delta s = 0$$

This quadratic equation can be solved for V_{i+1}. The following variables are used in the program,

$$FAC1 = \frac{f\Delta s}{\tau_V}$$

$$FAC2 = 1 + 4f_s\frac{\Delta s}{D_d}$$

$$U2 = U_{i+1}$$

$$UP1 = V_i$$

The equation for V_{i+1} becomes

$$V_{i+1} = \frac{-FAC1 + \sqrt{FAC1^2 + 2 * FAC2(FAC1 * U2 + g\Delta s + UP1^2/2)}}{FAC2}$$

The particle temperature equation, including radiative heat transfer, is

$$mc_d\frac{dT_d}{dt} = Nu\pi kD(T_c - T_d) - \sigma\epsilon\pi D^2(T_d^4 - T_w^4)$$

or

$$\frac{dT_d}{dt} = \frac{Nu}{2\tau_T}(T_c - T_d) - \frac{6\sigma\epsilon}{\rho_d c_d D}(T_d^4 - T_w^4)$$

The radiation term is nonlinear and can be linearized by

$$T_{d,i+1}^4 = T_{d,i+1}T_{d,i}^3$$

Using an implicit formulation, the finite difference equation becomes

$$\frac{T_{d,i+1} - T_{d,i}}{\Delta t} = \frac{Nu}{2\tau_T}(T_{c,i+1} - T_{d,i+1}) - \frac{6\sigma\epsilon}{\rho_d c_d D}(T_{d,i+1}T_{d,i}^3 - T_w^4)$$

or

$$T_{d,i+1}(1 + \frac{Nu\Delta t}{2\tau_T} + \frac{6\sigma\epsilon}{\rho_d c_d D}T_{d,i}^3) = T_{d,i} + \frac{Nu\Delta t}{2\tau_T}T_{c,i+1} + \frac{6\sigma\epsilon}{\rho_d c_d D}T_w^4$$

The time step is estimated from

$$\Delta t = \frac{2\Delta s}{V_i + V_{i+1}} = \frac{\Delta s}{V_{ave}}$$

The variables used in the program are

$$FAC3 = \frac{Nu\Delta s}{2\tau_T V_{ave}}$$

$$FAC4 = \frac{6\sigma\epsilon}{\rho_d c_d D}$$

$$T2 = T_{c,i+1}$$

The equation for particle temperature becomes

$$T_{d,i+1} = \frac{T_{d,i} + FAC3 * T2 + FAC4 * T_w^4}{1 + FAC3 + FAC4 * T_{d,i}^3}$$

For the situation where the temperature remains constant during a change in phase (fusion temperature), the particle energy equation becomes

$$mL\frac{d\theta}{dt} = Nu\pi kD(T_c - T_d) - \sigma\epsilon\pi D^2(T_d^4 - T_w^4)$$

where the variable θ is a progress variable for the change in phase. When the particle is in the solid state, θ is zero. When it is liquid, θ is unity. The variable PS stands for θ is the program. The change in PS over the cell is given by

$$\Delta PS = FAC3 * RLC * (T2 - TFUS) - FAC4 * RLC * (TFUS^4 - T_w^4)$$

where

$$RLC = \frac{c_d}{L}$$

and $TFUS$ is the fusion temperature.

D.3 Coupling parameters

The momentum coupling parameter is

$$S_{mom} = 3n\Delta s A\pi\mu f D(U - V)$$

The number density is related to the number flow rate by

$$\dot{n} = VAn$$

and the particle mass flow rate is

$$\dot{M}_d = \dot{n}m_d = nm_d AV$$

Therefore, the coupling term is

$$S_{mom} = 3\pi\mu f D\frac{\dot{M}_d}{m_d V} = Z\dot{M}_c f\frac{U - V}{\tau_V V}$$

The energy coupling for the total energy equation comes through convective heat transfer and work due to particle drag. The coupling term due to convective heat transfer is

$$S_{ener,con} = Nu\pi kD(T_d - T_c)n\Delta s A$$

which can be written as

$$S_{ener,con} = Z\dot{M}_c\frac{Nu}{2}\frac{T_d - T_c}{\tau_T V}$$

The work rate done due to particle drag is

$$S_{ener,drag} = S_{mom}V$$

Thus the energy coupling term becomes

$$S_{ener} = Z\dot{M}_c\frac{Nu}{2}\frac{T_d - T_c}{\tau_T V} + S_{mom}V$$

```
C*****************************************************************
C                                                               *
C        PROGRAM FOR QUASI-ONE-DIMENSIONAL COMPRESSIBLE FLOW IN A  *
C        VARIABLE DUCT USING CONSERVATIVE VARIABLES. IT IS ASSUMED *
C        THAT THE GAS IS IDEAL AND THAT THE VISCOSITY VARIES WITH THE *
C        SQUARE ROOT OF THE TEMPERATURE.                        *
C                                                               *
C*****************************************************************
C
        DIMENSION XL(101),DIA(101),AR(101),U(101),P(101),T(101),IC(101)
        DIMENSION DEN(101),UP(101),TP(101)
        COMMON TAUF,VISC,AG,FRICP,DX,DIAAV,PRN,SHR,UAVE,DENAV,DP,TMELT,
     1        TW,RLC,RADF,X1,CPS,TAVE,ZL
C
C---------------------------------------------------------------
C
C        INPUT PARAMETERS
C---------------------------------------------------------------
        OPEN(UNIT=1,FILE='nozzpar.dat',STATUS='UNKNOWN')
C--INLET DIAMETER (M)
        DIA1=0.05
C--EXIT DIAMETER (M)
        DIA2=0.025
C--NOZZLE LENGTH (M)
        AL=0.5
C--PIPE ROUGHNESS (M)
        ROUGH=0.003
C--INLET PRESSURE (PA)
        P0=101.E03
C--INLET TEMPERATURE (K)
        T0=293.0
C--INLET VELOCITY (M/S)
        U0=18.0
C--GAS CONSTANT (J/KG-K)
        RG=287
C--RATIO OF SPECIFIC HEATS
        AK=1.4
C--VISCOSITY (N-S/M2)
        VISCO=3.0E-05
C--PRANDTL NUMBER
        PRN=0.72
C--WALL TEMPERATURE (K)
        TW=293.0
C---------------------------------------------------------------
C
C        PARTICLE PARAMETERS
C---------------------------------------------------------------
C--PARTICLE DIAMETER (meters)
        DP=100.0E-06
C--PARTICLE DENSITY (kg/m3)
        RHOP=2500.0
C--LOADING
        ZL=1.0
C--PARTICLE WALL FRICTION COEFFICIENT
        FRICP=0.001
C--ACCELERATION DUE TO GRAVITY (m/s2)
        AG=0.0
C--INITIAL PARTICLE VELOCITY (m/s)
        UP0=18.0
C--INITIAL TEMPERATURE
        TP0=1000.0
C--PARTICLE SPECIFIC HEAT (J/kg/K)
        CPS=800.0
C--LATENT HEAT OF FUSION (J/kg)
        HLAT=10000.0
C--MELTING TEMPERATURE (K)
        TMELT=1200.0
C--EMISSIVITY OF PARTICLE SURFACE
        EPSILON=0.0
```

```
C---------------------------------------------------------------------
C
C         COMPUTATIONAL PARAMETERS
C---------------------------------------------------------------------
C--NUMBER OF GRID LINES
          IMAX=101
C--LINES SKIPPED IN PRINTOUT
          ISKIP=2
C--RESIDUAL FOR VELOCITY CONVERGENCE
          RESMAX=0.001
C---------------------------------------------------------------------
          PLANK=5.668E-08
          RES=RESMAX*U0
          XL(1)=0.0
          DX=AL/(IMAX-1)
          FAC1=AK/(AK+1)
          FAC2=2*(AK*AK-1)/AK/AK
          CPG=AK*RG/(AK-1)
          TAUF=RHOP*DP**2/18.0
          SHR=CPS/CPG
          RLC=CPS/HLAT
          RADF=6*PLANK*EPSILON/(RHOP*CPS*DP)
          IPHASE=0
          IF(TP0.LT.TMELT) PS=0.0
          IF(TP0.GT.TMELT) PS=1.0
          DEN0=P0/RG/T0
          IC(1)=0
C---------------------------------------------------------------------
C  CALL SUBROUTINE FOR DUCT GEOMETRY
C
          CALL DUCTGEO(AL,DX,DIA1,DIA2,IMAX,XL,DIA,AR)
C---------------------------------------------------------------------
C--INITIALIZE CONSERVATIVE VARIABLES
          X0=DEN0*AR(1)*U0
          Y0=X0*U0+P0*AR(1)
          Z0=X0*(CPG*T0+U0*U0/2)
          DEN1=DEN0
          U1=U0
          T1=T0
          P1=P0
          X1=X0
          Y1=Y0
          Z1=Z0
          UP1=UP0
          TP1=TP0
          U(1)=U0
          T(1)=T0
          P(1)=P0
          UP(1)=UP0
          TP(1)=TP0
          IC(1)=0
C---------------------------------------------------------------------
C         BEGIN DO LOOP
C---------------------------------------------------------------------
          DO 100 ISTEP=2,IMAX
C--ESTIMATE VELOCITY, PRESSURE, DENSITY AND TEMPERATURE AT NEXT STATION
          U2=U1*AR(ISTEP)/AR(ISTEP-1)
          P2=P1-DEN1*(U2*U2-U1*U1)/2
          DEN2=DEN1
          T2=T1
          IC(ISTEP)=0
          XL(ISTEP)=XL(ISTEP-1)+DX
C---------------------------------------------------------------------
C--BEGINNING ITERATION FOR PRESSURE AND VELOCITY AT NEXT STATION
C---------------------------------------------------------------------
    20    CONTINUE
          IC(ISTEP)=IC(ISTEP)+1
          U2E=U2
          UAVE=(U2+U1)*0.5
          DENAV=(DEN1+DEN2)*0.5
          TAVE=(T1+T2)*0.5
          VISC=VISCO*SQRT(TAVE/T0)
          DIAAV=(DIA(ISTEP)+DIA(ISTEP-1))*0.5
          RE=DENAV*UAVE*DIAAV/VISC
          RR=ROUGH/DIAAV
          FF=0.25/(0.434*ALOG(RR/3.7)+5.74/RE**0.9)**2
```

```
C-------------------------------------------------------------------
C--CALL THE PARTICLE SUBROUTINE TO EVALUATE PARTICLE VELOCITY AND
C  TEMPERATURE AND THE SOURCE TERMS
C
        CALL PARTICLE(UP1,TP1,U2,T2,UP2,TP2,PS,SMOMP,SENERP,IPHASE)
C-------------------------------------------------------------------
C       EVALUATING SOURCE TERMS
C-------------------------------------------------------------------
C
C--MASS SOURCE
        SMASS=0.0
C--MOMENTUM SOURCE
        SMOM=0.0
C------SOURCE TERM FOR AREA CHANGE
        SMOM=SMOM+(P1+P2)*(AR(ISTEP)-AR(ISTEP-1))*0.5
C------SOURCE TERM FOR FRICTION
        SMOM=SMOM-FF*DX*DENAV*UAVE*UAVE*3.1416*DIAAV
C------MOMENTUM SOURCE TERM FOR PARTICLES
        SMOM=SMOM+SMOMP
C--ENERGY SOURCE
        SENER=0.0
C------ENERGY SOURCE TERM FOR PARTICLES
        SENER=SENER+SENERP
C-------------------------------------------------------------------
C       UPDATING SOURCE TERMS
C-------------------------------------------------------------------
        X2=X1+SMASS
        Y2=Y1+SMOM
        Z2=Z1+SENER
        SONIC=FAC2*X2*Z2/Y2/Y2
        WRITE(*,2004) X2,Y2,Z2,SONIC,ISTEP
        IF(SONIC.GT.1.0) THEN
            WRITE(*,1001)
            STOP
        END IF
C-------------------------------------------------------------------
C       EVALUATING PRIMITIVE VARIABLES AT NEXT STATION
C-------------------------------------------------------------------
        U2=FAC1*Y2*(1.0-SQRT(1.0-SONIC))/X2
        DEN2=X2/AR(ISTEP)/U2
        P2=(Y2-X2*U2)/AR(ISTEP)
        T2=(Z2/X2-U2*U2/2)/CPG
        IF(IC(ISTEP).GT.20) THEN
            WRITE(*,1002)
            STOP
        END IF
        IF(ABS(U2E-U2).GT.RES) GOTO 20
C-------------------------------------------------------------------
C       END OF ITERATION LOOP AND INITIALIZING FOR NEXT STEP
C-------------------------------------------------------------------
        U1=U2
        U(ISTEP)=U1
        T1=T2
        T(ISTEP)=T2
        UP1=UP2
        UP(ISTEP)=UP2
        TP1=TP2
        TP(ISTEP)=TP2
        P1=P2
        P(ISTEP)=P2
        DEN1=DEN2
        DEN(ISTEP)=DEN2
C--CHECKING ON CHANGE OF PHASE
        IF(PS.EQ.0.0.AND.TP2.GT.TMELT) IPHASE=1
        IF(PS.EQ.1.0.AND.TP2.LT.TMELT) IPHASE=1
        IF(PS.GT.1.0) THEN
                PS=1.0
                IPHASE=0
        ENDIF
        IF(PS.LT.0.0) THEN
                PS=0.0
                IPHASE=0
        ENDIF
```

```
C------------------------------------------------------------------
        X1=X2
        Y1=Y2
        Z1=Z2
  100   CONTINUE
C------------------------------------------------------------------
C       PRINTOUT OF FINAL VALUES
C------------------------------------------------------------------
        WRITE(1,2000) P0,T0,U0
        WRITE(1,2002)
        DO 200 I=1,IMAX,ISKIP
        WRITE(1,2003) XL(I),U(I),T(I),P(I),UP(I),TP(I),IC(I)
  200   CONTINUE
        STOP
 1001   FORMAT(' INITIAL VELOCITY TOO HIGH')
 1002   FORMAT(' TOO MANY ITERATIONS')
 2000   FORMAT(' GAS FLOW IN A DUCT USING CONSERVATIVE VARIABLES'
     1  /' PRESS=',E12.4,4X,'TEMP=',F10.2,4X,'VEL=',F8.4)
 2002   FORMAT(/'       DISTANCE     VELOCITY     GAS TEMP    PRESSURE   PART
     1  VEL    PART TEMP    ITER')
 2003   FORMAT(3F12.4,1E12.4,2F12.4,I3)
 2004   FORMAT(4E12.4,I5)
        END
C------------------------------------------------------------------
C
C       SUBROUTINE FOR DUCT GEOMETRY
C
C       THIS SUBROUTINE CREATES A COSINE CURVE FOR THE NOZZLE GEOMETRY
C
C------------------------------------------------------------------
        SUBROUTINE DUCTGEO(AL,DX,DIA1,DIA2,IT,XL,DIAX,ARX)
        DIMENSION XL(IT),DIAX(IT),ARX(IT)
        DIAM=(DIA1+DIA2)*0.5
        DIAA=(DIA1-DIA2)*0.5
        PIL=2*3.1416/AL
        DO 10 I=1,IT
        XL(I)=(I-1)*DX
        DIAX(I)=DIAM+DIAA*COS(XL(I)*PIL)
        ARX(I)=0.785*DIAX(I)**2
  10    CONTINUE
        RETURN
        END
C------------------------------------------------------------------
C------------------------------------------------------------------
C
C       SUBROUTINE FOR CALCULATING PARTICLE VELOCITY, TEMPERATURE AND
C       COUPLING TERMS
C
        SUBROUTINE PARTICLE(UP1,TP1,U2,T2,UP2,TP2,PS,SMOMP,SENERP,IPHASE)
C--CALCULATE VELOCITY USING THE IMPLICIT QUADRATIC APPROACH. THIS SCHEME
C       WILL NOT WORK IF VELOCITY GOES NEGATIVE!
C
        COMMON TAUF,VISC,AG,FRICP,DX,DIAAV,PRN,SHR,UAVE,DENAV,DP,TMELT,
     1       TW,RLC,RADF,X1,CPS,TAVE,ZL
C------------------------------------------------------------------
C
        IF(UP1.LT.0.0) THEN
           WRITE(*,1003)
           STOP
        ENDIF
        RER=ABS(UAVE-UP1)*DENAV*DP/VISC
        IF(RER.LT.0.1) THEN
           DFAC=1.0
        ELSE
           DFAC=1.0+0.15*RER**0.67
        END IF
        TAU=TAUF/VISC
        FAC1=DFAC*DX/TAU
        FAC2=1+FRICP*DX*2/DIAAV
        UP2=(-FAC1+SQRT(FAC1**2+2*FAC2*(UP1**2/2+AG*DX+FAC1*U2)))/FAC2
```

```
C
C--CALCULATE TEMPERATURE USING IMPLICIT APPROACH
C
      UPAVE=(UP1+UP2)*0.5
      IF(RER.LT.0.1) THEN
        ANU=2.0
      ELSE
        ANU=2.0+0.6*RER**0.5*PRN**0.33
      END IF
      FAC3=ANU*DX/(3*PRN*SHR*TAU*UPAVE)
      FAC4=RADF*DX/UPAVE
      IF(IPHASE.EQ.1) THEN
        TP2=TMELT
        PS=PS+FAC3*RLC*(T2-TMELT)-FAC4*RLC*(TP2**4-TW**4)
      ELSE
        TP2=(TP1+FAC3*T2)/(1.0+FAC3)
      END IF
      TPAVE=(TP1+TP2)*0.5
C-----EVALUATION OF THE SOURCE TERM FOR MOMENTUM
      SMOMP=ZL*X1*FAC1*(UPAVE-UAVE)/UPAVE
C-----EVALUATION OF SOURCE TERM FOR ENERGY - HEAT TRANSFER AND WORK
      SENERP=ZL*X1*FAC3*(TPAVE-TAVE)*CPS+UPAVE*SMOMP
C
      RETURN
 1003 FORMAT(' VELOCITY IS NEGATIVE-PARTICLE SUBROUTINE INOPERATIVE')
      END
C--------------------------------------------------------------------
```

Index